Function and Evolution of Repeated DNA Sequences

SCIENCES

Biology, Field Director – Marie-Christine Maurel

Genetics, Epigenetics, Subject Head – Bernard Dujon

Function and Evolution of Repeated DNA Sequences

Coordinated by

Guy-Franck Richard

iSTE

WILEY

First published 2023 in Great Britain and the United States by ISTE Ltd and John Wiley & Sons, Inc.

ISTE Ltd
27-37 St George's Road
London SW19 4EU
UK

www.iste.co.uk

John Wiley & Sons, Inc.
111 River Street
Hoboken, NJ 07030
USA

www.wiley.com

Library of Congress Control Number: 2022949377

British Library Cataloguing-in-Publication Data
A CIP record for this book is available from the British Library
ISBN 978-1-78945-119-1

ERC code:
LS2 Genetics, 'Omics', Bioinformatics and Systems Biology
LS2_5 Epigenetics and gene regulation

Contents

Foreword

Our modern societies are too preoccupied with immediate performances to conceive of a world where the costs and efforts to achieve a result are not rationally minimized. And yet, life offers this image as soon as we take time to study it closely. The remarkable adaptation of different organisms to their living conditions revolves around genomes which are far from the products of what we may consider to be rational engineering. There is no such thing as a minimal genome: all of them are too large in comparison to the number of genes considered necessary to produce the organism hosting them. Often far too large, ours appears 50 times too large. They all contain identically (or almost identically) repeated sequences, sometimes they are repeated numerous times in the same genome; whereas random combinations of the four nucleotides make this phenomenon extremely unlikely, if not practically impossible.

This situation was observed as far back as the mid-20th century, long before the emergence of genomics, through the study of the renaturation kinetics of DNA molecules. The excessive amount of DNA and the abundance of repeated sequences remained a puzzle that some tended to quickly dismiss by referring to junk DNA, continuing to focus their studies only on what they already knew! Genome sequencing would come to solve this enigma by demonstrating just how incomplete our prior knowledge was. A new vision of genome organization and function is now provided, in which temporal dynamics combine with the present, because all genomes are simply imperfect copies of the genomes that preceded them and not new constructs. From now on, traces of the past mix with present events and together they lay the foundations of the future.

As the present work illustrates so remarkably, repeats found in genomes can result from major evolutionary accidents such as whole-genome duplications, which, in a singular phenomenon, tend to coincide with the transitions between major geological eras. But they may also come from repeated interactions with infectious elements – of the viral type – that eventually integrate into chromosomes and are transmitted to the offspring. Thus, there are both endogenous and exogenous causes for the existence of repeated sequences in genomes. In turn, repeats can form the basis of the formation of chromosome functional elements such as centromeres, telomeres and guanine quadruplexes. Copy number variation of long repeated sequences can play a critical role in phenotypes and in organism adaptation. Similarly, the instability of short-sequence repeats allows us to easily differentiate between individuals from the same population. However, this can sometimes lead to very serious syndromes. Finally, the different mobile elements, kinds of specialized molecular machines, present in various numbers in the different genomes cannot be ignored. By the mid-20th century, they had already been identified by their genetic effects – they are mutagenic – but we now have a much broader view of their diversity and of the consequences, sometimes considerable, of their activity.

If our knowledge of repeated genome sequences has only progressed belatedly, this is partly due to technical difficulties encountered in their sequencing. Until the recent advent of new technologies allowing longer reads, it was very difficult to correctly assemble repeated sequences and many so-called whole-genome sequences were in fact incomplete. For example, about 8% of the human genome, made up of highly repetitive sequences, remained unknown for two decades, until the application of special technologies this year. Similarly, the study of copy number variations due to segmental duplications, which have long been underestimated, is only just beginning. And let us not forget that our exploration of the living world is not only far from complete but is very much biased in favor of the already well-known groups of organisms. We can therefore expect new discoveries, or even surprises, in the study of this part of genomes, overlooked for too long, which demonstrates just how real long-term success differs from the illusion of immediate performance.

Gif-sur-Yvette

Bernard DUJON
Professor Emeritus at *Sorbonne Université*
and the *Institut Pasteur*
Member of the *Institut de France*
(Académie des Sciences)

May 2023

Introduction

About Repeated Genomes

Guy-Franck RICHARD
Instabilités naturelles & synthétiques des génomes,
Institut Pasteur, CNRS UMR3525, Paris, France

Genome [ˈdʒiːnəʊm] *nm Biol.*: Set of hereditary characteristics of a living being, of which a small part is composed of genes providing a function to the organism, and the majority is composed of repeated sequences for which it is unknown whether they have a function.

Taking matters a little further, this could be a modern definition of the word "genome", in the light of the knowledge garnered across three decades from sequencing the DNA content of living beings, in particular eukaryotic organisms, more complex than those of their bacterial and archaebacterial ancestors. Biologists were already aware back in the 1960s, long before the invention of the first DNA sequencing methods, that the content of genomes was difficult to comprehend. Denaturation-renaturation experiments highlighted that the speed of renaturation of the double-helix was proportional to its concentration. The C_0t parameter was the value at which renaturation of half the genomic DNA was complete, under controlled conditions. Each organism could then be defined by the C_0t value of its genome. In trying to establish the C_0t values of genomes of the simplest organisms – phages or bacteria – or of more complex organisms, such as vertebrates, it transpired that the latter contained three types of sequences presenting very different C_0t values (see Figure I.1).

Function and Evolution of Repeated DNA Sequences,
coordinated by Guy-Franck RICHARD.

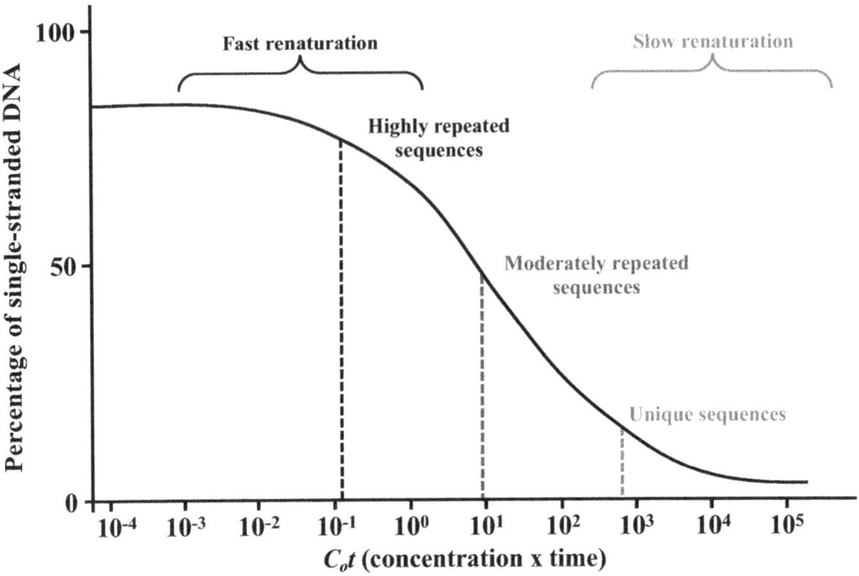

Figure I.1. *Example of* C_0t *curve*

It is thus possible to show that the mouse genome, for example, is composed of 70% unique sequences with slow renaturation, 20% moderately repeated sequences present in 1,000 to 100,000 copies per genome and 10% highly repeated sequences representing at least 1 million copies per genome and showing rapid renaturation (Britten and Kohne 1968). This approach, based on the physicochemical properties of DNA, slightly underestimated the quantity of repeated sequences because their renaturation rate depends on the identity between these sequences, divergent sequences (such as long terminal repeats (LTRs)) renaturing more slowly than identical sequences. Nowadays, $C_o t$ curves are still sometimes used to separate the highly repetitive fraction of a genome from its unique fraction in order to sequence specific DNA of either fraction (Peterson et al. 2008).

I.1. The "C-value" paradox

From the moment it was proven that DNA was the support of heredity, and theoretically contained all the genes necessary for the development of a living being, it seemed logical that the most sophisticated organisms had to contain more genes and therefore more DNA in their genome (the "C value") to encode these genes. This idea was to be questioned in the 1950s with the discovery that the nuclei of certain amphibians and fish contained 20 times more DNA than the nuclei of mammals. Given that the latter presented a greater developmental complexity, this appeared very much paradoxical, and was even used as an argument by the opponents of DNA being the sole support of heredity (Thomas 1971). This "C-value paradox" could finally be explained only decades later, when the first genomes were sequenced. It is now known that the number of genes in an organism has little to do with its size or level of complexity. The baker's yeast genome contains about 6,000 genes, that of fruit flies about 14,000 and the human genome (or those of its very close cousins, great apes) contents itself with 20,000 genes, with which it manages a very sophisticated level of developmental and behavioral complexity. But what about the paramecium with its 40,000 genes, twice as many as the human genome? Or *Trichomonas vaginalis*, a parasite of the genital tract, with its 60,000 genes? Or indeed wheat and its 124,000 genes, more than six times as many as our genes? Clearly, this so-called complexity could not be measured by the number of genes in an organism. Studies of comparative genomics[1] have shown that this high number of genes in certain organisms does in fact conceal ancestral events of partial or total genome duplication, followed by variable amounts of gene losses (Wolfe and Shields 1997; Jaillon et al. 2004). These events actively participate in the genetic redundancy and their identification as well as their underlying mechanisms will be addressed in Chapter 1.

If the complexity of an organism has nothing to do with the number of genes contained, the same is true of the amount of DNA. The human genome, with just over three times as many genes as brewer's yeast, contains 200 times more DNA. The genome of a rotifer – a small animal measuring just a few millimeters that lives in freshwaters – contains three times more genes than the human genome in 12 times less DNA! (see Figure I.2).

The genomic sequence of all these organisms showed that some of them had evolved a very compact genome, with high gene density, while others contained a multitude of repeated DNA sequences whose function did not appear obvious at first glance, and that some authors did not hesitate to call them "junk DNA" (Ohno 1972).

1 Comparative genomics is the field of genomics that focuses on the comparison of entire genomes with each other and not just of genes. Analysis tools have been specifically developed to compare the organization, structure, synteny (gene order along a chromosome) of genomes, considered as objects to be studied as a whole.

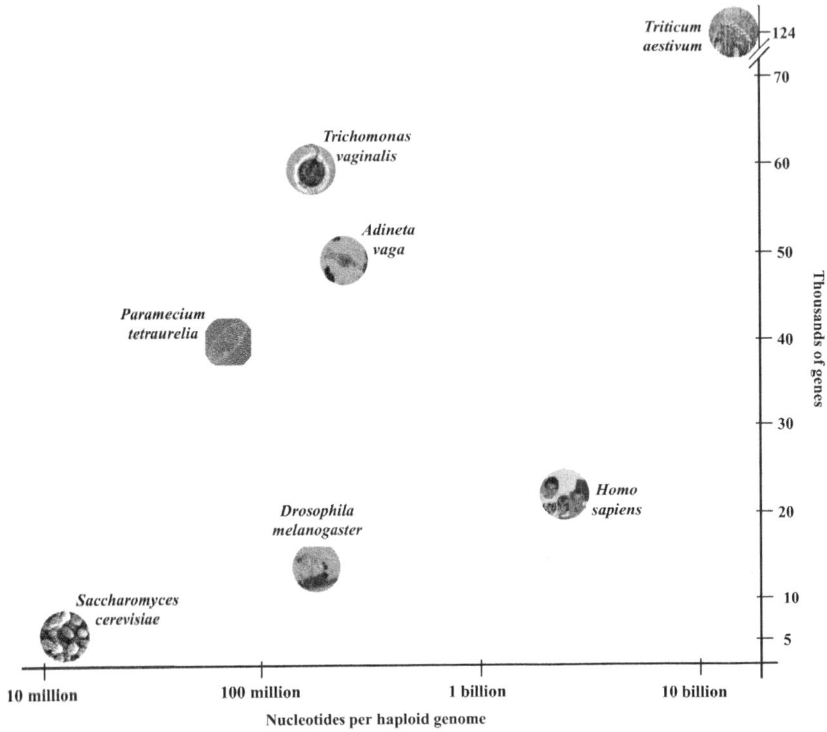

Figure I.2. *Comparison of genome sizes and gene numbers*

I.2. Recycling junk DNA

About 2% of the human genome is translated into proteins. Even by adding the untranslated genes (rRNA, tRNA, siRNA, snRNA, etc.), the percentage of "useful" DNA barely increases. So, what is the purpose of the 98% of DNA in our genome that has, apparently, no function? One conceivable answer is that it has none. The consortium led by Jeff Boeke, professor of genetics at Johns Hopkins University in Baltimore, set out to create the first synthetic yeast genome, using synthetic oligonucleotides. The brewer's yeast *Saccharomyces cerevisiae* is a eukaryotic organism whose genome contains 12.5 million nucleotides distributed across 16 chromosomes. The synthetic chromosomes were reconstructed one by one from 70 nucleotide-long sequences assembled in blocks of 750 base pairs, themselves assembled in mega-blocks of 2–4 kb, reintroduced one after the other in a hierarchical manner into the yeast genome in replacement of the natural sequences (Muller and Koszul 2015). When designing synthetic chromosomes, it was decided that all repeated sequences would be removed from the genome. All tRNA-encoding DNAs were grouped on a single circular chromosome, specifically built to carry them. Retrotransposons, microsatellites, minisatellites and other repeated elements inessential to life were removed from the new sequence. These synthetic chromosomes, with their junk DNA removed, are perfectly able to sustain life in yeast cells containing them, without any apparent phenotypic defect, at least under laboratory growth conditions (Dymond et al. 2011; Annaluru et al. 2014). One may conclude from the results of this project that junk DNA is useless. However that would be a mistake.

The human reference genome contains about 443,000 residual elements of past retroviral invasions, covering 8.3% of the total sequence (International Human Genome Sequencing Consortium 2001). These retroviral scars are the remains of successive invasions, occurring over the past hundred million years, of our mammalian ancestors by exogenous elements, which left the trace of their passage in the form of LTR[2]. These retroviral remains are therefore part of our junk DNA. Nevertheless, as we will see, their presence in our genome testifies to their distant but indispensable role in the existence of our lineage. Therian mammals, that is, those possessing a uterus within which the fertilized egg develops, are classified into two groups. Eutherians (or placentals) like humans and mice have a very elaborate placenta connecting the wall of the uterus to the embryo and allowing it to develop in complete safety throughout the entire gestation period. Marsupials (kangaroos and koalas) do not have placentas and the development of their young takes place mainly outside the uterus. Genome sequencing showed that the two human genes specifically expressed in the placenta, *syncytin-1* and *syncytin-2*, were derived from a gene encoding an ancestral viral protein, which infected the primate lineage 25–40 million years ago. Remarkably, the genome of the mouse, another placental mammal, also contains two viral genes having the same function as human genes but deriving from a slightly more recent viral infection than that of the human lineage. Thus, the placenta was invented twice, independently, in two lineages of mammals, by capture of genes of retroviral origin (Dupressoir et al. 2009). Another example is even more striking. Sexual reproduction was invented at the origin of the eukaryotic world. From the first primitive eukaryotic cells, a syngamy[3] system was developed that allowed the nuclei of two haploid cells to fuse to give birth to a diploid cell. The protein responsible for the fusion of male and female gametes is the same in plants and animals; it is the product of the *HAP2* gene. This protein is of viral origin and allows the envelope of a virus to fuse with the plasma membrane of its host's cells

2 LTR (long terminal repeat): repeated sequences typically found at the insertion sites of retroviruses.

3 Syngamy: nuclear fusion of two cells of opposite mating type.

(Fédry et al. 2017). Thus, a gene essential to sexual reproduction was captured from a virus by the genome of the very first eukaryotic cells about 1.5 billion years ago.

Other examples of the capture of a piece of transposable element exist, thus creating a new gene, a new function. Junk DNA is therefore regularly recycled during the course of evolution to bring diversity and novelty. As François Jacob (1977) said more than 40 years ago, evolution "tinkers", it makes new from old, reusing bits of genes, cutting them, splicing them and fusing them with others in order to create novelty. What appears today to the geneticists of the 21st century as junk DNA perhaps served in the past – or will serve in the future – to create diversity. The tremendous success of the eukaryotic world in invading all ecological niches under all climates and latitudes stems in part from the extraordinary flexibility of its genome and its ability to accumulate genetic elements that are seemingly useless but will be recycled in the long run to create novelty and enable the appearance of new living species.

I.3. The different repeat types

There are often several ways to classify genetic elements. Some authors have chosen to distinguish between dispersed repeated elements in contrast to tandem repeats, the latter being repeated at least twice in a row at the same genetic locus, unlike the former, which are repeated at different loci (Richard et al. 2008). But some dispersed repeats are so numerous in the genomes that they appear to be tandemly repeated. This is the case for *Alu* sequences in humans, which are frequently found grouped in introns or intergenic sequences. Repeated sequences of exogenous origin, that is to say originating from an organism other than the cell in which they are observed, could also be distinguished from repeated sequences of endogenous origin, manufactured by the cell in which they are observed. Transposable elements would belong to the first category, having invaded the genomes of eukaryotic (or prokaryotic) lineages, while the different satellite DNAs would belong to the second, being manufactured by molecular processes specific to the genomes that contain them. But other problems then arise. It is known, for example, that *Alu* elements, inactive retrotransposons that can be mobilized in *trans* by the machinery of other retroelements, are of endogenous origin. They result in fact from the duplication of the non-coding 7SL RNA, which is involved in the synthesis of excreted proteins. This duplication, prior to mammalian radiation, resulted in the fusion of two monomers of 130 nucleotides derived from 7SL RNA, separated by a short adenine-rich region (Ullu and Tschudi 1984). Achieving a coherent classification of the repeated sequences therefore proves a complicated task, particularly in the genomes of evolved plants and animals within which they are plethoric, both in structure and number.

We have therefore tried in the rest of this work to present the repeat elements in relation to their role (proven or assumed) in genomes, rather than according to their structure or their assumed origin (see Figure I.3).

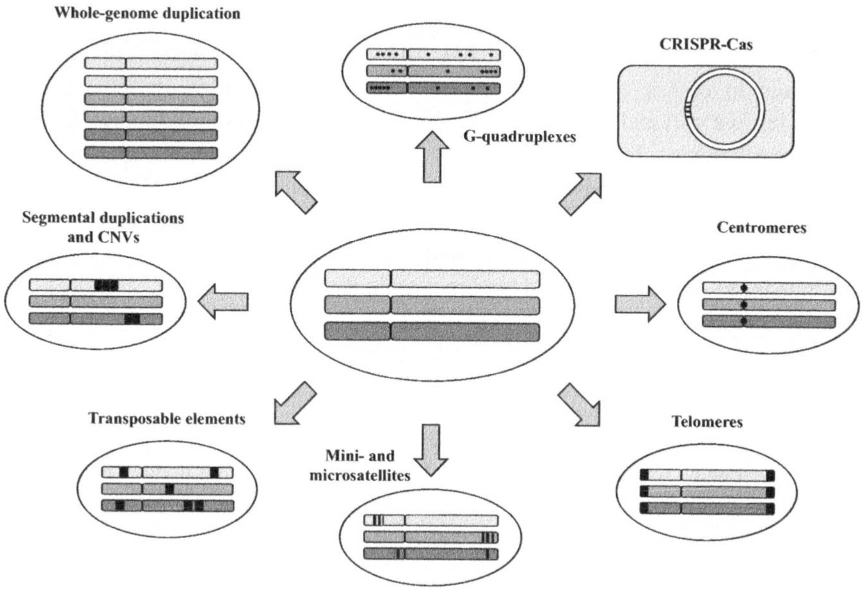

Figure I.3. *The different types of repeated DNA sequences*

After exploring total or partial genome duplications in Chapter 1, the duplications of large DNA segments, sometimes in multiple copies in tandem or dispersed within genomes, will be described in Chapter 2. These contribute significantly to the level of genetic redundancy and gene duplication and their study, although essential to understand the dynamics of complex genomes and the inheritability of certain traits is still in its infancy. Transposons and retrotransposons will be presented in Chapter 3, and their role in the generation of genetic novelties will be detailed. In most species, centromeres are present at a rate of one per chromosome. These very particular repeated elements are essential for the proper segregation of sister chromatids during cell divisions. They will be studied in Chapter 4 and as we will see, holocentric organisms depart from this rule by exhibiting several tens of centromeres per chromosome. Telomeres are highly repeated sequences found at the ends of chromosomes to prevent loss of genetic information. Their sequence and structure vary greatly from organism to organism, with some species having developed highly original telomeres that are made up of tandemly repeated elements. These concepts will be studied in Chapter 5. G-quadruplexes, these secondary DNA structures caused by the regular repeat of GC base pairs, are present in all eukaryotic genomes. Their distribution and their role in DNA transcription and replication will be discussed in Chapter 6. The different types of satellite DNA found in large numbers in eukaryotes, and whose precise function is not always clear, will be described in Chapter 7. As we will see, although prokaryotic genomes contain only few of them, some bacteria use them as camouflage to escape their host's immune system. Remaining in the world of prokaryotes, we will end in Chapter 8 with the fascinating study of another bacterial defense mechanism, directed against these other enemies that are plasmids and bacteriophages: the CRISPR-Cas system. The acquisition of small, tandemly repeated pieces of DNA from invaders foreign to the cell provides eubacteria and archaea with a robust line of defense. And it offers 21st-century geneticists myriad tools to manipulate their preferred genomes at their own convenience.

I.4. References

Annaluru, N., Muller, H., Mitchell, L.A., Ramalingam, S., Stracquadanio, G., Richardson, S.M., Dymond, J.S., Kuang, Z., Scheifele, L.Z., Cooper, E.M. et al. (2014). Total synthesis of a functional designer eukaryotic chromosome. *Science*, 344(6179), 55–58.

Britten, R.J. and Kohne, D.E. (1968). Repeated sequences in DNA. *Science*, 161, 529–540.

Dupressoir, A., Vernochet, C., Bawa, O., Harper, F., Pierron, G., Opolon, P., Heidmann, T. (2009). Syncytin-A knockout mice demonstrate the critical role in placentation of a fusogenic, endogenous retrovirus-derived, envelope gene. *Proceedings of the National Academy of Sciences*, 106(29), 12127–12132.

Dymond, J.S., Richardson, S.M., Coombes, C.E., Babatz, T., Muller, H., Annaluru, N., Blake, W.J., Schwerzmann, J.W., Dai, J., Lindstrom, D.L. et al. (2011). Synthetic chromosome arms function in yeast and generate phenotypic diversity by design. *Nature*, 477(7365), 471–476.

Fédry, J., Liu, Y., Péhau-Arnaudet, G., Pei, J., Li, W., Tortorici, M.A., Traincard, F., Meola, A., Bricogne, G., Grishin, N.V. et al. (2017). The ancient gamete fusogen HAP2 is a eukaryotic class II fusion protein. *Cell*, 168(5), 904–915.e10.

International Human Genome Sequencing Consortium (2001). Initial sequencing and analysis of the human genome. *Nature*, 409, 860–921.

Jacob, F. (1977). Evolution and tinkering. *Science*, 196(4295), 1161–1166.

Jaillon, O., Aury, J.-M., Brunet, F., Petit, J.-L., Stange-Thomann, N., Mauceli, E., Bouneau, L., Fischer, C., Ozouf-Costaz, C., Bernot, A. et al. (2004). Genome duplication in the teleost fish Tetraodon nigroviridis reveals the early vertebrate proto-karyotype. *Nature*, 431(7011), 946–957.

Muller, H. and Koszul, R. (2015). Conception et synthèse de néochromosomes. *Médecine thérapeutique/Médecine de la reproduction, gynécologie et endocrinologie*, 17(4), 228–236.

Ohno, S. (1972). So much "junk" DNA in our genome. *Evolution of Genetic Systems*, 23, 366–370.

Peterson, D.G., Schulze, S.R., Sciara, E.B., Lee, S.A., Bowers, J.E., Nagel, A., Jiang, N., Tibbitts, D.C., Wessler, S.R., Paterson, A.H. (2008). Integration of Cot analysis, DNA cloning, and high-throughput sequencing facilitates genome characterization and gene discovery. *Genome Research*, 12, 795–807.

Richard, G.-F., Kerrest, A., Dujon, B. (2008). Comparative genomics and molecular dynamics of DNA repeats in eukaryotes. *Microbiol. Mol. Biol. Rev.*, 72(4), 686–727.

Thomas Jr., C.A. (1971). The genetic organization of chromosomes. *Annu. Rev. Genet.*, 5, 237–256.

Ullu, E. and Tschudi, C. (1984). Alu sequences are processed 7SL RNA genes. *Nature*, 312, 171–172.

Wolfe, K.H. and Shields, D.C. (1997). Molecular evidence for an ancient duplication of the entire yeast genome. *Nature*, 387, 708–713.

1

Whole-Genome Duplications, a Source of Redundancy at the Entire-Genome Scale

Elise PAREY[1] and Camille BERTHELOT[1,2]

[1] *Institut de biologie de l'École Normale Supérieure, CNRS UMR8197, INSERM U1024, Paris, France*

[2] *Génomique fonctionnelle comparative, Institut Pasteur, CNRS UMR3525, Paris, France*

Whole-genome duplications, or polyploidizations, are high-impact mutational events, generating a copy of the entirety of a cell's chromosomes. Although genomes are made up of numerous locally repeated sequences (transposable elements, satellite DNA, duplicated genes, etc.), whole-genome duplications introduce a redundancy of the entire genome. Along with expansions of transposable elements, whole-genome duplications are the major contributors to the evolution of genome size.

This chapter presents the main examples of polyploids, the mechanisms at the origin of their formation and their consequences for genome functioning and organism evolution. Although this chapter focuses on polyploid organisms, we will also discuss the case of polyploid cells and cell populations within haploid or diploid individuals, and their implications in the normal and pathological physiology of organisms.

For a color version of all figures in this chapter, see http://www.iste.co.uk/richard/function.zip.

Function and Evolution of Repeated DNA Sequences,
coordinated by Guy-Franck RICHARD.

Glossary

Orthologs, paralogs and ohnologs: these terms correspond to the evolutionary origins of two genes (or sequences) observed in one or several different species. Orthologous genes are descended from a common ancestor by speciation ("same gene" in different species), whereas paralogs are descended from a duplication of the gene in an ancestor. Ohnologs are a subclass of paralogs and correspond specifically to paralogs derived from polyploidization.

Homeologs: duplicated chromosomes derived from polyploidization, as opposed to homologs, which correspond to pairs of chromosomes in a diploid genome. By extension, this term can also refer to duplicated chromosomal regions within a polyploid genome.

Synteny: the order and organization of genes, and more generally of sequences, along chromosomes. Synteny is said to be "conserved" when the gene order remains maintained during the course of evolution. It can be disrupted by genomic rearrangements and sequence deletions and insertions.

Paleopolyploids and neopolyploids: paleopolyploid species are descended from an ancient whole-genome duplication event. Their genome bears the traces of this duplication but no longer behaves like a polyploid genome. Neopolyploid species are recent polyploids whose genome is a clear assortment of two or more subgenomes, which have particular behaviors owing to their polyploid state.

Box 1.1. *Glossary of terms*

1.1. Prevalence of polyploids in the tree of life

1.1.1. *Whole duplications in eukaryotes*

Whole-genome duplications have profoundly marked the evolutionary history of eukaryotic genomes, and it is in this phylum that they are best documented (see Figure 1.1). We distinguish here between species descended from polyploid ancestors, known as paleopolyploids, and extant polyploids, known as neopolyploids. The "model" whole-duplication events, that is, those most studied in eukaryotes, are those of the yeast *Saccharomyces cerevisiae*: many examples in plants such as maize, wheat and rockcress; the two duplications at the stem of the phylogeny of vertebrates, known as 1R and 2R; and lastly, the duplication of the ancestor of teleost fish, to which several model organisms such as zebrafish and medaka belong. The most common polyploidizations in eukaryotes are *triploidizations*, at the origin of genomes with three sets of chromosomes (3n), tetraploidizations (4n) and hexaploidizations (6n). Here, we provide a quick, non-exhaustive overview of well-documented polyploidization events in major eukaryotic phyla.

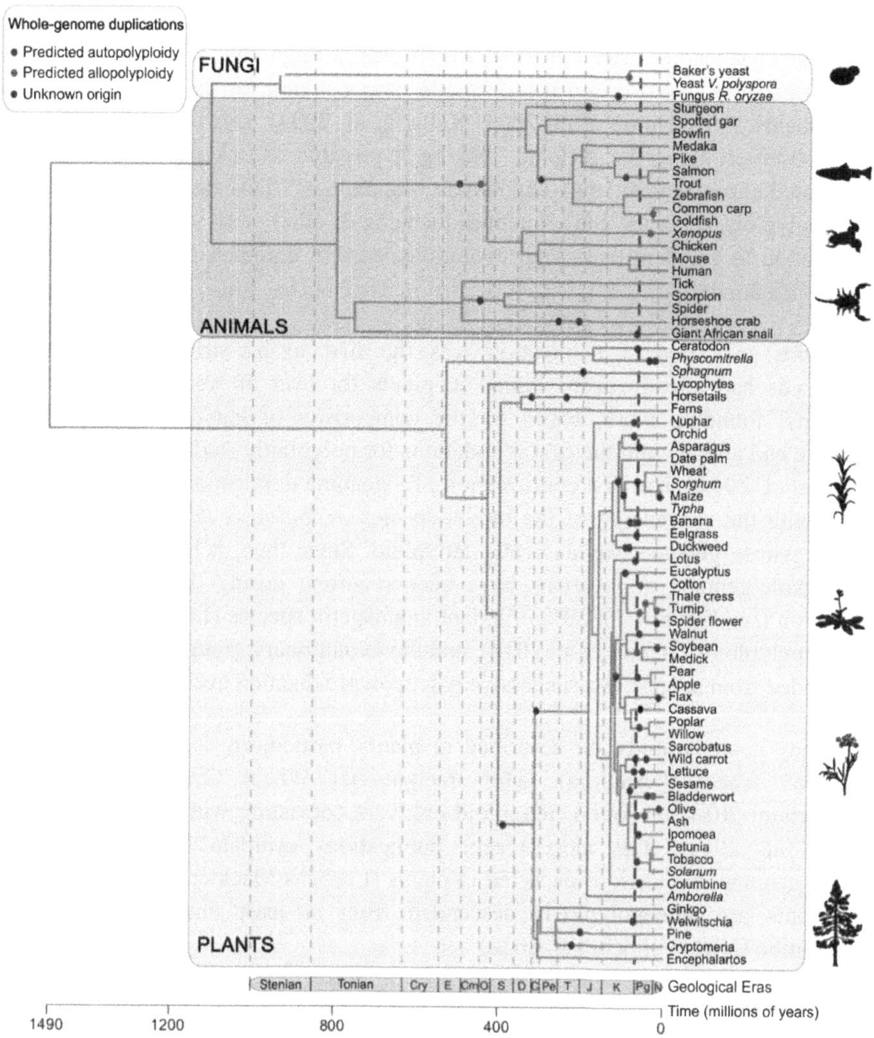

Figure 1.1. *Whole-genome duplications identified in the eukaryotic phylogenetic tree. Autopolyploidization events are shown in blue, allopolyploidizations in red and duplications with unidentified parental origins in black. Only paleopolyploidies are represented: neopolyploidies (<5 Mya, such as those of wheat) are not indicated on the tree for legibility reasons. The geological time scale is represented below the tree, highlighting how numerous polyploidization events in plants correspond to the Cretaceous-Tertiary crisis (K-Pg boundary). The majority of events presented correspond to the data cataloged by Van de Peer et al. (2017) and Clark and Donoghue (2018), updated to reflect the current literature*

Yeasts: baker's yeast *S. cerevisiae* was the first eukaryotic species to see its genome sequenced (Goffeau et al. 1996). It was through analysis of this sequence that its paleopolyploid origin was formally demonstrated for the first time (Wolfe and Shields 1997; Dujon et al. 2004; Kellis et al. 2004): sets of duplicated genes were observed along the genome, making it possible to link pairs of ancestrally duplicated chromosomes, today known as "homeologs". This signature is typical of paleopolyploid genomes. Subsequent sequencing of other yeast species enabled this duplication to be identified as an evolutionary event common to all yeasts of the genus *Saccharomyces*.

Plants: while yeast duplication was the first to be strictly demonstrated, polyploids had already been studied in plants for over 50 years. The geneticist Stebbins' founding work, based on the comparison of karyotypes of different species, had already laid the first foundations for polyploidy studies (Stebbins 1947; Soltis et al. 2014). The first confirmation of a genome duplication in plants came in 2000 with the sequencing of the model species *Arabidopsis thaliana* (Kaul et al. 2000), whose genome was anciently tetraploid. Since then, it has become known that whole-genome duplications have been recurrent during the course of plant evolution (see Figure 1.1): 20% of extant angiosperm species (flowering plants) are recent polyploids (Barker et al. 2016), and the evolutionary group of angiosperms is descended from at least two successive paleopolyploidization events.

Many neopolyploids are described in plants, particularly in cultivated species such as wheat (hexaploid) (The International Wheat Genome Sequencing Consortium 2014) or cotton (tetraploid cultivars coexisting with diploid cultivars) (Li et al. 2014). The most recent publications estimate that at least 250 whole-genome duplications occurred in plants (Leebens-Mack et al. 2019), and that all plants are paleopolyploids descended from at least one ancient genome duplication (Nieto Feliner et al. 2020).

Vertebrates: the hypothesis of total duplications in the vertebrate ancestral lineage originated in the work of Susumu Ohno in the 1960s and 1970s. After observing the existence of many families of duplicated genes distributed across chromosomes in vertebrate genomes, Ohno (1970) proposed the classically accepted hypothesis known as the "2R hypothesis", that is, the presence of two whole-genome duplications in the ancestor of the vertebrate lineage, denoted 1R and 2R. This hypothesis was then reinforced by the study of clusters of Hox genes, which exist as a single cluster in invertebrates, versus four in vertebrates (Popovici et al. 2001), then by studying multigene families on the whole genomes of several model vertebrates (Dehal and Boore 2005) as well as by reconstructions of ancestral genomes (Sacerdot et al. 2018). Similarly, the seven clusters of Hox genes present in the teleost fish genome have made it possible to propose a third duplication specific to this evolutionary group (known as "3R") (Amores et al. 1998), confirmed with the sequencing of the tetraodon genome (Jaillon et al. 2004). In general, whole duplications in vertebrates are ancient and therefore still poorly characterized compared with events occurring in plants and yeasts. Several duplications took place subsequent to events 1R and 2R, for example, in xenopes, including the well-documented duplication of the model species *Xenopus laevis* (Session et al. 2016), as well as in non-teleost fish, notably sturgeon (Du et al. 2020). Additional duplications are also cataloged in teleosts (after 3R), including one in carp (Chen et al. 2019; Xu et al. 2019) and one in salmonids (Berthelot et al. 2014; Lien et al. 2016). With no whole-genome duplication being cataloged in warm-blooded vertebrate species, some authors have proposed that polyploids are not stable or viable in these groups (Wertheim et al. 2013).

Paramecia: in addition to yeast, paramecia represent another unicellular model of interest having undergone several whole duplications of their genome. Three duplications were highlighted by sequencing of the genome of *Paramecium tetraurelia (*Aury et al. 2006). The most recent duplication is shared by 15 morphologically very similar species (the *P. aurelia* complex). The notable genomic feature of paramecia is the separation of the somatic genome (MAC: macronucleus) from the germline genome (MIC: micronucleus). Only the MIC is subject to meiosis and is transmitted to the descendant cell and can thus be the transmitter of whole-genome duplications; the MAC is systematically reformed from the MIC during the paramecium lifecycle. In addition to the genomic redundancy provided by total duplications, the MACs of paramecia contain a second form of redundancy: they correspond to a rearranged and extremely polyploid version of the MIC, which can contain up to 800 identical copies of the chromosomes. The implications of this nuclear dimorphism on the ability to tolerate total duplications, as well as their consequences, remain poorly understood in the absence of other documented examples in eukaryotes.

Invertebrates: polyploidizations have been relatively little studied, and therefore more rarely documented, in animal species outside the vertebrate phylum. Three whole-duplication events have been suggested in the horseshoe-crab lineage (Kenny et al. 2016), as well as tetraploidization in the ancestor of scorpions and spiders (Schwager et al. 2017). It should be noted, however, that uncertainties regarding the phylogeny of these species and the still low number of species sequenced do not permit accurate dating of these events (Ballesteros and Sharma 2019; Nong et al. 2021). In addition, in hexapods, transcriptomic data suggest several events of massive gene duplications (Li et al. 2018). Nevertheless, in the absence of genomes sequenced to reveal the signatures of homeologous regions characteristic of whole-genome duplications, the origin of these gene copies is debated (Roelofs et al. 2020). Lastly, in mollusks, a whole-duplication event in the lineage of the giant African snail has recently been demonstrated through the sequencing of its genome (Liu et al. 2021).

Many authors have questioned the overrepresentation of genome duplications in plants compared to animals. The hypothesis most commonly put forward proposes that the cellular disruptions involved in total duplication, to which we will return later, would not be supported by the constraints of the development of certain groups of animals, particularly mammals. In contrast, polyploidizations would be a common mechanism favored by natural selection in plant evolution. Some authors have also proposed that the large number of polyploids in plants may reflect a "ratchet" effect, in which genome duplications would be well tolerated but irreversible. Unable to be lost once acquired, the polyploidies would then be fated to progressive accumulation during the course of evolution in this phylum (Meyers and Levin 2006).

1.1.2. *Polyploidies in prokaryotic organisms*

The existence of polyploid organisms is now also well documented in bacteria and archaea. Many examples have been described, notably in the majority of halophilic and methanogenic *Euryarchaeota* clades, in Proteobacteria, Cyanobacteria, *Deinococci* and Gram-positive bacteria. In these groups of unicellular organisms, the ploidy of the organisms can reach beyond 50 copies of the genome per cell (50n) (Markov and Kaznacheev 2016), and even hundreds of thousands in the case of *Epulopiscium* spp., a group of Firmicutes known to be symbiotic with certain fishes (Mendell et al. 2008). This ploidy is rarely stable and is often higher when the cells are in the growth phase compared to the stationary phase. Thus, wide variations in ploidy can be observed within a species or even within a population.

The evolutionary origin and dating of these polyploidization events is often difficult to determine in the absence of exhaustive genomic and phylogenetic studies to trace these events. The functional and evolutionary consequences of genome duplications are also less studied in these groups, although several avenues have been proposed. While the consequences linked to genome redundancy, which we will discuss later, present points in common with eukaryotes, it can nevertheless be noted that other advantages have been proposed in prokaryotes, without any real link to the genetic information itself. For example, certain species such as *Haloferax volcanii*, a halophilic archaea, use the duplication of their genome as a phosphate storage mechanism, independently of any genetic consideration (Zerulla et al. 2014). In *Epulopiscium* spp., the very high ploidy leads to giant cells that escape predation by ciliates and can migrate into the digestive tract of their host (Mendell et al. 2008). As these consequences are unrelated to the presence of duplicated genomic sequences and the information that they carry, we will not discuss them any further in this chapter, but we invite readers to read Soppa's (2014) review for further examples. It should be noted, however, that prokaryotic polyploidies have been proposed as one of the original mechanisms involved in the evolution of mitosis, meiosis and sexual reproduction at the time of the appearance of eukaryotes (Markov and Kaznacheev 2016).

1.1.3. *Polyploid cells in normal and pathological physiology*

Genome duplications do not always concern entire organisms, and can also occur in the somatic tissues of diploid multicellular individuals where these cells can form polyploid clonal lineages, sometimes called endopolyploids. This phenomenon of polyploidization can be normal and constitutive, for example in certain cells in the liver or the placenta in humans (Ganem and Pellman 2007), in the intestine in *Drosophila* (Fox et al. 2010), or in many tissues in angiosperm plants, where endopolyploidy is frequent (D'Amato 1984). In the most extreme cases, endopolyploid cells can contain several hundred thousand copies of the genome, as in certain neural cells of *Aplysia* (Edgar et al. 2014). Endopolyploidy can also appear in a normal physiological context in response to stress factors or the environment, as is the case with wound healing in humans (Scholes and Paige 2015; Yant and Bomblies 2015).

However, it is in the context of tumorigenesis that polyploid cells are most often described in usually diploid organisms. It is estimated that more than 30% of human tumors contain cells with a duplicated genome and that endopolyploidy is part of the common processes of carcinogenesis, with karyotype rearrangements (Bielski et al. 2018). Thus, the presence of polyploid cells forms part of the classical progression markers in many types of cancers, including breast and cervical cancers.

1.2. Mechanisms for the appearance of whole-genome duplications

1.2.1. *Non-separation of chromosomes after replication*

Whether concerning polyploid individuals or somatic lineages, the appearance of whole-genome duplications involves similar mechanisms leading to the appearance of a first polyploid cell either during mitosis or meiosis. The main mechanism of polyploid cell formation occurs by successive replications of DNA content without cell division, through a mechanism called endoreplication (Edgar et al. 2014; Yant and Bomblies 2015) (see Figure 1.2). Endoreplication is evolutionarily well-conserved from plants to animals. It is reflected in a variation of the classical cell cycle: the G1 → S → G2 → M → G1 cycle is altered either by a reduction in the division phase (incomplete mitosis, without cytodieresis: endomyosis) or in G → S → G (without mitosis: endocycle). Programmed endoreplication of healthy polyploid cells is triggered by developmental or environmental cues, and is often related to stress response mechanisms, particularly in plants (Scholes and Paige 2015). These result in inhibition of the entry into division as well as activation of the cycle progression and exit signals (see the review by Fox and Duronio (2013) for details of the molecular mechanisms that may be involved). Constitutive endopolyploidy thus leads to the formation of specialized, differentiated cells, which lose their ability to divide.

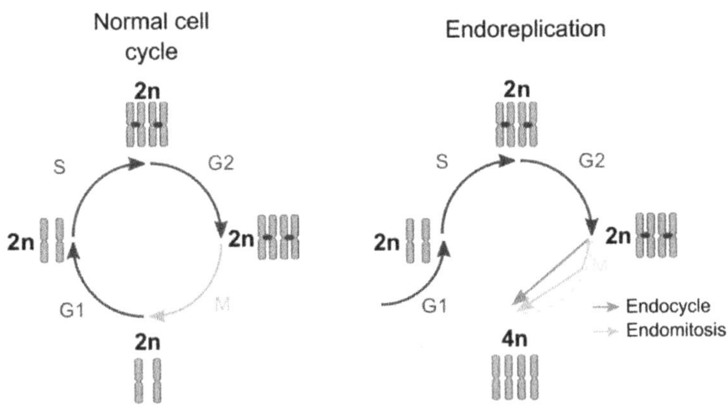

Figure 1.2. *Tetraploidization by endoreplication. In the case of a normal cell cycle, the genome is replicated in the S phase, then the sister chromatids separate during mitosis in the M phase. During an endoreplication cycle, chromatids can separate without entering the M phase (endocycle, in red), or the cell can initiate the mitosis phase but not complete it until separation of the daughter cells, producing a single tetraploid cell (endomitosis, in yellow). The mechanism is essentially identical for endoreplications during meiosis. If the chromatids do not separate, this mechanism can give rise to polytene chromosomes*

In the pathological case of cancer cells, endoreplication is aberrantly triggered. When these polyploid cells regain their ability to proliferate, their unstable genome and mitoses make them particularly prone to the accumulation of mutations, a characteristic that facilitates the progression of cancers (Yant and Bomblies 2015).

In the particular case of meiosis, endoreplication leads to unreduced gametes, most frequently diploid. The frequency of unreduced meioses is variable according to eukaryotic clades and increased in hybrid species: it is estimated to be 0.73% in humans (Egozcue et al. 2002) and 0.56% on average in non-hybrid plants compared with 27.52% in hybrids (Ramsey and Schemske 1998). It is also well documented that unreduced gametes increase in response to environmental conditions, notably changes in temperature, across many species groups among plants, fish and amphibians.

1.2.2. *Autopolyploidization, a perfect genome redundancy*

Individuals carrying whole-genome duplications occur mostly following the fusion of unreduced gametes. In the example of a diploid species, the fusion of two unreduced 2n gametes results in tetraploidization: the transition to a 4n state. More rarely, polyploid formation can also occur after somatic doubling during the early stages of development. Autopolyploidy results from the fusion of genomes of individuals of the same species, in contrast to allopolyploidy (developed in the following section), where the parental genomes come from different species (Stebbins 1947). In plants, autopolyploids can result from the duplication of the genome of the same individual by self-fertilization, in which case genetic redundancy can be total.

As a result, autopolyploid genomes initially contain almost-identical genome copies: the only differences correspond to the polymorphisms carried by the individual(s) at the origin of the polyploidization event. Many examples of natural neo-autopolyploids are known in plants, such as the common *Biscutella* (Parisod et al. 2010) or *Dactylis glomerata* (cat grass) (Lumaret et al. 1989). Several paleopolyploids are also suggested to be of autopolyploid origin: the most recent duplication of the banana genome (so-called “alpha” duplication) (D’Hont et al. 2012), that of the soybean genome (Schmutz et al. 2010), or that of the poplar genome (Garsmeur et al. 2014). In vertebrates, the parental origins of the majority of whole-genome duplications are not known, but two events are generally accepted as probable cases of autopolyploidization: the tetraploidization at the origin of the salmonid lineage (Berthelot et al. 2014; Lien et al. 2016) and that in the sturgeon lineage (Du et al. 2020).

1.2.3. *Allopolyploidization, an overlapping of genomes of similar species*

Allopolyploidy can result, similarly to autopolyploidy, from the fusion of unreduced gametes but originating from individuals of two genetically similar species. Nevertheless, in the case of allopolyploidy, duplication generally occurs after an initial hybridization stage between the two species leading to the emergence of a polyploid population. Indeed, meioses of hybrids are often unstable, with a higher probability of forming unreduced gametes (Ramsey and Schemske 1998).

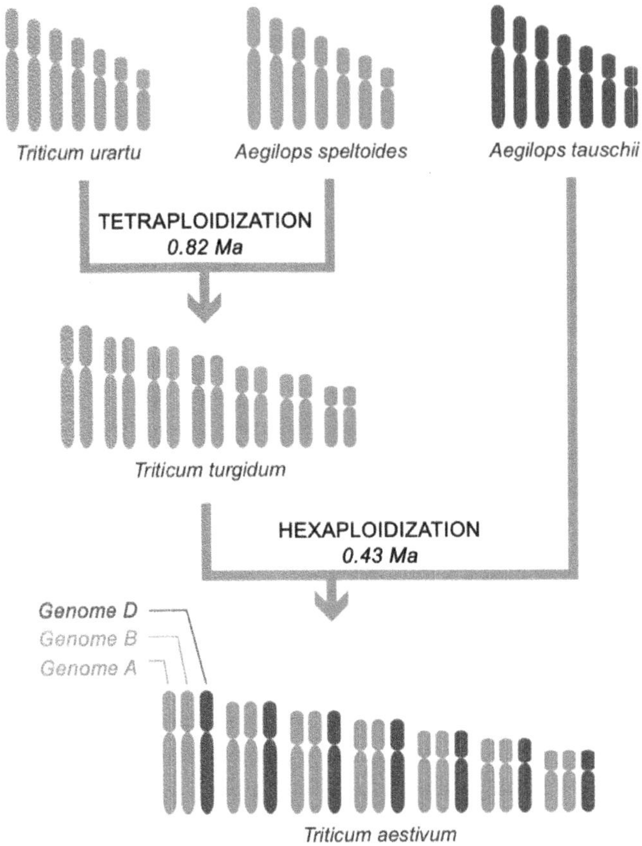

Figure 1.3. *Allopolyploidizations in the lineage of wheat (Triticum aestivum). Common wheat is a hexaploid species (2n = 42) resulting from two recent allopolyploidization events, dated to 820 ka and 430 ka ago. Wheat is descended from the hybridization of still-existing species, einkorn wheat (Triticum urartu), Aegilops speltoides, and a species of wild goatgrass (Aegilops tauschii). These species each have a haploid karyotype with seven chromosomes, which are preserved in the current wheat genome in the form of subgenomes clasically annotated A, B and D. The karyotypes shown are the haploid karyotypes*

In comparison with autopolyploids, allopolyploids contain two copies of more divergent genomes, where the different alleles can present an assortment of more varied polymorphisms. Depending on the distance between the parental species, the two subgenomes may have been subject to different evolutionary dynamics for a relatively long time and accumulated differences in terms of transposable element repertoires, genomic composition (GC, codon usage) and a divergence in coding and non-coding sequences. Indeed, it is the observation of these divergences between the two subgenomes that makes it possible to classify a polyploidization event as corresponding to an auto- or allopolyploidy. Again, many examples exist of allopolyploidizations in plants, such as the ancient maize genome duplication (Garsmeur et al. 2014), and more recent polyploidizations in wheat (The International Wheat Genome Sequencing Consortium 2014) and cotton lineages (Li et al. 2015). In certain cases, the parental genomes that have hybridized and duplicated are known, as is the case for common wheat, as a hexaploid species derived from two successive allopolyploidization events between still-extant wild cereal species (see Figure 1.3). In vertebrates, the best-documented allopolyploidies are the duplication of the genome of carp, of which at least one of the progenitors is derived from the barb clade, the other remaining unidentified (Xu et al. 2019), as well as that of the *Xenopus laevis* genome, derived from the hybridization of two ancestral xenopes, approximately 17 Mya (Session et al. 2016). Lastly, whole-genome duplication of the yeast *S. cerevisiae* is also an allopolyploidy (Marcet-Houben and Gabaldón 2015).

The relative frequency of auto- and allopolyploidies in eukaryotes is poorly known. Indeed, for many ancient events, the ancestral species are not known and the question of their origin has not been resolved. In plants, allopolyploidies were initially thought to constitute the majority. Nevertheless, it would seem that this observation is the result of an overrepresentation bias with respect to allopolyploids in published studies, whereas the two mechanisms are in fact now considered to occur in equal measure (Barker et al. 2016).

1.3. Cellular consequences of whole-genome duplications

1.3.1. *Disruption of cell and nucleus organization*

At the cellular level, one of the most obvious and universal consequences of genome duplications is the increase in cell size. Polyploid cells are typically larger than those of normal ploidy, and this is observed across all species groups from prokaryotes to vertebrates, as well as in normal or pathological polyploid cell lineages (Yant and Bomblies 2015). The increase in the amount of DNA hosted logically leads to an increase in the nuclear volume (Cavalier-Smith 1978), as well as an increase in the duration of the cell cycle, where the replication phase in particular is lengthened (Mable 2001). In plants, it has been proposed that the consequences of this nucleotypic effect reach as far as the developmental duration and generation time of polyploid and paleopolyploid species (Levin 1983), although the effect is less documented in other groups of the tree of life. In general, neopolyploid species are larger in adulthood than their diploid parents, with an allometric effect established between genome size, cell size and the size of the individual (Gregory et al. 2000; Otto 2007).

The doubling of genetic material also disrupts the structural organization of the nucleus: the presence of additional copies of the genome requires a reorganization of the nuclear territories. In cotton, where the existence of diploid and tetraploid cultivars allows detailed comparisons of the nuclear organization, one study notably showed that tetraploidization was accompanied by modifications in the topological arrangement of the chromosomes in the nucleus and by an inversion of open and closed genomic compartments (Wang et al. 2018). These modifications also extend to higher resolution topological structures, with a reorganization of certain chromatin domains (TADs[1]), whose structure is important for the regulation of gene expression. Also in cotton, new TADs appeared after polyploidization, preferentially in the open-chromatin regions. These TADs maintain the homeologous regions in a spatially closed environment, suggesting that these regions could remain co-regulated (Wang et al. 2018). In Arabidopsis, whose genome is not organized into well-defined TADs, similar modifications have nevertheless been observed with artificial induction of tetraploidization, with an increase in nuclear contacts between interchromosomal regions and a restriction of short-range interactions (Zhang et al. 2019).

Given that the techniques allowing detailed investigation of the three-dimensional organization of the nucleus are relatively recent, studies that analyze the effects of genome duplications are currently few in number and are focused on the phylum of plants, where the numerous neopolyploids allow a comparison between parental and polyploid genomes. Thus, the effects on nuclear organization in the other clades remain poorly known.

1 TADs (topologically associated domains) are genomic regions presenting strong physical interactions in the cell nucleus. They correspond to three-dimensional folding units of chromatin, separated from each other by insulator proteins. It has been proposed that these domains participate in gene regulation by spatially delimiting co-expressed genes and their regulatory elements.

1.3.2. *Modifications in the expression of genes and transposons*

It is now well documented that genome duplications are accompanied by profound and immediate changes in gene expression (Adams and Wendel 2005). These effects have been clearly demonstrated by experiments with artificial tetraploidization, particularly in *Arabidopsis* and cotton, where hundreds of genes are differentially expressed between the parent individual and the polyploid individual (Adams et al. 2003; Wang et al. 2004; Zhang et al. 2019). According to a hypothesis widely developed in the literature, some of these modifications of gene expression are related to dosage balance (Guo et al. 1996; Papp et al. 2003). Indeed, the expression dosage of certain genes is important to their function, either because their stoichiometry with respect to other components of the cell is important for the formation of complexes, or because the action of the proteins they encode is dose dependent. By increasing gene redundancy, genome duplications modify dosage balance and disrupt gene expression networks – a phenomenon also observed in smaller scale duplications and copy number variable genomic regions (Gout and Lynch 2015). Beyond the dosage balance hypothesis, many factors appear to contribute to the modification of gene expression immediately after a genome duplication, like the reorganization of the nucleus discussed above, and changes in the epigenetic marks carried by the genome. Thus, profound modifications in methylation (Salmon et al. 2005) and histone marks (Zhang et al. 2019) have been observed in natural and artificial polyploid plants, suggesting that these upheavals are at the origin of at least some of the many transcriptional changes observed.

In the particular case of allopolyploids, one of the additional consequences of whole-genome duplication is the merging of two different subgenomes in a common nuclear environment. This situation has consequences on cells and nuclei also observed in the case of hybrids, notably trans effects on the expression of genes, whose regulatory sequences can now recruit transcription factors and alleles from the other subgenome (Shi et al. 2012; Bottani et al. 2018). These trans-regulatory effects can in some cases lead to polyploids that are more stable and vigorous than their parents, as is the case in certain hybrids (Ni et al. 2009; Shi et al. 2012). In other cases, this pooling of two genomes can lead to a phenomenon of dominance, where one of the parental subgenomes is overexpressed relative to the other. This is the case with maize, which is descended from an allopolyploidization event estimated at 10 Mya and whose parental species are not known. The two subgenomes, *maize1* and *maize2*, contribute unequally to the maize transcriptome: ohnologous genes are expressed at a significantly higher level in the *maize1* subgenome (Renny-Byfield et al. 2017).

The mechanisms by which the dominance of one subgenome is exerted over the other are not fully understood, but could be related to the repression of transposable elements. Indeed, when the two subgenomes have accumulated different populations of transposable elements against which they have developed appropriate domestication strategies, bringing them together in a common nuclear environment can lead to a failure to suppress these transposable elements and to their activation – this is the "genomic shock" hypothesis, first proposed by Barbara McClintock (1984). Thus, new peaks of transposable element invasions have been observed after polyploidization events in natural allotetraploid coffee (Lopes et al. 2013), allotetraploid synthetic rapeseed (An et al. 2014), allotetraploid *Xenopus* (Session et al. 2016) and even in some autopolyploids such as rice (Zhang et al. 2015). These invasions are often resolved evolutionarily by hypermethylation of the transposable elements. These regulations of transposable elements can have a ricochet effect on the expression of neighboring genes by modifying their epigenetic context. It has been proposed that it is this defense mechanism against transposable elements that leads one of the parental subgenomes to become dominant over the other by reducing its expression by epigenetic regulation (Freeling et al. 2012). Note that although the two concepts are often grouped together under the generic term "genome dominance", subgenome dominance (also known as homeolog expression bias) should be distinguished from expression-level dominance (Grover et al. 2012). Expression-level dominance refers to cases where the transcriptomic profile of the genes of a polyploid more closely resembles that of one of its two parents. In contrast to subgenome dominance, which evaluates the bias in the contribution of the two subgenomes to gene expression in the polyploid, it is necessary to know the transcription profiles in the parent species of a polyploid in order to characterize the dominance of the expression level. The links between subgenome dominance and expression-level dominance are still poorly understood and may not follow the same direction. For example, cotton presents no subgenome dominance bias, but an expression-level dominance bias (Rapp et al. 2009; Yoo et al. 2013).

Lastly, allopolyploidizations can be the cause of conflicts between nuclear genomes and genomes of other organelles, chloroplasts and mitochondria. This effect has been observed in rapeseed, groundnut, tobacco and cotton (Gong et al. 2014), presumably driving the preferential expression of certain genes of the maternal genome or the conversion of paternal genes by the maternal copy in order to resolve these genetic conflicts.

1.3.3. *Unstable meiosis*

Whole-genome duplications pose a real challenge for sexually reproducing eukaryotic organisms at the moment of meiosis: indeed, the almost-complete redundancy of the genome means that chromosomes can establish crossing-overs[2] by homologous recombination with not one but three other chromosomes in the case of a tetraploidy, and more when the ploidy is higher (Stebbins 1947). These crossing-overs give rise to multivalents bringing together several homeologous chromosomes during the metaphase, instead of the bivalents normally observed (see Figure 1.4). The formation of these multivalents makes meiosis unstable: they prevent the normal, balanced segregation of chromosomes toward the cell poles at anaphase, inducing an increase in gametes carrying an abnormal chromosomal assortment (aneuploids), and leading to a decrease in fertility (Comai 2005).

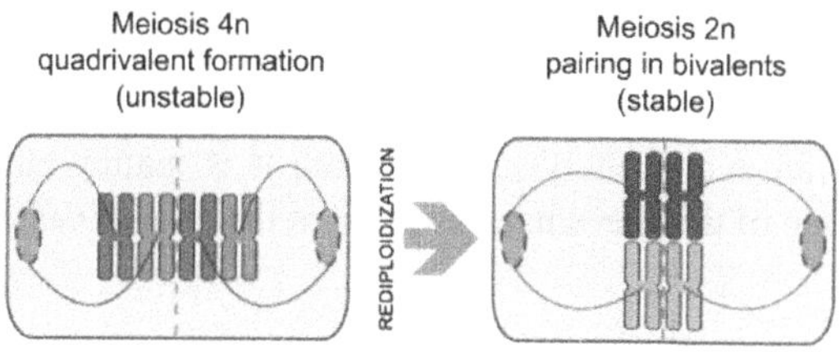

Figure 1.4. *Restoration of meiosis in polyploids. In polyploid species, homologous and homeologous chromosomes pair as multivalents on the metaphase plate during meiosis. These pairings are unstable and can lead to poor chromosome separation in the daughter cells, generating unbalanced gametes. During rediploidization, the divergence of homeologous chromosomes increases, once more allowing stable pairing of homologous chromosome pairs*

These difficulties are particularly marked in autopolyploids, where genetic redundancy is almost perfect. In allopolyploids, the existing genetic differences between the two parental subgenomes, derived from distinct species, in some cases enable the problem of meiosis to be circumvented: if the genetic divergence is sufficient, it allows for the preferential pairing of chromosomes derived from the same subgenome and the formation of normal bivalents. In certain cases of recent allopolyploids, the correct pairing of homologous chromosomes is controlled by specific genetic loci: this is the case in wheat, where the homeologs differ at the *Th1*

2 Crossing-overs are genetic structures established during meiosis, where homologous chromosomes exchange a pair of sister chromatids. These structures are essential for meiosis to occur correctly and for the separation of chromosomes during gametogenesis.

locus, which ensures almost-diploid meiosis despite the great similarity of the subgenomes (Griffiths et al. 2006). These difficulties in passing the meiosis stage have been proposed to be one of the major obstacles to the establishment of polyploid populations (Ramsey and Schemske 1998; Otto 2007), particularly in clades where the frequency of unreduced gametes is high and where the accidental formation of neo-polyploid individuals is non-negligible, as is the case in plants.

1.4. Rediploidization: evolutionary reduction in genetic redundancy

It is important to note that the complete, or almost complete, redundancy of the genome after a polyploidization event is not evolutionarily stable. When a polyploid population appears, its genome undergoes a phenomenon called “rediploidization” that begins soon after the whole-duplication event and continues for tens, if not hundreds, of millions of years. Rediploidization leads to the evolutionary divergence of the two subgenomes, and to the resolution of the majority of conflicts encountered by the neopolyploid individuals discussed above. Because rediploidization is an evolutionary phenomenon, it is mainly studied in polyploid individuals, but some of these events also apply in the case of cancer lineages.

1.4.1. *Resolution of meiosis by karyotype rearrangement*

As mentioned above, meiosis is a major obstacle for polyploid individuals. The restoration of normal diploid meiosis is considered to be one of the fundamental steps in the stabilization and longevity of individuals with a duplicated genome. One of the mechanisms involved in this resolution is the reorganization of the karyotype. This reorganization usually involves a wave of chromosomal rearrangements shortly after whole-genome duplication, including many chromosome fusions and translocations. Several studies have shown that the rate of genomic rearrangements increases after whole-genome duplication (Sémon and Wolfe 2007; Rajeh et al. 2018; Du et al. 2020) and suggested that this karyotype reorganization could be evolutionarily favored. By producing a reassortment of the duplicated regions on chromosomes, this reorganization will indeed enable the restoration of normal pairing of the chromosomes as bivalents at the moment of meiosis. The rainbow-trout genome is a good example for visualizing the effects of this reorganization: after the whole-genome duplication that appeared in the ancestor of salmonids about 100 million years ago, the genome reverted to a number of chromosomes almost identical to the pre-duplication ancestor by chromosome fusions, with homeologous chromosomes derived from the duplication each fusing with different chromosomes (Berthelot et al. 2014). These ancient chromosomes have thus become chromosome arms, whose structure is still visible in the genome of rainbow trout today (see Figure 1.5). This mechanism of chromosome fusions, which reduces the number of chromosomes present in the karyotype, is called descending dysploidy and has been frequently observed in ancient polyploids, suggesting that it is a general mechanism of rediploidization (Mandáková and Lysak 2018).

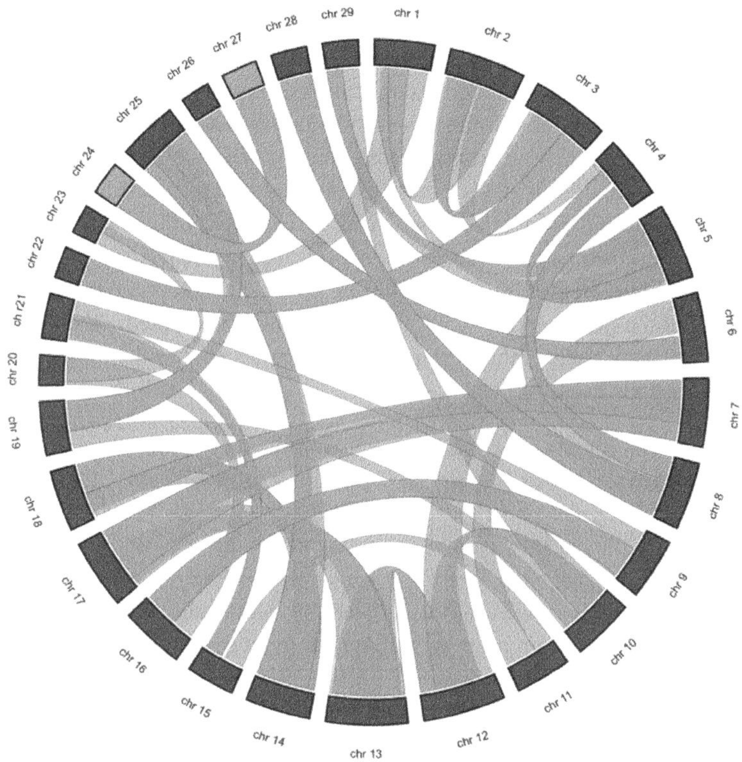

Figure 1.5. *Organization of homeologous regions in the genome of rainbow trout (Oncorhynchus mykiss). Each block around the outer edge of the circle represents a chromosome, and the links connect the homeologous regions derived from whole-genome duplication in salmonids, dated to about 100 Mya. The karyotype has undergone many rearrangements since whole-genome duplication and the majority of chromosomes share homeologous regions with several other chromosomes. An example of unrearranged homeologous chromosomes, chromosomes 24 and 27, is highlighted in pink*

These karyotypic reorganizations give the anciently duplicated genomes a typical structure, where, although the duplicated chromosomes are degraded, each region of the genome can be linked to a sister region with which it shares duplicated genes and homologous regions. This structure today remains among the most commonly used and definitive proof to establish the existence of an ancestral polyploidization: for example, it was decisive in confirming the existence of whole-genome duplication in the phylum of the yeast *Saccharomyces cerevisiae* (Kellis et al. 2004) and in that of teleost fish (Jaillon et al. 2004) during the first whole-genome sequencings. The presence of traces of such regions sharing duplicated gene families is also drawn upon by the main studies confirming that vertebrates are descended from two very ancient whole-genome duplications. Indeed, these duplications are so ancient (dated to about 550 million years ago) that the duplicated chromosome structure is almost entirely degraded in the genomes of extant vertebrates. Several works have demonstrated that vertebrate chromosomes can be grouped into tetrads sharing an unexpected number of duplicated genes and multigene families, which represent the remnants of the ancestral organization of the octoploid vertebrate genome (Dehal and Boore 2005; Sacerdot et al. 2018). Although the existence of these two whole-genome duplications is generally accepted, the mode and temporality of their appearance in the vertebrate clade remain the subject of lively debate (Smith et al. 2018).

1.4.2. *Evolutionary divergence of duplicated sequences*

As stable polyploid genomes evolve, duplicated sequences will also accumulate mutations, which will increase their divergence. This phenomenon has mainly been studied with respect to genes duplicated by polyploidization, called ohnologs in homage to Susumu Ohno, in order to distinguish them from paralogous genes in the general sense, derived from all types of gene duplications. In conjunction with the karyotype reorganization, the duplicated genome will generally be subject to a massive loss of genes: only part of the duplicated genes is maintained over long evolutionary distances (Sankoff et al. 2010; Inoue et al. 2015; Wendel et al. 2018). Two explanations have been put forward for this phenomenon, called gene fractionation: on the one hand, both copies of each gene produced by duplication are functionally redundant, and for the majority of genes, one of the copies is dispensable. One of the copies will therefore be able to accumulate deleterious mutations with no selective effect on the organism, since the function of the gene continues to be ensured by the other copy. Ultimately, the mutated copy is pseudogenized and lost, and the other copy remains the functional copy of the paleopolyploid genome, under selection pressure. This interpretation of neutral divergence draws on the fact that many polyploid genomes contain a high number of pseudogenes corresponding to ancient ohnologs deactivated by mutation (Dehal and Boore 2005; Berthelot et al. 2014; Elurbe et al. 2017; Mascagni et al. 2021).

On the other hand, it also appears that polyploid genomes rapidly lose part of their duplicated genes through deletions of genes or groups of genes along certain chromosomes (Byrnes et al. 2006; Sankoff et al. 2010), sometimes even through loss of entire chromosomal segments, as is the case in sturgeons (Du et al. 2020). It has been proposed that these losses of long sequence fragments contribute to the divergence of homeologous chromosomes, limiting the possibilities of homologous recombination between duplicated regions in addition to the karyotypic rearrangements discussed above. These deletions could therefore be favored by selection in favoring a return to normal meiosis in polyploid individuals.

Like karyotype rearrangements during rediploidization, gene fractionation results in a structure typical of anciently duplicated genomes, known as double-conserved synteny. Indeed, the genes of each ancestral region are found distributed over two genomic regions post-duplication: compared to an external species that has not undergone duplication, we observe a conservation of the synteny between species, that is, of the order and organization of the genes, but this syntenic organization is duplicated in polyploid species with a duplicated genome (see Figure 1.6). Only part of the genes are maintained in two ohnolog copies, the others having reverted to a single-copy state, or singleton, on one or the other of the homeologous regions.

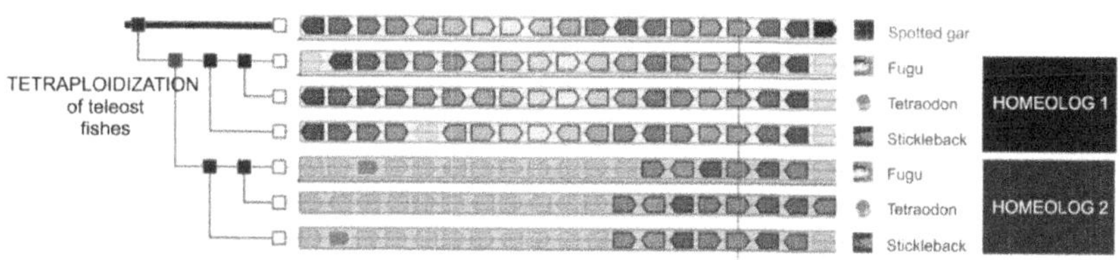

Figure 1.6. *Homeologous regions with double-conserved synteny in teleost fish genomes. Teleost fish underwent a very ancient genome tetraploidization, dated 320 million years ago, after their divergence from the group of holostean fish to which the spotted gar belongs. Each region of the teleost genome is therefore duplicated in relation to the genome of the gar. The duplication of the genome is schematically represented by a red square on the phylogenetic tree on the left, and the speciations of the different species are represented by blue squares. The colored boxes represent genes that are homologous between species. It can be noted that in this region, the majority of duplicated genes in teleosts were retained on homeologous chromosome 1 and lost on homeologous chromosome 2*

The mutational mechanisms of gene fractionation are still poorly known, in part because the majority of well-studied polyploidization events are ancient, and their rediploidization is almost complete. Several studies based on synthetic polyploids in plants have, however, demonstrated that mutations and deletions begin from the first generations, in connection with the unstable meiosis of these genomes (Song et al. 1995; Gaeta et al. 2007). This phenomenon then slows down but continues over a long evolutionary period: thus, in rainbow trout (duplication age: 100 million years), more than 50% of the genes are still present in two ohnolog copies (Berthelot et al. 2014). In zebrafish (duplication: 320 million years), it is estimated that about 25% of the genes are still duplicated (Howe et al. 2013).

1.4.3. *Bias and dominance during rediploidization*

Rediploidization and gene fractionation are not always balanced phenomena, equitably distributing the genes over one or the other of the duplicated genomic regions derived from polyploidization. An observation that can be made in many paleopolyploids is that gene fractionation tends to favor retention on one or the other of the homeologous chromosomes, sometimes still derived from one of the parental subgenomes: this is frequently the case in plants, where such bias has been observed in many examples such as *Arabidopsis suecica*, an allotetraploid derived from the hybridization of two species of rockcress (Chang et al. 2010) or maize (Schnable et al. 2011; Renny-Byfield et al. 2017). This phenomenon is also well documented in certain vertebrates such as carp (Xu et al. 2019), goldfish (Chen et al. 2019) and *Xenopus* (Session et al. 2016). The origin of this bias is poorly understood, but unequal gene fractionation appears more common in the case of allopolyploids, whose two subgenomes come from different species. In combination with the fact that one subgenome is often favored for gene expression in the case of neoallopolyploids, it has been proposed that this biased rediploidization reflects the dominance exerted by one of the subgenomes over the other. The molecular mechanisms of this dominance are poorly understood, but could again be related to the suppression of the expression of transposable elements that can create conflicts in allopolyploids, as mentioned above (Bottani et al. 2018). The most commonly accepted explanation is that reducing the expression of one of the subgenomes by epigenetic mechanisms and defending against the transposable elements that it comprises, allows this dominated subgenome to accumulate a high number of mutations without being subjected to negative selection pressure, including at gene sequence level. Because these gene copies are less expressed, the deleterious effects of mutations would be masked by the dominant copy expressed at a high level. This dynamic is potentially double-edged in its consequences: the dominated subgenome is likely to have more evolutionary plasticity and evolvability by relieving selection pressure on its sequences; but, in the long term, the accumulation of deleterious mutations can lead to the loss of genes and functional regions on this subgenome (Van de Peer et al. 2017).

It should be noted, however, that this phenomenon of dominance is not universal: several examples of allopolyploidy without dominance exist, particularly in squash (Sun et al. 2017) and plants of the genus *Ephedra* (Wu et al. 2021), where no gene expression or retention bias has been observed. While the presence of a retention bias is often used as proof that a paleopolyploidy event corresponds to an allopolyploidization, the absence of bias does not allow any definitive conclusion to be drawn as to the origin of parents of a whole-genome duplication.

1.4.4. *Incomplete and lineage-specific rediploidizations*

Lastly, rediploidization is a phenomenon that is not only continuous over time after the genome duplication event, but it is not necessarily always homogeneous and synchronous over the entire genome. Thus, certain regions of the genome can revert to an essentially diploid state very rapidly, while others remain almost entirely duplicated in a stable manner. An example of such stable regions are the HOX clusters in vertebrate genomes, where the four gene clusters derived from the two whole-duplication events that occurred 550 million years ago are maintained in an almost-whole state with the majority of genes conserved in order for the majority of vertebrates (Garcia-Fernàndez and Holland 1994). In certain clades, it is well established that rediploidization remains incomplete in certain regions because these regions maintain (or have long maintained) chromosome exchanges during meiosis: this is the case in salmon, where seven pairs of chromosomal regions derived from the salmonid whole-genome duplication (100 Mya) still recombine at a high frequency (Lien et al. 2011; Lien et al. 2016). These regions conserve a strong sequence identity and essentially still behave like tetraploid alleles in the genome of salmon and many other salmonids (Allendorf et al. 2015).

The prevalence of these phenomena of incomplete rediploidization in paleopolyploids is not known: it is possible that they are frequent, but apart from the case of salmonids, they remain little studied today. These observations have nevertheless made it possible to propose a more general model of rediploidization as a progressive process, which can occur in parallel with species evolution and in particular speciation. Thus, rediploidization and sequence divergence can occur independently in different subclades derived from a polyploidization event, if it is not already advanced when speciation occurs. This phenomenon, known as delayed rediploidization or lineage-specific ohnolog rediploidization events (LORe) (Robertson et al. 2017), introduces discrepancies in the evolutionary history of the sequences with respect to that of the genomic loci that carry them, since the sequences have continued to recombine and to behave as a single locus long after the duplication event (see Figure 1.7). It should be noted that ancestral and delayed rediploidizations can coexist within the same genome, with certain regions of the genome rediploidized rapidly and others later in evolution, as is the case in salmonids.

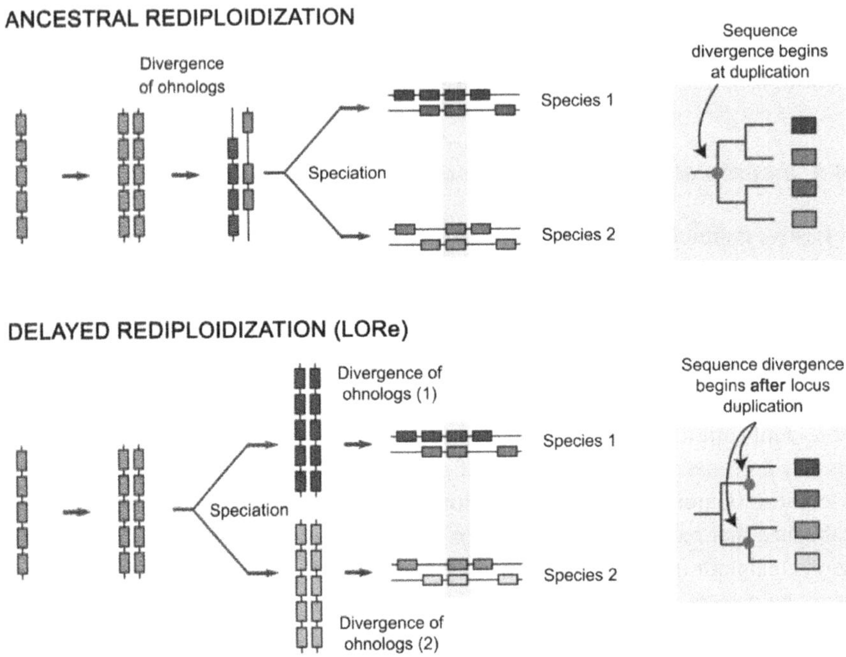

Figure 1.7. *Ancestral and delayed rediploidization. In the most classical case, rediploidization starts immediately after the polyploidization event, and the homeologous sequences begin to diverge rapidly. The descendant species therefore share a unique rediploidization. In the case of delayed rediploidization, the homeologous regions recombine and diverge little: speciation events can occur before rediploidization begins. Homeolog divergence then occurs independently in the descendant lineages. There is then a discrepancy between the history of the locus (duplicated at the moment of polyploidization) and of the sequence (which behaves as a duplicated sequence from rediploidization only), such that strictly speaking there are no paralogous or orthologous genes between the species of the different descendant clades*

1.5. Functions and evolution of duplicated genes

Despite rediploidization and the loss of part of the redundancy induced by whole-genome duplications, many genes remain conserved in duplicated copies over long evolutionary distances in ancient polyploids. These ohnologous genes present functional characteristics that distinguish them both from genes that have reverted to the singleton state, but also from paralogous genes inherited from other types of duplication events. They are crucial contributors to the complexification of the gene repertoire and to the appearance of evolutionary innovations in paleopolyploid clades.

1.5.1. *Redundancy and subfunctionalization*

At the moment of the polyploidization event, the copies of duplicated genes carried by the homeologous chromosomes are generally functionally redundant: they fulfill the same functions and are expressed in the same tissues, aside from divergences that may have been adopted by the genomes of two parents in the case of allopolyploidy. This functional redundancy can be selectively advantageous: this is notably the case for many essential genes, where one of the ohnologs can function as a backup copy preserving the ancestral function in the case where the other copy accumulates deleterious alleles (Makino et al. 2009). This redundancy can also be selected where increasing the gene expression dosage is advantageous because adding the expression of the two copies can provide an immediate evolutionary advantage to the organism. For example, in yeasts of the genus *Saccharomyces*, many genes involved in glycolysis were maintained in two copies, giving them faster growth on a glucose-rich medium (Conant and Wolfe 2007). Similarly, in *Arabidopsis thaliana,* many genes maintained duplicated after the "beta" paleopolyploidy (about 80 Mya) are involved in high-flux metabolic reactions, for which having multiple copies enables this flow to be accelerated and represents a selective advantage (Bekaert et al. 2011).

However, the preferred pathway when the two ohnolog copies are maintained appears to be that of subfunctionalization (Lynch and Force 2000). This is a generally neutral evolutionary process in which both copies of the gene are subject to functional divergence. Indeed, the majority of genes fulfill several functions, either by participating in several cellular processes or by being expressed in several tissues. The ohnolog copies can evolve in a specialized manner, each retaining only a part of the ancestral functions. This evolutionary model has been formalized by Force et al. (1999) and Stoltzfus (1999) as a process of duplication–degeneration–complementation (see Figure 1.8): under this model, each gene copy can tolerate mutations in its protein sequence (Henikoff et al. 1997) or in its regulatory sequences (Kleinjan et al. 2008), as long as the entirety of the ancestral functions remain assured by the sum of the two copies. This is the case, for example, of the genes *pax6a* and *pax6b* in zebrafish, two copies derived from the teleost-specific whole-genome duplication, expressed in different embryonic structures after accumulation of mutations in their regulatory sequences, until reaching functional complementation, where the sum of the two copies recapitulates the expression of the ancestral gene (Kleinjan et al. 2008). This joint degeneration of the two duplicated genes leads to a situation where both copies become necessary in order to fulfill the entirety of the functions of the gene (Gout and Lynch 2015; De Smet et al. 2017). This complementation can extend to complete specialization, where the functions of the two copies are no longer redundant, and where, for example, each have their own territory of expression separate from that of the other gene.

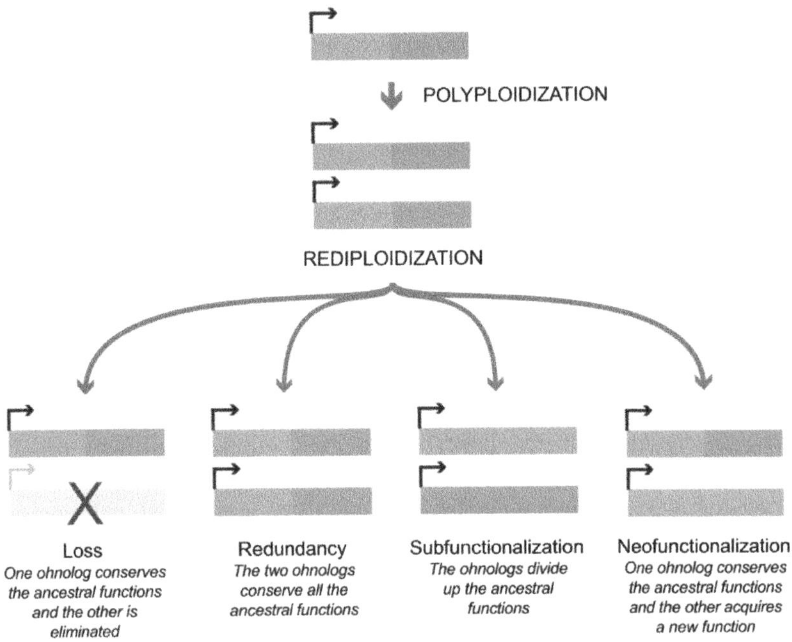

Figure 1.8. *Outcome of duplicated genes after a polyploidization event. During rediploidization, ohnologous genes can either revert to a simple copy state, where one of the ohnologs is lost, or be maintained in two copies with different functional fates*

1.5.2. *Neofunctionalization and evolutionary innovations*

Another possible evolutionary fate of ohnologous genes is the adoption by one of the copies of a new function that did not exist in the ancestor or neofunctionalization (Ohno 1970). Indeed, as we have already mentioned, the functional redundancy induced by duplication has the consequence of relieving the selection pressure on one or the other of the copies. In certain cases, one of the copies of the gene will thus be free to accumulate mutations and thus increase the evolvability of this gene, which will be able to explore new combinations of mutations and ultimately evolve toward new functions beneficial to the organism (Byrne and Wolfe 2007; Tirosh and Barkai 2007; Kassahn et al. 2009). Unlike subfunctionalization, neofunctionalization begins with a phase of neutral evolution but is then an adaptive phenomenon, where positive selection for the new gene function plays an important role in the establishment of this function. In the case of alopolyploid genomes with genomic dominance, it has been proposed that neofunctionalization primarily affects the dominated subgenome, whose expression is repressed, because the mutations would have less functional impact on this subgenome (Wendel et al. 2018). The acquisition of new adaptive functions in the dominated subgenome could be one of the important mechanisms for maintaining paleopolyploidy in species where the contribution of the two subgenomes to gene expression is biased.

Neofunctionalization processes increase the evolutionary potentialities of polyploids and are the source of significant evolutionary innovations in these clades, which have undoubtedly been decisive for their evolutionary success despite the transmission difficulties posed by highly redundant genomes. Thus, several whole-duplication events at the evolutionary origin of flowering plants were decisive in expanding the repertoire of *MADS-box*-type transcription factors (Airoldi and Davies 2012; Panchy et al. 2016), which control the development of the different flower elements and are one of the best-described examples of neofunctionalization. Other notable examples include the diversification of developmental gene repertoires in vertebrates, such as the *Hox*, *Cdx ParaHox* and *Pax* genes (Amores et al. 1998; Kleinjan et al. 2008; Marlétaz et al. 2018), which control the establishment of the organizational plan of the body, and for which the acquisition of new copies during the whole-genome duplications 1R and 2R likely played a crucial role in the appearance of new vertebrate-specific structures during embryonic development.

The relative contributions of neofunctionalization and subfunctionalization to the maintenance and specialization of duplicated genes derived from polyploidization events remain hotly debated (Sandve et al. 2018). Indeed, for many genes derived from very ancient duplications, ancestral functions are not known and it remains difficult to establish whether the functions and territories of expression observed in paleopolyploid species come under sub- or neo-functionalization (see Figure 1.9). It is likely, however, that the evolution of the majority of ohnologs is in between the two, and some authors have proposed that neofunctionalization is the inevitable fate of ohnologous genes maintained over long evolutionary distances, with subfunctionalization only a transitional state along the way (Rastogi and Liberles 2005).

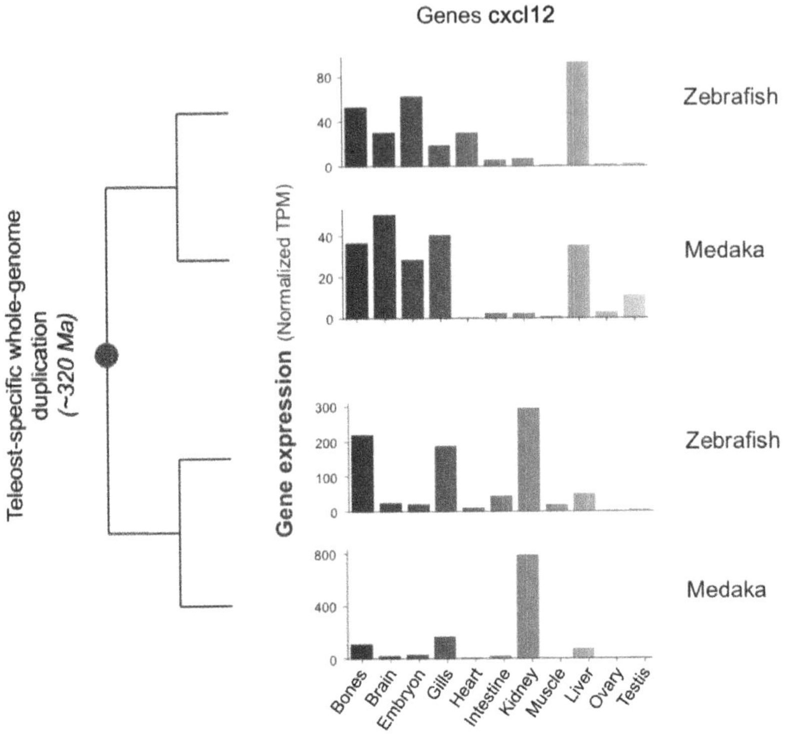

Figure 1.9. *Example of ohnologous genes with divergent territories of expression. The genes of the Cxcl12 family are present in two ohnolog copies derived from tetraploidization of teleost fish, dated to 320 Mya, such as zebrafish and medaka. The two ohnolog copies of the gene diverged in expression and acquired different territories of expression, with one copy expressed in the bones, brain, embryo, gills and liver, and the other copy strongly expressed in the bones, gills and kidney. In the absence of data on the ancestral territory of expression of Cxcl12, it is not possible to say whether this pattern of expression is a case of sub- or neofunctionalization*

1.5.3. *Gene repertoire bias*

In paleopolyploid genomes, many biases are observed among genes whose ohnolog copies are retained. Compared to genes that have returned to single-copy state, ohnologs are notably enriched in certain functional categories such as transcription factors, genes involved in development, involved in diseases, and linked to the response to the environment, in particular immunity (Papp et al. 2003; Blanc and Wolfe 2004; Makino et al. 2009; Makino and McLysaght 2010). Genes for which one of the copies was lost during fractionation, on the other hand, are enriched with housekeeping genes, basal metabolic genes and deeply conserved cellular functions. These functional biases are observed fairly homogeneously both in plants and vertebrates, and over the course of whole-genome duplications when they follow one another in the same clades. Thus, certain functional families are amplified and retained recurrently by successive polyploidizations, as is the case in vertebrates, where we observe a significant excess of gene families that have retained the two ohnolog copies after each of the duplication events 1R, 2R (at the stem of vertebrates, around 550 Mya) and 3R (specific to teleost fish, around 320 Mya) (Howe et al. 2013). In the human genome, biased retentions have participated in the expansion of many multigene families such as *Fox* factors, which are involved in many aspects of development, and MHC-like genes, involved in immunity (Blomme et al. 2006).

Ohnolog retention biases do not stop at the functions fulfilled by genes. It has been observed that ohnologs are more frequently involved in protein complexes and interactions than genes that have reverted to the singleton state during rediploidization, but also more than the paralogous genes derived from other duplication mechanisms such as segmental duplications or gene retrocopies mediated by reverse transcriptases. The most commonly accepted explanation is that these genes involved in protein complexes are frequently subject to dosage balance (Makino and McLysaght 2010; McLysaght et al. 2014). As mentioned above, many genes are subject to selection pressure on their level of expression, which must respect the stoichiometry of other cellular compounds or other genes. In the case of protein complexes, the balance of expression of the different subparts of the complex is important: if one of the elements of the complex is over- or under-expressed in relation to the others, part of the protein complexes formed is incomplete through non-availability (see Figure 1.10). This dosage imbalance is generally deleterious and therefore counter-selected during evolution (Rice and McLysaght 2017). In the case of polyploidization, all genes involved in the formation of the complex are duplicated simultaneously, so that the dosage balance between the subparts of the complex is respected. Conversely, segmental and retrocopy duplications amplify only one gene in the complex at a time and are typically lost by negative selection. Whole-genome duplications would therefore be a preferred mechanism for amplification of genes that produce protein complexes, which can be recruited for new functions by neofunctionalization. This model predicts that when one of the genes involved in the complex returns to the singleton state by fractionation, a selection pressure is exerted on all of its partners to also lose one of their ohnolog copies and restore the dosage balance.

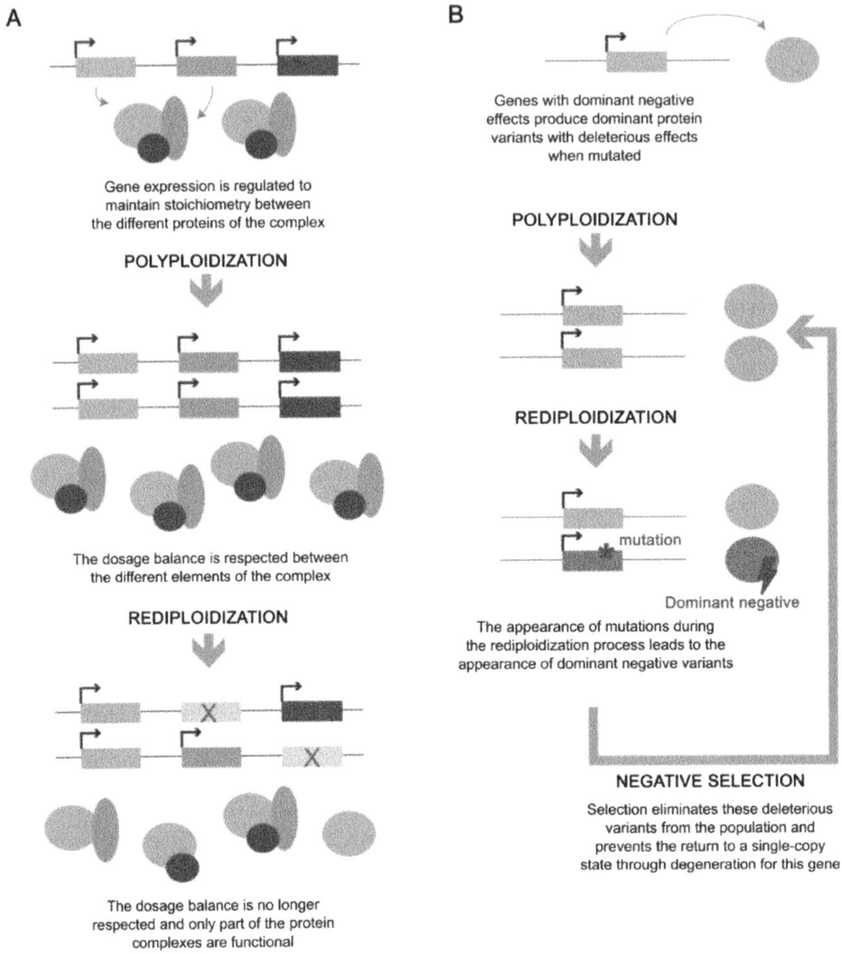

Figure 1.10. *Preferential retention mechanisms of ohnolog copies of genes. The figure details two examples of ohnologous gene retention: genes subject to dosage balance because they are involved in protein complexes and a gene with a dominant negative mutation*

Furthermore, the retained ohnologous genes are generally biased in terms of expression levels, but these effects differ according to clades, whole-duplication events and paleopolyploidization age. In the paramecium *Paramecium tetraurelia*, highly expressed genes were more frequently retained in ohnologous pairs than the genes with lower expression over the three whole-genome duplication events that affected the evolution of this species (Gout et al. 2010). In salmonids, on the other hand, on average ohnologs have lower expression levels and have frequently undergone downward modifications in their expression compared to the ancestral expression level inferred from data from external groups (Lien et al. 2016; Varadharajan et al. 2018; Gillard et al. 2020). All of these trends have also been interpreted as diverse repercussions of dosage balance: because the consequences of a sudden imbalance in expression levels are more profound for highly expressed genes, these genes would tend to be retained in two copies, and also to be regulated in order to decrease the sum of their expression toward a value closer to the ancestral value.

Lastly, it has also been observed that genes involved in human diseases, genes with dominant negative effects (dominant when mutated), and especially oncogenes, are overrepresented among genes retained in several ohnolog copies since duplications 1R and 2R of vertebrates (Singh et al. 2012). This maintaining is surprising, because it leads to the amplification of genes likely to have deleterious effects by simple mutation in the human gene repertoire, whereas evolutionary theories would instead predict their elimination. Although dosage balance has been put forward as an explanation for this maintaining, a second hypothesis has been proposed that is based on the so-called “dangerous gene” theory originally offered by Gibson and Spring (1998). This hypothesis suggests that these genes are maintained because the accumulation of mutations in their sequences is not tolerable by evolution, as their deleterious effects are so strong. These genes can therefore only be eliminated by the whole deletion of their sequence, and cannot be inactivated by mutations and transition to the pseudogene state, as is the case with a large part of the genes returning to the singleton state. Thus, paradoxically, these genes remain conserved over large evolutionary distances despite their cost to the organism.

1.5.4. *Regulatory blocks and splitting of regulatory regions*

While many studies have looked at the evolution of ohnologous genes after whole-genome duplication, the evolution of the non-coding genome has been far less studied. In particular, the regions regulating gene expression are also duplicated with their target gene during polyploidization, but their evolutionary mechanisms are less well understood. Many works have suggested that these sequences generally evolve faster than the genes themselves (Schmidt et al. 2010; Villar et al. 2015). After a genome duplication, it appears that the *cis*-regulatory regions of genes actively participate in the functional divergence of copies, by accumulating mutations that modify gene expression patterns, but also by recruiting new regulatory sequences around sub- and neo-functionalized genes (Kleinjan et al. 2008; Shi et al. 2012; Hu and Wendel 2019). Thus, a comparison of the genomes of several vertebrates with the amphioxus, a chordate not having been subject to the whole-genome duplications specific to vertebrates, revealed that genes having diverged strongly in expression after polyploidization concentrated a greater number of active non-coding regions in their vicinity, suggesting that the adoption of new regulatory sequences played a role in the functional evolution of these genes (Marlétaz et al. 2018).

A notable consequence of gene rediploidization is the unnesting of the regulatory networks of certain genes with complex regulatory landscapes. Indeed, it is now well demonstrated that certain genes, notably involved in development, are surrounded by many non-coding regulatory elements, whose sequence is often evolutionarily conserved, and whose contribution is important for the fine regulation of their expression (Engström et al. 2007; Kikuta et al. 2007). These regulatory elements can be located several hundred or even thousands of kilobases from their target gene, often in the introns of neighboring genes, and can be separated from their target by several genes, called “bystander genes”. The selection pressure exerted to maintain these regulatory sequences in the vicinity of their target gene preserves the configuration and order of the set of bystander genes and of the target gene, as the chromosomal rearrangements that would separate these genes would disrupt the regulatory landscape of the target gene by distancing its regulatory sequences (Irimia et al. 2013). These sets of genes thus remain conserved in synteny over large evolutionary distances and are described as genomic regulatory blocks. After a whole-duplication event, the presence of two copies of the regulatory block allows relaxation of the stresses on the bystander genes, which can be rearranged or subjected to an alternate rediploidization, where one of the homeologous regions retains the highly regulated gene and its regulatory sequences, while the other retains only bystander genes (see Figure 1.11) (Kikuta et al. 2007; Akalin et al. 2009). This characteristic of duplicated genomes has notably been used to document developmental gene regulatory landscapes in humans by studying how non-coding elements were retained on the zebrafish genome, which was duplicated about 320 My ago (Naville et al. 2015; Clément et al. 2020). Indeed, many non-coding elements with the sequence conserved but with poorly defined functions in vertebrates are retained only on one of the duplicated regions in teleost fish: the joint retention of surrounding genes, often involved in embryonic development, suggests these genes are likely targets of these conserved non-coding elements. It has also been proposed that these mechanisms for unnesting gene regulatory blocks are at the origin of large regions devoid of genes observed around developmental genes in vertebrate genomes, known as "gene deserts", which can reach several megabases in length and whose evolutionary origin remains debated (Touceda-Suárez et al. 2020).

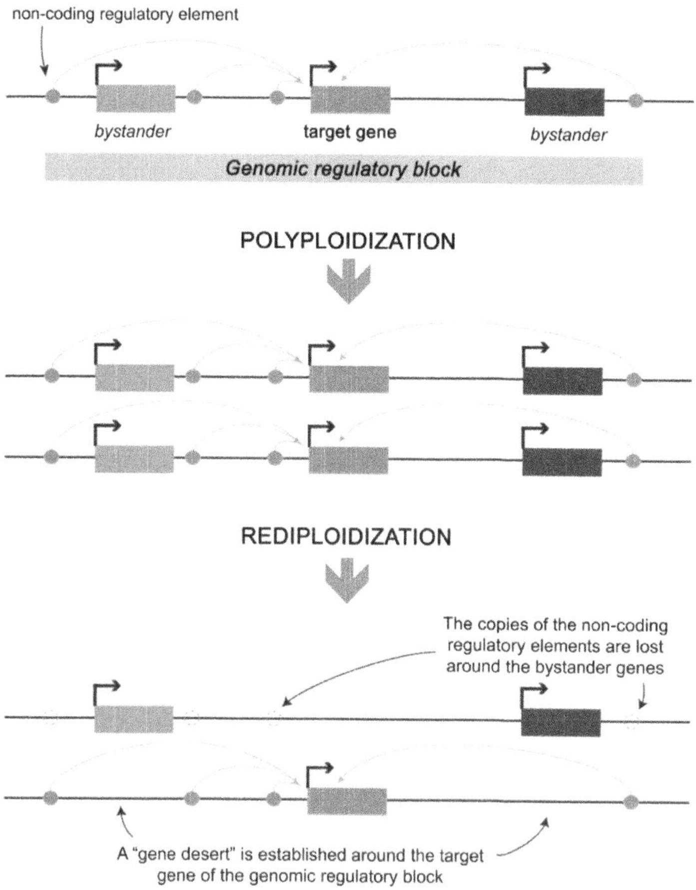

Figure 1.11. *Disentangling of a gene regulatory block during rediploidization. Gene regulatory blocks correspond to a set of genes that are maintained in order, conserved between species because of the presence of regulatory elements nested in this gene set. These non-coding elements typically regulate a single gene in the block, known as the "target gene". The other genes ("bystanders") are simply maintained in the block by negative selection against rearrangements between the regulatory elements and the target gene. After polyploidization, the selection pressure is lifted on one of the homeologous regions, which can promote the separation of bystander genes on a homeolog free to rearrange, and the conservation of the target gene and its regulatory elements on the other homeolog. In this case, the homeolog carrying the target gene can become a gene desert*

1.6. Whole-genome duplications and evolutionary diversification

Due to their contribution to genome diversification, polyploidizations have long been perceived as a major mechanism for diversifying phenotypes and species. But the link between genomic modifications and their consequences for adaptation, the acquisition of evolutionary innovations and speciation nevertheless often remains difficult to test. Here, we discuss the main avenues raised by the literature, indicating that despite the challenges that whole-genome duplications pose to organisms, they have undoubtedly been crucial for the evolutionary success of the anciently polyploid clades in which they occurred.

1.6.1. *Association with geological crises*

Beyond the taxonomic bias of whole-genome duplications in the eukaryotic tree of life, a temporal bias has also been noted and discussed in the literature (see Figure 1.1). Many polyploidization events appear to correlate with major geological crises. This observation is mainly based on the distribution of genome duplications in plants, where the number of events is sufficient to observe trends. In particular, a high number of duplications are clustered at the Cretaceous-Paleogene boundary (Fawcett et al. 2009; Vanneste et al. 2014), a period of mass extinction dated to 66 million years ago and at the origin of the extinction of the dinosaurs; at the Paleocene-Eocene boundary (Cai et al. 2019), an environmental crisis dated to 55 million years ago marked by the most sudden and significant global warming of the Cenozoic era; and at the end of Miocene 5–10 million years ago (Estep et al. 2014), a period characterized by a drop in the atmospheric CO_2 level. Different arguments relating to the environmental and mutational robustness of polyploid organisms propose a causative link between duplication and increased chances of survival under unstable conditions.

As developed above, genome duplications can generate functional innovations through the neofunctionalization of one of the copies of a duplicated gene (Ohno 1970). These additional gene copies would allow the polyploids to build up an "adaptive reserve", offering them greater environmental robustness. The high adaptive potential of polyploids is supported by the theory of population genetics. Put simply, the rate of appearance of beneficial mutations in a population can be considered to be proportional to the rate of beneficial mutations per gene (i), to the number of genes (ii), and to the size of the population (iii). Thus, with twice as many copies, the rate of appearance of beneficial mutations would be roughly twice as high in a tetraploid population as in a diploid population (Otto 2007).

In addition, the allelic redundancy generated by duplication gives the polyploids greater mutational robustness. Indeed, the presence of a higher number of alleles can dilute the expression of deleterious alleles. This effect of masking the expression of deleterious alleles is particularly decisive in the case of cancers, upon the appearance of "mutator" phenotypes, where the accumulation of aberrant mutations should theoretically jeopardize cell viability (López et al. 2020). The masking of recessive mutations also favors the appearance and fixation of dominant mutations increasing fitness, and in particular, in the case of cancers, those involved in cell proliferation. In polyploid species, the high copy number decreases the probability of homozygous deleterious alleles and thus their expression, which could contribute to the evolutionary success of these individuals (Gu et al. 2003; Crow and Wagner 2006). In a period of major environmental changes, this robustness can be a double-edged sword: it theoretically allows organisms to tolerate higher rates of mutations, but it can also mask the effect of beneficial mutations and thus slow or prevent their fixation (Otto 2007).

The hypothesis of a greater adaptive potential of polyploids is supported by in silico evolution experiments. The advantage of such evolutionary simulations is that they make it possible to test hypotheses by generating a large amount of evolutionary trajectories that are otherwise difficult to obtain accurately from real data. These simulations showed that digital organisms were predominantly diploid under stable conditions and that more and more polyploids appeared when confronted with major environmental changes (Yao et al. 2019).

1.6.2. *Evolutionary speciations and radiations*

Certain clades marked by a whole-genome duplication at their stem have experienced remarkable evolutionary success and diversification: this is the case, in particular, of vertebrates, but also of teleost fish, which account for more than 50% of vertebrates, and angiosperms, which form 90% of terrestrial plants. Conversely, certain episodes of duplication do not seem to have led to a diversification of descendants, such as those of horseshoe crabs (Kenny et al. 2016), represented by only four extant species. In addition, it is estimated that the number of neopolyploidations in plants is significantly greater than the number of paleopolyploidy events (Van de Peer et al. 2017). This result suggests that in most cases, whole-genome duplications do not allow for long-term evolutionary success. These different observations raise the question of the link between whole-genome duplication and diversification of species.

No demonstrated direct correlation exists between whole-genome duplication and increased rates of species diversification. Even when they become established in the long term, whole-genome duplications do not seem to induce sudden evolutionary radiation. Nevertheless, it is proposed that the effect of duplications on diversification rates would be separated by a latency time, which can last for up to tens or even hundreds of millions of years. This decoupling between duplication and diversification, formalized in the "radiation time-lag model" (Schranz et al. 2012), explains the evolutionary trajectories of species after different duplications in plants (Tank et al. 2015; Soltis and Soltis 2016; Walden et al. 2020). Importantly, lineage-specific rediploidization represents a mechanism that can explain the observed lag between whole-duplications events and the evolutionary success of descendant clades. By maintaining genomic regions in tetraploid state over long evolutionary periods, adaptive potential is "stored" in genomes and can be expressed as a function of environmental conditions. This mechanism has been proposed to explain the evolutionary radiation of salmonids, delayed to 50 million years after their duplication (Robertson et al. 2017).

The appearance of genetic incompatibilities in polyploid populations during the genome rediploidization process could also promote speciation. Losses of differential genes, also called reciprocal losses, correspond to the loss of distinct homeologous copies of a duplicated gene in different individuals. Studies in *Arabidopsis* (Bikard et al. 2009) and in experimentally evolved yeast populations (Maclean and Greig 2011) have shown that these losses can lead to the establishment of a reproductive barrier within a polyploid population. Indeed, while two subpopulations have each kept a single copy of the gene on different homeologous chromosomes, a quarter of the hybrid descendants within these populations inherit no copy of the gene in question. If the gene is essential, then the hybrid formed is not viable. In yeasts, this phenomenon has been linked to speciation events after their whole-genome duplication (Scannell et al. 2006). However, it is estimated that their role in the diversification of species remains minor (Muir and Hahn 2015).

1.7. Perspectives and conclusions

With the advent of the genome sequencing era, whole-genome duplications have moved in a few years from an evolutionary phenomenon considered dramatic but rare, to a common, central mechanism of species evolution. While it was initially well described in plants, we now know that this mechanism affects all phyla and is at the origin of major evolutionary innovations. Polyploids remain difficult organisms to study from a genomic perspective. When these polyploids are recent, the high redundancy of their genomic content makes their sequencing difficult: homeologous regions then resemble alleles and are often difficult to differentiate from true homologs, leading to a collapse during sequence assembly, where two distinct genomic regions are assembled into a single sequence. Conversely, when these polyploids are ancient, it is often difficult to trace the evolutionary history of homeologous regions because of the numerous genomic rearrangements and the sometimes rapid divergence of these sequences after polyploidization. As a result, many aspects of the evolution of polyploid genomes still elude us, and the study of whole-genome duplications remains an active area of research. The recent development of major international biodiversity sequencing projects, such as the Earth Biogenome Project and the Darwin Tree of Life, will contribute significantly to our understanding of polyploid species, their prevalence in less-documented living groups, and the diversity of modes of evolution, speciation and radiation resulting from the tremendous genomic variability of which whole-genome duplications are the source.

1.8. References

Adams, K.L. and Wendel, J.F. (2005). Novel patterns of gene expression in polyploid plants. *Trends in Genetics*, 21(10), 539–543.

Adams, K.L., Cronn, R., Percifield, R., Wendel, J.F. (2003). Genes duplicated by polyploidy show unequal contributions to the transcriptome and organ-specific reciprocal silencing. *Proceedings of the National Academy of Sciences of the United States of America*, 100(8), 4649–4654.

Airoldi, C.A. and Davies, B. (2012). Gene duplication and the evolution of plant MADS-box transcription factors. *Journal of Genetics and Genomics*, 39(4), 157–165.

Akalin, A., Fredman, D., Arner, E., Dong, X., Bryne, J.C., Suzuki, H., Daub, C.O., Hayashizaki, Y., Lenhard, B. (2009). Transcriptional features of genomic regulatory blocks. *Genome Biology*, 10(4), R38.

Allendorf, F.W., Bassham, S., Cresko, W.A., Limborg, M.T., Seeb, L.W., Seeb, J.E. (2015). Effects of crossovers between homeologs on inheritance and population genomics in polyploid-derived salmonid fishes. *Journal of Heredity*, 106(3), 217–227.

Amores, A., Force, A., Yan, Y.-L., Joly, L., Amemiya, C., Fritz, A., Ho, R.K., Langeland, J., Prince, V., Wang, Y.-L., et al. (1998). Zebrafish hox clusters and vertebrate genome evolution. *Science*, 282(5394), 1711–1714.

An, Z., Tang, Z., Ma, B., Mason, A.S., Guo, Y., Yin, J., Gao, C., Wei, L., Li, J., Fu, D. (2014). Transposon variation by order during allopolyploidisation between *Brassica oleracea* and *Brassica rapa*. *Plant Biology (Stuttgart, Germany)*, 16(4), 825–835.

Aury, J.-M., Jaillon, O., Duret, L., Noel, B., Jubin, C., Porcel, B.M., Ségurens, B., Daubin, V., Anthouard, V., Aiach, N., et al. (2006). Global trends of whole-genome duplications revealed by the ciliate *Paramecium tetraurelia*. *Nature*, 444(7116), 171–178. DOI: 10.1038/nature05230.

Ballesteros, J.A. and Sharma, P.P. (2019). A critical appraisal of the placement of Xiphosura (Chelicerata) with account of known sources of phylogenetic error. *Systematic Biology*, 68(6), 896–917.

Barker, M.S., Arrigo, N., Baniaga, A.E., Li, Z., Levin, D.A. (2016). On the relative abundance of autopolyploids and allopolyploids. *New Phytologist*, 210(2), 391–398.

Bekaert, M., Edger, P.P., Pires, J.C., Conant, G.C. (2011). Two-phase resolution of polyploidy in the arabidopsis metabolic network gives rise to relative and absolute dosage constraints. *The Plant Cell*, 23(5), 1719–1728.

Berthelot, C., Brunet, F., Chalopin, D., Juanchich, A., Bernard, M., Noël, B., Bento, P., Da Silva, C., Labadie, K., Alberti, A., et al. (2014). The rainbow trout genome provides novel insights into evolution after whole-genome duplication in vertebrates. *Nat. Commun.*, 5(1), 1–10. DOI: 10.1038/ncomms4657.

Bielski, C.M., Zehir, A., Penson, A.V., Donoghue, M.T.A., Chatila, W., Armenia, J., Chang, M.T., Schram, A.M., Jonsson, P., Bandlamudi, C., et al. (2018). Genome doubling shapes the evolution and prognosis of advanced cancers. *Nat. Genet.*, 50, 1189–1195. DOI: 10.1038/s41588-018-0165-1.

Bikard, D., Patel, D., Metté, C.L., Giorgi, V., Camilleri, C., Bennett, M.J., Loudet, O. (2009). Divergent evolution of duplicate genes leads to genetic incompatibilities within *A. thaliana*. *Science*, 323(5914), 623–626.

Blanc, G. and Wolfe, K.H. (2004). Functional divergence of duplicated genes formed by polyploidy during arabidopsis evolution. *The Plant Cell*, 16(7), 1679–1691.

Blomme, T., Vandepoele, K., Bodt, S.D., Simillion, C., Maere, S., de Peer, Y.V. (2006). The gain and loss of genes during 600 million years of vertebrate evolution. *Genome Biology*, 7(5), R43.

Bottani, S., Zabet, N.R., Wendel, J.F., Veitia, R.A. (2018). Gene expression dominance in allopolyploids: Hypotheses and models. *Trends in Plant Science*, 23(5), 393–402.

Byrne, K.P. and Wolfe, K.H. (2007). Consistent patterns of rate asymmetry and gene loss indicate widespread neofunctionalization of yeast genes after whole-genome duplication. *Genetics*, 175(3), 1341–1350.

Byrnes, J.K., Morris, G.P., Li, W.-H. (2006). Reorganization of adjacent gene relationships in yeast genomes by whole-genome duplication and gene deletion. *Molecular Biology and Evolution*, 23(6), 1136–1143.

Cai, L., Xi, Z., Amorim, A.M., Sugumaran, M., Rest, J.S., Liu, L., Davis, C.C. (2019). Widespread ancient whole-genome duplications in Malpighiales coincide with Eocene global climatic upheaval. *New Phytologist*, 221(1), 565–576.

Cavalier-Smith, T. (1978). Nuclear volume control by nucleoskeletal DNA, selection for cell volume and cell growth rate, and the solution of the DNA C-value paradox. *Journal of Cell Science*, 34(1), 247–278.

Chang, P.L., Dilkes, B.P., McMahon, M., Comai, L., Nuzhdin, S.V. (2010). Homoeolog-specific retention and use in allotetraploid *Arabidopsis suecica* depends on parent of origin and network partners. *Genome Biology*, 11(12), R125.

Chen, Z., Omori, Y., Koren, S., Shirokiya, T., Kuroda, T., Miyamoto, A., Wada, H., Fujiyama, A., Toyoda, A., Zhang, S., et al. (2019). De novo assembly of the goldfish (*Carassius auratus*) genome and the evolution of genes after whole-genome duplication. *Science Advances*, 5(6), eaav0547.

Clark, J.W. and Donoghue, P.C.J. (2018). Whole-genome duplication and plant macroevolution. *Trends in Plant Science*, 23(10), 933–945.

Clément, Y., Torbey, P., Gilardi-Hebenstreit, P., Crollius, H.R. (2020). Enhancer–gene maps in the human and zebrafish genomes using evolutionary linkage conservation. *Nucleic Acids Research*, 48(5), 2357–2371.

Comai, L. (2005). The advantages and disadvantages of being polyploid. *Nature Reviews Genetics*, 6(11), 836–846.

Conant, G.C. and Wolfe, K.H. (2007). Increased glycolytic flux as an outcome of whole-genome duplication in yeast. *Molecular Systems Biology*, 3 [Online]. Available at: https://www.ncbi.nlm.nih.gov/pmc/articles/PMC1943425/ [Accessed 2 April 2021].

Crow, K.D. and Wagner, G.P. (2006). What is the role of genome duplication in the evolution of complexity and diversity? *Molecular Biology and Evolution*, 23(5), 887–892.

D'Amato, F. (1984). Role of polyploidy in reproductive organs and tissues. In *Embryology of Angiosperms*, Johri, B.M. (ed.). Springer, Berlin, Heidelberg. DOI.org/10.1007/978-3-642-69302-1_11.

De Smet, R., Sabaghian, E., Li, Z., Saeys, Y., Van de Peer, Y. (2017). Coordinated functional divergence of genes after genome duplication in *Arabidopsis thaliana*. *The Plant Cell*, 29(11), 2786–2800.

Dehal, P. and Boore, J.L. (2005). Two rounds of whole genome duplication in the ancestral vertebrate. *PLOS Biology*, 3(10), e314.

D'Hont, A., Denoeud, F., Aury, J.-M., Baurens, F.-C., Carreel, F., Garsmeur, O., Noel, B., Bocs, S., Droc, G., Rouard, M., et al. (2012). The banana (*Musa acuminata*) genome and the evolution of monocotyledonous plants. *Nature*, 488, 213–217. DOI: 10.1038/nature11241.

Du, K., Stöck, M., Kneitz, S., Klopp, C., Woltering, J.M., Adolfi, M.C., Feron, R., Prokopov, D., Makunin, A., Kichigin, I., et al. (2020). The sterlet sturgeon genome sequence and the mechanisms of segmental rediploidization. *Nat. Ecol. Evol.*, 1–12. DOI: 10.1038/s41559-020-1166-x.

Dujon, B., Sherman, D., Fischer, G., Durrens, P., Casaregola, S., Lafontaine, I., de Montigny, J., Marck, C., Neuvéglise, C., Talla, E., et al. (2004). Genome evolution in yeasts. *Nature*, 430, 35–44. DOI: 10.1038/nature02579.

Edgar, B.A., Zielke, N., Gutierrez, C. (2014). Endocycles: A recurrent evolutionary innovation for post-mitotic cell growth. *Nature Reviews Molecular Cell Biology*, 15(3), 197–210.

Egozcue, S., Blanco, J., Vidal, F., Egozcue, J. (2002). Diploid sperm and the origin of triploidy. *Human Reproduction (Oxford, England)*, 17(1), 5–7.

Elurbe, D.M., Paranjpe, S.S., Georgiou, G., van Kruijsbergen, I., Bogdanovic, O., Gibeaux, R., Heald, R., Lister, R., Huynen, M.A., van Heeringen, S.J., et al. (2017). Regulatory remodeling in the allo-tetraploid frog *Xenopus laevis*. *Genome Biology*, 18(1), 198.

Engström, P.G., Sui, S.J.H., Drivenes, Ø., Becker, T.S., Lenhard, B. (2007). Genomic regulatory blocks underlie extensive microsynteny conservation in insects. *Genome Research*, 17(12), 1898–1908.

Estep, M.C., McKain, M.R., Diaz, D.V., Zhong, J., Hodge, J.G., Hodkinson, T.R., Layton, D.J., Malcomber, S.T., Pasquet, R., Kellogg, E.A. (2014). Allopolyploidy, diversification, and the Miocene grassland expansion. *Proceedings of the National Academy of Sciences*, 111(42), 15149–15154.

Fawcett, J.A., Maere, S., de Peer, Y.V. (2009). Plants with double genomes might have had a better chance to survive the Cretaceous–Tertiary extinction event. *Proceedings of the National Academy of Sciences*, 106(14), 5737–5742.

Force, A., Lynch, M., Pickett, F.B., Amores, A., Yan, Y.L., Postlethwait, J. (1999). Preservation of duplicate genes by complementary, degenerative mutations. *Genetics*, 151(4), 1531–1545.

Fox, D.T. and Duronio, R.J. (2013). Endoreplication and polyploidy: Insights into development and disease. *Development*, 140(1), 3–12.

Fox, D.T., Gall, J.G., Spradling, A.C. (2010). Error-prone polyploid mitosis during normal Drosophila development. *Genes & Development*, 24(20), 2294–2302.

Freeling, M., Woodhouse, M.R., Subramaniam, S., Turco, G., Lisch, D., Schnable, J.C. (2012). Fractionation mutagenesis and similar consequences of mechanisms removing dispensable or less-expressed DNA in plants. *Current Opinion in Plant Biology*, 15(2), 131–139.

Gaeta, R.T., Pires, J.C., Iniguez-Luy, F., Leon, E., Osborn, T.C. (2007). Genomic changes in resynthesized *Brassica napus* and their effect on gene expression and phenotype. *The Plant Cell*, 19(11), 3403–3417.

Ganem, N.J. and Pellman, D. (2007). Limiting the proliferation of polyploid cells. *Cell*, 131(3), 437–440.

Garcia-Fernàndez, J. and Holland, P.W.H. (1994). Archetypal organization of the amphioxus Hox gene cluster. *Nature*, 370(6490), 563–566.

Garsmeur, O., Schnable, J.C., Almeida, A., Jourda, C., D'Hont, A., Freeling, M. (2014). Two evolutionarily distinct classes of paleopolyploidy. *Molecular Biology and Evolution*, 31(2), 448–454.

Gibson, T.J. and Spring, J. (1998). Genetic redundancy in vertebrates: Polyploidy and persistence of genes encoding multidomain proteins. *Trends in Genetics*, 14(2), 46–49.

Gillard, G.B., Grønvold, L., Røsæg, L.L., Holen, M.M., Monsen, Ø., Koop, B.F., Rondeau, E.B., Gundappa, M.K., Mendoza, J., Macqueen, D.J., et al. (2020). Comparative regulomics reveals pervasive selection on gene dosage following whole genome duplication. *bioRxiv*. DOI: 2020.07.20.212316.

Goffeau, A., Barrell, B.G., Bussey, H., Davis, R.W., Dujon, B., Feldmann, H., Galibert, F., Hoheisel, J.D., Jacq, C., Johnston, M., et al. (1996). Life with 6000 genes. *Science*, 274(5287), 546–567.

Gong, L., Olson, M., Wendel, J.F. (2014). Cytonuclear evolution of rubisco in four allopolyploid lineages. *Molecular Biology and Evolution*, 31(10), 2624–2636.

Gout, J.-F. and Lynch, M. (2015). Maintenance and loss of duplicated genes by dosage subfunctionalization. *Molecular Biology and Evolution*, 32(8), 2141–2148.

Gout, J.-F., Kahn, D., Duret, L., Paramecium Post-Genomics Consortium (2010). The relationship among gene expression, the evolution of gene dosage, and the rate of protein evolution. *PLoS Genetics*, 6(5), e1000944.

Gregory, T.R., Hebert, P.D.N., Kolasa, J. (2000). Evolutionary implications of the relationship between genome size and body size in flatworms and copepods. *Heredity*, 84(2), 201–208.

Griffiths, S., Sharp, R., Foote, T.N., Bertin, I., Wanous, M., Reader, S., Colas, I., Moore, G. (2006). Molecular characterization of Ph1 as a major chromosome pairing locus in polyploid wheat. *Nature*, 439(7077), 749–752.

Grover, C.E., Gallagher, J.P., Szadkowski, E.P., Yoo, M.J., Flagel, L.E., Wendel, J.F. (2012). Homoeolog expression bias and expression level dominance in allopolyploids. *New Phytologist*, 196(4), 966–971.

Gu, Z., Steinmetz, L.M., Gu, X., Scharfe, C., Davis, R.W., Li, W.-H. (2003). Role of duplicate genes in genetic robustness against null mutations. *Nature*, 421(6918), 63–66.

Guo, M., Davis, D., Birchler, J.A. (1996). Dosage effects on gene expression in a maize ploidy series. *Genetics*, 142(4), 1349–1355.

Henikoff, S., Greene, E.A., Pietrokovski, S., Bork, P., Attwood, T.K., Hood, L. (1997). Gene families: The taxonomy of protein paralogs and chimeras. *Science*, 278(5338), 609–614.

Howe, K., Clark, M.D., Torroja, C.F., Torrance, J., Berthelot, C., Muffato, M., Collins, J.E., Humphray, S., McLaren, K., Matthews, L., et al. (2013). The zebrafish reference genome sequence and its relationship to the human genome. *Nature*, 496, 498–503. DOI: 10.1038/nature12111.

Hu, G. and Wendel, J.F. (2019). Cis–trans controls and regulatory novelty accompanying allopolyploidization. *New Phytologist*, 221(4), 1691–1700.

Inoue, J., Sato, Y., Sinclair, R., Tsukamoto, K., Nishida, M. (2015). Rapid genome reshaping by multiple-gene loss after whole-genome duplication in teleost fish suggested by mathematical modeling. *Proceedings of the National Academy of Sciences of the United States of America*, 112(48), 14918–14923.

International Wheat Genome Sequencing Consortium (2014). A chromosome-based draft sequence of the hexaploid bread wheat (*Triticum aestivum*) genome. *Science*, 345(6194) [Online]. Available at: http://science.sciencemag.org/content/345/6194/1251788 [Accessed 29 March 2021].

Irimia, M., Maeso, I., Roy, S.W., Fraser, H.B. (2013). Ancient cis-regulatory constraints and the evolution of genome architecture. *Trends in Genetics*, 29(9), 521–528.

Jaillon, O., Aury, J.-M., Brunet, F., Petit, J.-L., Stange-Thomann, N., Mauceli, E., Bouneau, L., Fischer, C., Ozouf-Costaz, C., Bernot, A., et al. (2004). Genome duplication in the teleost fish Tetraodon nigroviridis reveals the early vertebrate proto-karyotype. *Nature*, 431, 946–957. DOI: 10.1038/nature03025.

Kassahn, K.S., Dang, V.T., Wilkins, S.J., Perkins, A.C., Ragan, M.A. (2009). Evolution of gene function and regulatory control after whole-genome duplication: Comparative analyses in vertebrates. *Genome Research*, 19(8), 1404–1418.

Kellis, M., Birren, B.W., Lander, E.S. (2004). Proof and evolutionary analysis of ancient genome duplication in the yeast *Saccharomyces cerevisiae*. *Nature*, 428(6983), 617–624.

Kenny, N.J., Chan, K.W., Nong, W., Qu, Z., Maeso, I., Yip, H.Y., Chan, T.F., Kwan, H.S., Holland, P.W.H., Chu, K.H., et al. (2016). Ancestral whole-genome duplication in the marine chelicerate horseshoe crabs. *Heredity*, 116(2), 190–199.

Kikuta, H., Laplante, M., Navratilova, P., Komisarczuk, A.Z., Engström, P.G., Fredman, D., Akalin, A., Caccamo, M., Sealy, I., Howe, K. (2007). Genomic regulatory blocks encompass multiple neighboring genes and maintain conserved synteny in vertebrates. *Genome Research*, 17(5), 545–555.

Kleinjan, D.A., Bancewicz, R.M., Gautier, P., Dahm, R., Schonthaler, H.B., Damante, G., Seawright, A., Hever, A.M., Yeyati, P.L., van Heyningen, V., Coutinho, P. (2008). Subfunctionalization of duplicated zebrafish pax6 genes by cis-regulatory divergence. *PLoS Genetics*, 4(2) [Online]. Available at: https://www.ncbi.nlm.nih.gov/pmc/articles/PMC2242813/ [Accessed 8 April 2021].

Leebens-Mack, J.H., Barker, M.S., Carpenter, E.J., Deyholos, M.K., Gitzendanner, M.A., Graham, S.W., Grosse, I., Li, Z., Melkonian, M., Mirarab, S., et al. (2019). One thousand plant transcriptomes and the phylogenomics of green plants. *Nature*, 574, 679–685. DOI: 10.1038/s41586-019-1693-2.

Levin, D.A. (1983). Polyploidy and novelty in flowering plants. *The American Naturalist*, 122(1), 1–25.

Li, F., Fan, G., Wang, K., Sun, F., Yuan, Y., Song, G., Li, Q., Ma, Z., Lu, C., Zou, C., et al. (2014). Genome sequence of the cultivated cotton Gossypium arboreum. *Nat. Genet.*, 46, 567–572. DOI: 10.1038/ng.2987.

Li, F., Fan, G., Lu, C., Xiao, G., Zou, C., Kohel, R.J., Ma, Z., Shang, H., Ma, X., Wu, J., et al. (2015). Genome sequence of cultivated Upland cotton (Gossypium hirsutum TM-1) provides insights into genome evolution. *Nat. Biotechnol*, 33, 524–530. DOI: 10.1038/nbt.3208.

Li, Z., Tiley, G.P., Galuska, S.R., Reardon, C.R., Kidder, T.I., Rundell, R.J., Barker, M.S. (2018). Multiple large-scale gene and genome duplications during the evolution of hexapods. *Proceedings of the National Academy of Sciences*, 115(18), 4713–4718.

Lien, S., Gidskehaug, L., Moen, T., Hayes, B.J., Berg, P.R., Davidson, W.S., Omholt, S.W., Kent, M.P. (2011). A dense SNP-based linkage map for Atlantic salmon (*Salmo salar*) reveals extended chromosome homeologies and striking differences in sex-specific recombination patterns. *BMC Genomics*, 12(1), 615.

Lien, S., Koop, B.F., Sandve, S.R., Miller, J.R., Kent, M.P., Nome, T., Hvidsten, T.R., Leong, J.S., Minkley, D.R., Zimin, A., et al. (2016). The Atlantic salmon genome provides insights into rediploidization. *Nature*, 533, 200–205. DOI: 10.1038/nature17164.

Liu, C., Ren, Y., Li, Z., Hu, Q., Yin, L., Wang, H., Qiao, X., Zhang, Y., Xing, L., Xi, Y., et al. (2021). Giant African snail genomes provide insights into molluscan whole-genome duplication and aquatic–terrestrial transition. *Molecular Ecology Resources*, 21(2), 478–494.

Lopes, F.R., Jjingo, D., da Silva, C.R.M., Andrade, A.C., Marraccini, P., Teixeira, J.B., Carazzolle, M.F., Pereira, G.A.G., Pereira, L.F.P., Vanzela, A.L.L., et al. (2013). Transcriptional activity, chromosomal distribution and expression effects of transposable elements in *Coffea* genomes. *PLoS One*, 8(11), e78931.

López, S., Lim, E.L., Horswell, S., Haase, K., Huebner, A., Dietzen, M., Mourikis, T.P., Watkins, T.B.K., Rowan, A., Dewhurst, S.M., et al. (2020). Interplay between whole genome doubling and the accumulation of deleterious alterations in cancer evolution. *Nature Genetics*, 52(3), 283–293.

Lumaret, R., Bowman, C.M., Dyer, T.A. (1989). Autopolyploidy in *Dactylis glomerata* L.: Further evidence from studies of chloroplast DNA variation. *Theoretical and Applied Genetics*, 78(3), 393–399.

Lynch, M. and Force, A. (2000). The probability of duplicate gene preservation by subfunctionalization. *Genetics*, 154(1), 459–473.

Mable, B.K. (2001). Ploidy evolution in the yeast *Saccharomyces cerevisiae*: A test of the nutrient limitation hypothesis. *Journal of Evolutionary Biology*, 14(1), 157–170.

Maclean, C.J. and Greig, D. (2011). Reciprocal gene loss following experimental whole-genome duplication causes reproductive isolation in yeast. *Evolution*, 65(4), 932–945.

Makino, T. and McLysaght, A. (2010). Ohnologs in the human genome are dosage balanced and frequently associated with disease. *Proceedings of the National Academy of Sciences of the United States of America*, 107(20), 9270–9274.

Makino, T., Hokamp, K., McLysaght, A. (2009). The complex relationship of gene duplication and essentiality. *Trends in Genetics*, 25(4), 152–155.

Mandáková, T. and Lysak, M.A. (2018). Post-polyploid diploidization and diversification through dysploid changes. *Current Opinion in Plant Biology*, 42, 55–65.

Marcet-Houben, M. and Gabaldón, T. (2015). Beyond the whole-genome duplication: Phylogenetic evidence for an ancient interspecies hybridization in the Baker's yeast lineage. *PLOS Biology*, 13(8), e1002220.

Markov, A.V. and Kaznacheev, I.S. (2016). Evolutionary consequences of polyploidy in prokaryotes and the origin of mitosis and meiosis. *Biology Direct*, 11(1), 28.

Marlétaz, F., Firbas, P.N., Maeso, I., Tena, J.J., Bogdanovic, O., Perry, M., Wyatt, C.D.R., de la Calle-Mustienes, E., Bertrand, S., Burguera, D., et al. (2018). Amphioxus functional genomics and the origins of vertebrate gene regulation. *Nature*, 564, 64–70. DOI: 10.1038/s41586-018-0734-6.

Mascagni, F., Usai, G., Cavallini, A., Porceddu, A. (2021). Structural characterization and duplication modes of pseudogenes in plants. *Scientific Reports*, 11(1), 5292.

McClintock, B. (1984). The significance of responses of the genome to challenge. *Science*, 226(4676), 792–801.

McLysaght, A., Makino, T., Grayton, H.M., Tropeano, M., Mitchell, K.J., Vassos, E., Collier, D.A. (2014). Ohnologs are overrepresented in pathogenic copy number mutations. *Proceedings of the National Academy of Sciences*, 111(1), 361–366.

Mendell, J.E., Clements, K.D., Choat, J.H., Angert, E.R. (2008). Extreme polyploidy in a large bacterium. *Proceedings of the National Academy of Sciences of the United States of America*, 105(18), 6730–6734.

Meyers, L.A. and Levin, D.A. (2006). On the abundance of polyploids in flowering plants. *Evolution: International Journal of Organic Evolution*, 60(6), 1198–1206.

Muir, C.D. and Hahn, M.W. (2015). The limited contribution of reciprocal gene loss to increased speciation rates following whole-genome duplication. *The American Naturalist*, 185(1), 70–86.

Naville, M., Ishibashi, M., Ferg, M., Bengani, H., Rinkwitz, S., Krecsmarik, M., Hawkins, T.A., Wilson, S.W., Manning, E., Chilamakuri, C.S.R., et al. (2015). Long-range evolutionary constraints reveal cis-regulatory interactions on the human X chromosome. *Nat. Commun.*, 6. DOI: 10.1038/ncomms7904.

Ni, Z., Kim, E.-D., Ha, M., Lackey, E., Liu, J., Zhang, Y., Sun, Q., Chen, Z.J. (2009). Altered circadian rhythms regulate growth vigor in hybrids and allopolyploids. *Nature*, 457(7227), 327–331.

Nieto Feliner, G., Casacuberta, J., Wendel, J.F. (2020). Genomics of evolutionary novelty in hybrids and polyploids. *Frontiers in Genetics*, 11 [Online]. Available at: https://www.frontiersin.org/articles/10.3389/fgene.2020.00792/full [Accessed 31 August 2020].

Nong, W., Qu, Z., Li, Y., Barton-Owen, T., Wong, A.Y.P., Yip, H.Y., Lee, H.T., Narayana, S., Baril, T., Swale, T., et al. (2021). Horseshoe crab genomes reveal the evolution of genes and microRNAs after three rounds of whole genome duplication. *Communications Biology*, 4, 1–11. DOI: 10.1038/s42003-020-01637-2.

Ohno, S. (1970). *Evolution by Gene Duplication*. Springer Berlin, Heidelberg.

Otto, S.P. (2007). The evolutionary consequences of polyploidy. *Cell*, 131(3), 452–462.

Panchy, N., Lehti-Shiu, M., Shiu, S.-H. (2016). Evolution of gene duplication in plants. *Plant Physiology*, 171(4), 2294–2316.

Papp, B., Pál, C., Hurst, L.D. (2003). Dosage sensitivity and the evolution of gene families in yeast. *Nature*, 424(6945), 194–197.

Parisod, C., Holderegger, R., Brochmann, C. (2010). Evolutionary consequences of autopolyploidy. *New Phytologist*, 186(1), 5–17.

Popovici, C., Leveugle, M., Birnbaum, D., Coulier, F. (2001). Homeobox gene clusters and the human paralogy map. *FEBS Letters*, 491(3), 237–242.

Rajeh, A., Lv, J., Lin, Z. (2018). Heterogeneous rates of genome rearrangement contributed to the disparity of species richness in Ascomycota. *BMC Genomics*, 19(1), 282.

Ramsey, J. and Schemske, D.W. (1998). Pathways, mechanisms, and rates of polyploid formation in flowering plants. *Annual Review of Ecology and Systematics*, 29(1), 467–501.

Rapp, R.A., Udall, J.A., Wendel, J.F. (2009). Genomic expression dominance in allopolyploids. *BMC Biology*, 7(1), 18.

Rastogi, S. and Liberles, D.A. (2005). Subfunctionalization of duplicated genes as a transition state to neofunctionalization. *BMC Evolutionary Biology*, 5, 28.

Renny-Byfield, S., Rodgers-Melnick, E., Ross-Ibarra, J. (2017). Gene fractionation and function in the ancient subgenomes of maize. *Molecular Biology and Evolution*, 34(8), 1825–1832.

Rice, A.M. and McLysaght, A. (2017). Dosage sensitivity is a major determinant of human copy number variant pathogenicity. *Nature Communications*, 8(1), 14366.

Robertson, F.M., Gundappa, M.K., Grammes, F., Hvidsten, T.R., Redmond, A.K., Lien, S., Martin, S.A.M., Holland, P.W.H., Sandve, S.R., Macqueen, D.J. (2017). Lineage-specific rediploidization is a mechanism to explain time-lags between genome duplication and evolutionary diversification. *Genome Biology*, 18, 111.

Roelofs, D., Zwaenepoel, A., Sistermans, T., Nap, J., Kampfraath, A.A., Van de Peer, Y., Ellers, J., Kraaijeveld, K. (2020). Multi-faceted analysis provides little evidence for recurrent whole-genome duplications during hexapod evolution. *BMC Biology*, 18(1), 57.

Sacerdot, C., Louis, A., Bon, C., Berthelot, C., Crollius, H.R. (2018). Chromosome evolution at the origin of the ancestral vertebrate genome. *Genome Biology*, 19(1), 166.

Salmon, A., Ainouche, M.L., Wendel, J.F. (2005). Genetic and epigenetic consequences of recent hybridization and polyploidy in Spartina (Poaceae). *Molecular Ecology*, 14(4), 1163–1175.

Sandve, S.R., Rohlfs, R.V., Hvidsten, T.R. (2018). Subfunctionalization versus neofunctionalization after whole-genome duplication. *Nature Genetics*, 50(7), 908–909.

Sankoff, D., Zheng, C., Zhu, Q. (2010). The collapse of gene complement following whole genome duplication. *BMC Genomics*, 11(1), 313.

Scannell, D.R., Byrne, K.P., Gordon, J.L., Wong, S., Wolfe, K.H. (2006). Multiple rounds of speciation associated with reciprocal gene loss in polyploid yeasts. *Nature*, 440(7082), 341–345.

Schmidt, D., Wilson, M.D., Ballester, B., Schwalie, P.C., Brown, G.D., Marshall, A., Kutter, C., Watt, S., Martinez-Jimenez, C.P., Mackay, S., et al. (2010). Five-vertebrate ChIP-seq reveals the evolutionary dynamics of transcription factor binding. *Science*, 328(5981), 1036–1040.

Schmutz, J., Cannon, S.B., Schlueter, J., Ma, J., Mitros, T., Nelson, W., Hyten, D.L., Song, Q., Thelen, J.J., Cheng, J., et al. (2010). Genome sequence of the palaeopolyploid soybean. *Nature*, 463, 178–183. DOI: 10.1038/nature08670.

Schnable, J.C., Springer, N.M., Freeling, M. (2011). Differentiation of the maize subgenomes by genome dominance and both ancient and ongoing gene loss. *Proceedings of the National Academy of Sciences*, 108(10), 4069–4074.

Scholes, D.R. and Paige, K.N. (2015). Plasticity in ploidy: A generalized response to stress. *Trends in Plant Science*, 20(3), 165–175.

Schranz, M.E., Mohammadin, S., Edger, P.P. (2012). Ancient whole genome duplications, novelty and diversification: The WGD radiation lag-time model. *Current Opinion in Plant Biology*, 15(2), 147–153.

Schwager, E.E., Sharma, P.P., Clarke, T., Leite, D.J., Wierschin, T., Pechmann, M., Akiyama-Oda, Y., Esposito, L., Bechsgaard, J., Bilde, T., et al. (2017). The house spider genome reveals an ancient whole-genome duplication during arachnid evolution. *BMC Biology*, 15, 62. DOI: 10.1186/s12915-017-0399-x.

Sémon, M. and Wolfe, K.H. (2007). Rearrangement rate following the whole-genome duplication in teleosts. *Molecular Biology and Evolution*, 24(3), 860–867.

Session, A.M., Uno, Y., Kwon, T., Chapman, J.A., Toyoda, A., Takahashi, S., Fukui, A., Hikosaka, A., Suzuki, A., Kondo, M., et al. (2016). Genome evolution in the allotetraploid frog Xenopus laevis. *Nature*, 538, 336–343. DOI: 10.1038/nature19840.

Shi, X., Ng, D.W.-K., Zhang, C., Comai, L., Ye, W., Jeffrey Chen, Z. (2012). Cis- and trans-regulatory divergence between progenitor species determines gene-expression novelty in *Arabidopsis allopolyploids*. *Nature Communications*, 3(1), 950.

Singh, P.P., Affeldt, S., Cascone, I., Selimoglu, R., Camonis, J., Isambert, H. (2012). On the expansion of "dangerous" gene repertoires by whole-genome duplications in early vertebrates. *Cell Reports*, 2(5), 1387–1398.

Smith, J.J., Timoshevskaya, N., Ye, C., Holt, C., Keinath, M.C., Parker, H.J., Cook, M.E., Hess, J.E., Narum, S.R., Lamanna, F., et al. (2018). The sea lamprey germline genome provides insights into programmed genome rearrangement and vertebrate evolution. *Nature Genetics*, 50, 270–277. DOI: 10.1038/s41588-017-0036-1.

Soltis, P.S. and Soltis, D.E. (2016). Ancient WGD events as drivers of key innovations in angiosperms. *Current Opinion in Plant Biology*, 30, 159–165.

Soltis, D.E., Visger, C.J., Soltis, P.S. (2014). The polyploidy revolution then… and now: Stebbins revisited. *American Journal of Botany*, 101(7), 1057–1078.

Song, K., Lu, P., Tang, K., Osborn, T.C. (1995). Rapid genome change in synthetic polyploids of Brassica and its implications for polyploid evolution. *Proceedings of the National Academy of Sciences*, 92(17), 7719–7723.

Soppa, J. (2014). Polyploidy in archaea and bacteria: About desiccation resistance, giant cell size, long-term survival, enforcement by a eukaryotic host and additional aspects. *Microbial Physiology*, 24(5–6), 409–419.

Stebbins, G.L. (1947). Types of polyploids: Their classification and significance. In *Advances in Genetics*, Demerec, M. (ed.). Elsevier, Amsterdam [Online]. Available at: http://www.sciencedirect.com/science/article/pii/S0065266008604903 [Accessed 14 October 2020].

Stoltzfus, A. (1999). On the possibility of constructive neutral evolution. *Journal of Molecular Evolution*, 49(2), 169–181.

Sun, H., Wu, S., Zhang, G., Jiao, C., Guo, S., Ren, Y., Zhang, J., Zhang, H., Gong, G., Jia, Z., et al. (2017). Karyotype stability and unbiased fractionation in the paleo-allotetraploid cucurbita genomes. *Molecular Plant*, 10(10), 1293–1306.

Tank, D.C., Eastman, J.M., Pennell, M.W., Soltis, P.S., Soltis, D.E., Hinchliff, C.E., Brown, J.W., Sessa, E.B., Harmon, L.J. (2015). Nested radiations and the pulse of angiosperm diversification: Increased diversification rates often follow whole genome duplications. *The New Phytologist*, 207(2), 454–467.

The Arabidopsis Genome Initiative (2000). Analysis of the genome sequence of the flowering plant Arabidopsis thaliana. *Nature*, 408, 796–815. DOI: 10.1038/35048692.

Tirosh, I. and Barkai, N. (2007). Comparative analysis indicates regulatory neofunctionalization of yeast duplicates. *Genome Biology*, 8(4), R50.

Touceda-Suárez, M., Kita, E.M., Acemel, R.D., Firbas, P.N., Magri, M.S., Naranjo, S., Tena, J.J., Gómez-Skarmeta, J.L., Maeso, I., Irimia, M. (2020). Ancient genomic regulatory blocks are a source for regulatory gene deserts in vertebrates after whole-genome duplications. *Molecular Biology and Evolution*, 37(10), 2857–2864.

Van de Peer, Y., Mizrachi, E., Marchal, K. (2017). The evolutionary significance of polyploidy. *Nature Reviews Genetics*, 18(7), 411–424.

Vanneste, K., Baele, G., Maere, S., de Peer, Y.V. (2014). Analysis of 41 plant genomes supports a wave of successful genome duplications in association with the Cretaceous–Paleogene boundary. *Genome Research*, 24(8), 1334–1347.

Varadharajan, S., Sandve, S.R., Gillard, G.B., Tørresen, O.K., Mulugeta, T.D., Hvidsten, T.R., Lien, S., Asbjørn Vøllestad, L., Jentoft, S., Nederbragt, A.J., et al. (2018). The grayling genome reveals selection on gene expression regulation after whole-genome duplication. *Genome Biology and Evolution*, 10(10), 2785–2800.

Villar, D., Berthelot, C., Aldridge, S., Rayner, T.F., Lukk, M., Pignatelli, M., Park, T.J., Deaville, R., Erichsen, J.T., Jasinska, A.J., et al. (2015). Enhancer evolution across 20 mammalian species. *Cell*, 160(3), 554–566.

Walden, N., German, D.A., Wolf, E.M., Kiefer, M., Rigault, P., Huang, X.-C., Kiefer, C., Schmickl, R., Franzke, A., Neuffer, B., et al. (2020). Nested whole-genome duplications coincide with diversification and high morphological disparity in Brassicaceae. *Nature Communications*, 11(1), 3795.

Wang, J., Tian, L., Madlung, A., Lee, H.-S., Chen, M., Lee, J.J., Watson, B., Kagochi, T., Comai, L., Chen, Z.J. (2004). Stochastic and epigenetic changes of gene expression in *Arabidopsis polyploids*. *Genetics*, 167(4), 1961–1973.

Wang, M., Wang, P., Lin, M., Ye, Z., Li, G., Tu, L., Shen, C., Li, J., Yang, Q., Zhang, X. (2018). Evolutionary dynamics of 3D genome architecture following polyploidization in cotton. *Nature Plants*, 4(2), 90–97.

Wendel, J.F., Lisch, D., Hu, G., Mason, A.S. (2018). The long and short of doubling down: Polyploidy, epigenetics, and the temporal dynamics of genome fractionation. *Current Opinion in Genetics & Development*, 49, 1–7.

Wertheim, B., Beukeboom, L.W., van de Zande, L. (2013). Polyploidy in animals: Effects of gene expression on sex determination, evolution and ecology. *Cytogenetic and Genome Research*, 140(2–4), 256–269.

Wolfe, K.H. and Shields, D.C. (1997). Molecular evidence for an ancient duplication of the entire yeast genome. *Nature*, 387(6634), 708–713.

Wu, H., Yu, Q., Ran, J.-H., Wang, X.-Q. (2021). Unbiased subgenome evolution in allotetraploid species of Ephedra and its implications for the evolution of large genomes in gymnosperms. *Genome Biology and Evolution*, 13(evaa236). DOI: gbe/evaa236.

Xu, P., Xu, J., Liu, G., Chen, L., Zhou, Z., Peng, W., Jiang, Y., Zhao, Z., Jia, Z., Sun, Y., et al. (2019). The allotetraploid origin and asymmetrical genome evolution of the common carp Cyprinus carpio. *Nature Communications*, 10, 4625. DOI: 10.1038/s41467-019-12644-1.

Yant, L. and Bomblies, K. (2015). Genome management and mismanagement–cell-level opportunities and challenges of whole-genome duplication. *Genes & Development*, 29(23), 2405–2419.

Yao, Y., Carretero-Paulet, L., de Peer, Y.V. (2019). Using digital organisms to study the evolutionary consequences of whole genome duplication and polyploidy. *PLoS One*, 14(7), e0220257.

Yoo, M.-J., Szadkowski, E., Wendel, J.F. (2013). Homoeolog expression bias and expression level dominance in allopolyploid cotton. *Heredity*, 110(2), 171–180.

Zerulla, K., Chimileski, S., Näther, D., Gophna, U., Papke, R.T., Soppa, J. (2014). DNA as a phosphate storage polymer and the alternative advantages of polyploidy for growth or survival. *PLoS One*, 9(4), e94819.

Zhang, J., Liu, Y., Xia, E.-H., Yao, Q.-Y., Liu, X.-D., Gao, L.-Z. (2015). Autotetraploid rice methylome analysis reveals methylation variation of transposable elements and their effects on gene expression. *Proceedings of the National Academy of Sciences*, 112(50), E7022–E7029.

Zhang, H., Zheng, R., Wang, Y., Zhang, Y., Hong, P., Fang, Y., Li, G., Fang, Y. (2019). The effects of Arabidopsis genome duplication on the chromatin organization and transcriptional regulation. *Nucleic Acids Research*, 47(15), 7857–7869.

2

Segmental Duplications and CNVs: Adaptive Potential of Structural Polymorphism

Patricia BALARESQUE[1] and Franklin DELEHELLE[2]

[1] *Diversification écologique et évolution des populations, Laboratoire d'Evolution et Diversité Biologique, Université Paul Sabatier, CNRS UMR5174, Toulouse, France*

[2] *Institut de Biologie de l'École Normale Supérieure, CNRS UMR8197, Université Paris Sciences & Lettres, France*

Structural polymorphism is a concept that describes the variations in composition and organization of genetic material within the genomes of several individuals of the same species, but also between species. Segmental duplications, defined as large sequences longer than 1 kbp, are present in several copies with a pairwise percentage identity of at least 90%; they are the preferential substrate for CNV-type structural variations. Through the action of different molecular mechanisms (homologous and non-homologous recombination), they induce insertions, deletions and inversions resulting in a local modification of the molecular structure, but also of the relative gene quantity in an organism. These differences in the gene copy number play a crucial role in genome complexity and species adaptability. With these copies responsible for dose effects, their fate will depend on several factors, notably the action of selection and drift effect, depending on the selective advantage and the

For a color version of all figures in this chapter, see http://www.iste.co.uk/richard/function.zip.

Function and Evolution of Repeated DNA Sequences,
coordinated by Guy-Franck RICHARD.

fitness[1] associated with this structural modification. From their adaptive potential to the role they play in many pathologies, we invite you to enter the world of redundancy and its consequences. While duplication is an undisputed source of innovation, the mechanisms involved are still far from being elucidated. But the rapid pace of structural changes and their potential impact on genomes make them particularly interesting polymorphisms in responding to a rapidly changing environment in which organisms will need to adapt. Let us shed light on a part of the genome that is little studied, but chock full of dynamism.

2.1. The multiple facets of genetic polymorphism

The genome is a highly dynamic structure. Many sources of variability exist from one genome to another, explaining the existence of as many genomes as individuals. The multiple forms that these genomes take correspond to what is called "polymorphism". Several sources of polymorphism exist: (1) a single mutation affecting the nucleotide, the basic unit of DNA; (2) the addition or removal of one or more bases at a given locus, called insertions/deletions; (3) a structural modification of the genome by varying the number and/or position of highly homologous DNA segments (such as SD, CNV, microsatellites, minisatellites, satellites and TE) (4) an epigenetic modification (Bernstein et al. 2007), including the modulation of gene expression by methylation (Li and Zhang 2014), modification of histones (Bannister and Kouzarides 2011) or through non-coding RNAs (Rinn and Chang 2012); and (5) modifications that affect the 3D structure of DNA (Spielmann et al. 2018). These different variations do not affect our genomes to the same extent: some will modify the nucleoid sequence (i.e. DNA), others will modify its 2D/3D topological structure – without there being a direct relationship between the size of the alteration and the severity of the impact. The density and distribution of segmental duplications most certainly give genomes a unique adaptability, although consequences are still difficult to fully grasp.

In this chapter, we focus on the structural polymorphism of eukaryotic genomes, where large duplicated DNA fragments (>1kbp) modify the genome structure. These fragments with high levels of identity (>90%) are substrates favorable to non-allelic recombination or gene conversion. The resulting CNV, existing from one individual to another, can be associated with variations in the number of genes or regulatory elements involved in phenotypic variations between individuals (Zhang et al. 2009). After introducing a few elements of terminology, we will discuss the characteristics of Segmental Duplications (SDs), their distribution and the molecular mechanisms responsible for their polymorphism. We will present an original analysis of Segmental Duplications (SDs) for organisms rarely studied from this angle. Lastly,

1 The selective value, or fitness, measures the suitability of a species with respect to its environment.

we will discuss the repercussions of polymorphism, which can take different forms: a selectively neutral modification, a novel positively selected gene favoring an individual adaptation to its environment, or a very severe pathology.

2.2. From Segmental Duplications to Copy Number Variants: terminology

Initial studies focusing on the duplication process as a key factor of genome evolution were conducted by Ohno (1970) and regularly updated (Zhang et al. 2009). The duplication process implies that a sequence is copied several times in the genome. These different copies present high percentage identities between them and can be located either on the same chromosome (intrachromosomal duplications), or on several (interchromosomal duplications), and can originate from different mechanisms described later in this review.

The terms to describe large duplicated sequences have multiplied since they were first described, varying according to the period, the molecular techniques used, the organisms, and also the purpose of the study. Among these, the terms low copy repeat, segmental duplications and duplicons are frequently found. The term gene amplification (MAJ) is used when it comes to bacteria and multigene families to emphasize the gene content of a fragment found in several copies. It should first be noted that genomic characteristics (size, content, percentage of identity between copies, location) vary from one study to another, rendering the synthesis exercise complex. This lack of a precise shared definition represents the first limitation for our review.

Regions containing SDs can exhibit copy number polymorphism, both between individuals and between species. On an interspecific scale, the variability in the number of duplications is referred to as copy number difference (CND); on an intraspecific scale, we speak of copy number variant (CNV), a term that highlights the variable copy number observed from one individual to another – because while a region can contain SD-type duplicated sequences, it does not necessarily present intraspecific copy number variability since the number of SDs can be fixed. The term CNV is therefore more commonly used in studies to analyze the quantitative nature of variability (copy number), more than the qualitative nature (size, percentage GC, gene content, etc.). All of these terms will be used in this review.

2.3. SDs: a general overview

2.3.1. *Background*

The duplication process was first characterized in 1925 in *Drosophila melanogaster* by the study of the *bar* gene and estimated at 10^{-3} events per generation (Sturtevant 1925). As early as 1936, duplication was considered an effective means of increasing the length of a genome and providing the organism with new substrates necessary for biological innovation (Bridges 1936), whose evolutionary potential was discussed by numerous scientists (Haldane 1933, Muller and Prokofyeva 1935, Huxley et al. 1944). Indeed, the duplication and divergence of copies is largely responsible for the evolution of genomes and their complexity (Ohno 1970): more than 50% of prokaryotic genes (Brenner et al. 1995, Teichmann et al. 1998) and 90% of eukaryotic genes (Gough et al. 2001) will have arisen from duplications. Studies conducted in recent years have shown that the copy number variation (of genes, but also regulatory and repeated elements) is much more abundant than initially estimated (Lipinski et al. 2011, Schrider et al. 2013, Lynch and Wagner 2008). The high frequency of rearrangement rate and variants of CNV has very significant consequences, whether on the evolution, speciation or the inheritability of certain pathologies.

2.3.2. *SDs: more than a category of sequences, superstructures*

Within the landscape of the human genome's sequences (Figure 2.1), segmental duplications occupy a unique place. They are defined fairly arbitrarily based on a length and percentage of identity between the minimal number of copies (1 kb and 90%). SDs are large sequences that possess an average mutation rate in humans ranging from 10^{-5} to 10^{-2} events per locus and per generation. While they are studied on the same basis as other repeated sequences such as SINEs or microsatellites, they are nonetheless very different. SDs are not a type of repeated sequences as such because they do not have a characteristic molecular structure. They are composed of several types of sequences – themselves single-copy or repeated (microsatellites, minisatellites, SINE) – and vary in terms of length, composition and location. They are more superstructures than a category of sequence per se: their compositions and structures are not easily definable, and the mutational mechanisms responsible for their dynamics are currently only partly understood.

In the current state of knowledge, it is difficult to say exactly to which mechanisms the SDs are subject since these can vary, as we will see, depending on their size and location, content or different types of stress to which organisms are subject. In order to legitimize the study of SDs per se, it is necessary to define, at best, their genomic characteristics such as their size, location, gene content, GC content, the associated CNV polymorphism and many other factors. These elements can explain the variability of the SDs observed within and between genomes.

Because of their repeated nature, SDs constitute veritable mutational hotspots, capable of generating extreme local variability, conducive to the emergence of new chromosomal structures, functions and phenotypes. SDs also play a significant role in the emergence of many pathologies due to their ability to generate duplications and deletions on large segments, thereby altering the dosage of one or several genes of an organism. It is interesting to note that ultra-conserved elements (UCEs) of genomes that play a decisive role at cellular level (DNA binding, DNA/RNA transcription, transcription regulation and development) are devoid of SDs and CNVs; duplications and deletions of UCEs, which are a source of instability, would therefore be eliminated at the cell, organism or species level, through lethality, distortion of segregation or a decrease in fitness, in order to maintain the stability of these regions (Derti et al. 2006). Maximum intolerance to CNV at the UCE level could confer a fitness advantage on individuals, protecting them from new pathologies induced by structural variation, as is the case in many cancers (Derti et al. 2006). Similarly, genes with ubiquitous expression specialized in "housekeeping" functions would be less prone to exhibit strong structural polymorphism than genes presenting tissue-specific expressions, their conservation being paramount for the essential functioning of an organism.

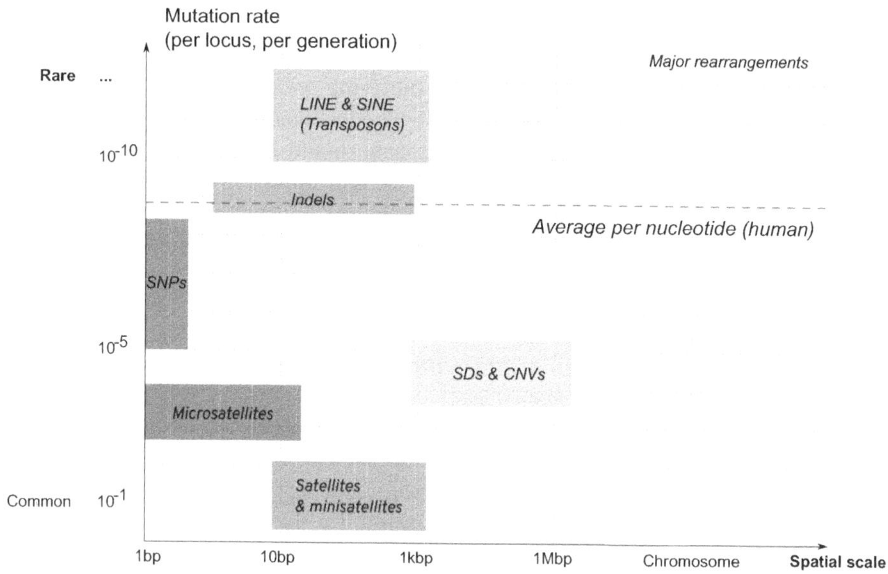

Figure 2.1. *General view of mutations affecting eukaryotic genomes, their impact and their appearance rate*

2.3.3. *SD and CNV: study biases related to the attractiveness of subjects as well as to the technological developments of the moment*

The literature on SDs is mainly documented in humans and great apes, as well as for some model species. Otherwise, the information remains partial and only covers certain multigene families, or the presence of CNV in certain species. At the time of writing it is difficult to obtain an overview of the mechanisms that affect SDs in most species, and it would not be conceivable to simply propose general conclusions. To go a little further than the data present in the literature, we will also propose an analysis of SDs for 12 other species (section 2.9).

The imbalance of information between humans and other species is rooted not only in a bias linked to the attractiveness of the subjects studied (humans being systematically favored), but also technical reasons related to sequencing technologies: short-fragment sequencing techniques have constituted the vast majority of genomes produced until very recently, with long-fragment techniques having only gained a foothold in more than one year (De Coster et al. 2021). This sequencing, which has the advantage of simplicity and a reasonable cost, is mainly oriented, by its inherent limitations, toward an understanding of the mechanisms affecting regions in single copies, introducing a bias toward potentially more stable/conserved fractions (including UCEs). This technological bias, which certainly induces a fundamental bias in our understanding of the genome and the living, could result in an over-estimation of the events of purifying selection, to the detriment of those of adaptive or balanced selection. The emergence and current democratization of sequencing techniques based on long fragments should facilitate the collection of information on SDs in the future. With these, we will improve our overall vision of genome evolution and mechanisms for adaptation, response to environmental pressures and stress. This methodological bias is the second identified limit of this review.

2.3.4. *SD: characteristics in human and non-human primates*

Segmental duplications are most often defined as DNA blocks larger than 1 kbp, separated by more than 1 Mbp, present in the form of multiple copies at several locations in the genome (up to 5 Mbp), exhibiting very high pairwise levels of identity (from 90% to 100%). SDs are responsible for the structural polymorphism observed in genomes: this polymorphism concerns DNA segments from 500 bp to 5 Mbp or more, which constitute substrates of choice for homologous and non-homologous recombination mechanisms. The proportion of SDs is estimated to be 5.2% in the human genome (Bailey et al. 2002), and the proportion of fragments larger than 5 Mbp is estimated to be 3.5% of the genome (Cheung et al. 2003). The average size of human SDs is about 16 kbp (Zhang et al. 2005). They can be located on one or several chromosomes and are preferentially observed in subtelomeric (40% (Linardopoulou et al. 2005)) and pericentromeric regions (33% (She et al. 2004)), where the activity of duplicative transposition appears to be significantly greater than elsewhere on the genome (Mefford and Trask 2002, She et al. 2004, Linardopoulou et al. 2005). SDs are heterogeneously distributed between chromosomes, in rates ranging from 1% to 27% (Samonte and Eichler 2002, Zhang et al. 2005). It should be noted that the sex chromosomes appear to be particularly rich in SDs (from 5% to 27.5%), especially the Y chromosome, whose SDs occupy more than 30% of the available long arm (Skaletsky et al. 2003).

Both human and chimpanzee genomes consist of 5% SDs (Samonte and Eichler 2002), a high proportion compared to 1.18% in orangutans and 1.55% in macaques (Gibbs et al. 2007). This highlights the explosion of SDs in African great apes compared to their Asian counterparts (Orangutans) (Marques-Bonet et al. 2009a). The accumulation of SDs in African great apes appears to be concomitant with the decline of other mutational processes, such as point mutation or retrotransposition (Marques-Bonet et al. 2009b). These apparent changes in the mutational dynamics of the global genome are not explained to date (Rogers and Gibbs 2014).

In humans, SDs are concentrated in 400 distinct regions that constitute true blocks of duplications (She et al. 2004). Within these blocks, SDs are organized according to a complex mosaic including SDs of several origins; with each new copy wave, it is an even more complex structure that emerges if it is not eliminated by natural selection: "from duplication comes duplication". These structures constitute an organizational characteristic of the Segmental Duplications (SDs) of *hominoidea* (Johnson et al. 2006): around 14 very-gene-rich sequences from 5 to 30 kbp are associated with most intrachromosomal duplications. These observations support the hypothesis that these sequences could be the catalysts for recent, independent expansions of duplications within these lineages.

2.4. Methodologies for detecting structural variation in genomes

Studies of SDs as biological objects depend on the discovery of their positions and on the extraction of their sequences from the relevant genomes. These operations, which are inherently dependent on the genome sequencing method considered, can take place at three moments relative to it:

– upstream, directly within the DNA molecule through biochemical processes;

– between the sequencing itself and assembly, using intermediate results;

– lastly, at the end of the chain, directly on the assembled genome.

The first family of methods, known as in vitro, typically revolves around two phases: first, a chemical labeling of the DNA areas of interest, then a microscopy image analysis. The last two, known as in silico, are purely digital and use a digitized form of DNA as the data source.

2.4.1. *In vitro methods*

They have the advantage of being based directly on the DNA molecule, and thus of ignoring any bias or error that may be introduced by the sequencing process. However, they only have this single advantage: they require heavy infrastructure, are costly, and are difficult to reproduce. As a result, they have been rendered obsolete through the reduction in the cost of sequencing methods and the increasing quality of the genomes thus digitized.

2.4.2. *Methods on reads*

Genome sequencing, especially when using so-called high-throughput methods (Goodwin et al. 2016), undergoes the following process: after a succession of chemical pre-treatments aiming to extract, purify and concentrate DNA molecules, a machine produces millions of partial scans of the genome, or reads, which vary in size from several dozen to several tens of thousands of bases according to the technology used. These reads are then chained by their overlapping parts into sequences longer the scaffolds by dedicated software.

The highly similar different copies of an Segmental duplication (SD) are typically fused during the operation: the reads belonging to the different copies are practically indistinguishable, and their divergences are attributed to sequencing errors. The only traces of these fused copies are then the peaks in the average coverage of the genome, which presents the number of reads covering each base of the assembled genome. This is supposed to remain relatively constant along the genome, so sudden variations are indicative of assembly errors, where reads from similar but distinct regions have been erroneously fused into a single region by the assembly software.

Ad hoc software programs using statistical anomaly detection methods have been developed to detect these unexpected peaks and deduce the underlying SDs. These are imprecise, with uncertainty in the order of around a hundred bases, and have difficulty placing ghost copies back in their original positions. But in the absence of additional data or auxiliary sequencing, they remain indispensable.

2.4.3. *Post-assembly methods*

Advances in long-read sequencing methods developed and improved over the past decade, supported by assembly software that uses deep learning (Boža et al. 2017), enable reads of several tens of thousands of bases to be obtained with increasing accuracy, even in the most difficult genomic areas to sequence (Wenger et al. 2019). Combined with pre-segmentation or placement methods (Schwartz et al. 1993), it becomes possible to assemble genomes with their duplicated regions intact (Miga et al. 2020).

> The algorithmic complexity of a program estimates the relationship between the size of the input data and the resources required to process them – memory, computation time, storage, etc. The most commonly used measurement is the Landau big $\mathcal{O}$ notation, which describes the decline of asymptotic behavior between two functions.
>
> Thus, for example, an algorithm that has the complexity of $\mathcal{O}(n^2)$ would use, for sufficiently large inputs, a number of operations proportional to the square of its size.

Box 2.1. *Algorithmic complexity*

This is of course the simplest scenario for searching for SDs, the availability of genomes assumed to be practically exact then reducing it to a purely algorithmic problem of searching for similarities at a given threshold within a character string. Exact solutions to this problem exist, but their algorithmic complexity, especially in terms of computational time, makes them de facto unusable at the scale of a whole genome, typically several billion bases long – at least in relation to today's computers.

This requires us to turn to heuristic methods which, in exchange for a degree of uncertainty as to the quality of the solution found, offer vastly improved performance. A first family is based on the analysis of alignments of a whole genome against itself, the matches then being gradually chained together to find the families of SDs. These methods are typically fairly accurate, but are extremely dependent on the whole-genome alignment stage, which is costly and potentially difficult to calibrate.

A second family circumvents whole alignment by relying on the fact that, for a given homology rate and minimal copy length, the existence of exact match areas of a certain length between copies of a family of SD is guaranteed (note that differences between copies of the same family tend to be grouped together rather than homogeneously distributed). These algorithms will first look for exact matches, potentially belonging to copies of the same family. This phase can be conducted much more efficiently, especially at the genomic scale, than the search for imperfect matches, for example with the use of suffix trees (Weiner 1973). Then, in a second stage, these anchors will be gradually extended in parallel, tolerating a certain rate of errors representative of the level of homology between the duplicons produced, until the members of the family under consideration are found.

Lastly, while a significant breakthrough has not yet been made, other proposals explore novel alternative approaches, based on frequency (Sharma et al. 2004) or statistical analyses (Girgis 2015) of genomic sequences.

Classes	Data	Advantages	Disadvantages	Examples
In vitro	DNA molecule	Directly accesses genetic information	Costly, slow, logistically cumbersome	FISH
Coverage profiling	Reads	Does not require any assembly	Not very accurate in location and in cardinal	WSSD (Bailey et al. 2002), WGAC (Bailey et al. 2001), DuplicationDetector (Djedatin et al. 2017)
Alignment chaining	Whole-genome alignment	Fairly accurate, easy to use	Costly and non-optimal	ReD (Audemard et al. 2012), DAGChainer (Haas et al. 2004), RECON (Bao and Eddy 2002), Cinteny (Sinha and Meller 2007)
Dedicated algorithms	Sequenced and assembled genome	Practically very accurate	Dependent on sequencing quality	ASGART (Delehelle et al. 2018), SEDEF (Numanagić et al. 2018), VMatch (unpublished)

Table 2.1. *Summary of SD detection methods*

2.5. The molecular mechanisms at the origin of structural variation

The genesis and dispersal mechanisms of SDs in genomes remain poorly understood (Marques-Bonet et al. 2009b). Changes in chromosome structure, whether by adding, removing or inverting genetic material, would occur primarily through two major types of mechanisms: homologous recombination (HR) and non-homologous recombination (NHR). These two mechanisms, very well described and summarized by Hastings et al. (2009), do not require the same degree of homology between the sequences involved: homologous recombination requires almost-perfect homology between sequences, whereas non-homologous recombination requires little or no homology.

2.5.1. *Homologous recombination mechanisms*

Homologous recombination is the molecular process by which a sequence presenting altered nucleotides will be repaired by converting them to those of a healthy homolog sequence, which most often happens to be the sister chromatid (Hastings et al. 2009). Among the repair mechanisms, the most-studied mechanism is that of homologous recombination induced by double-strand break (DSB).

If both ends of the double strand are present, the repair can be performed via the SDSA (synthesis-dependant strand annealing) route, which does not generate any crossover[2], but can produce CNVs. Otherwise, it will occur by resolving a double Holliday junction, which can result in crossovers potentially combined with gene conversion events. It is all the more likely that a crossover is formed, the larger the repeats, favoring the creation of a double Holliday junction. During the crossover, if the target sequence is not repaired using the dedicated homolog sequence, but another homolog sequence oriented in the same direction positioned elsewhere on the same chromosome (as can be the case in a region enriched with SD), a non-allelic recombination (NAHR), or ectopic recombination occurs (Stankiewicz and Lupski 2002). It will generate a duplication or a deletion, and if the target sequence is oriented in the opposite direction, an inversion (Hastings et al. 2009).

If only one end is present, homologous recombination also occurs through the BIR (break-induced replication) repair mechanism. As previously mentioned, if the damaged end pairs with a homologous sequence located elsewhere on the chromosome rather than with the sister molecule, the same consequences will occur (duplication, deletion and inversion), resulting in the appearance of CNV-type structural polymorphism.

The change in the structure of a chromosome can therefore occur through homologous recombination, more particularly during the repair of a DSB. As discussed by Hastings et al. (2009), it occurs primarily, not because the repair mechanism is not optimal, but because in a given region including SDs, the target sequence is not necessarily repaired using the dedicated homologous sequence. The cell should normally regulate the choice of target through the mismatch repair mechanism. When this mechanism fails, especially in the absence of cohesion between the broken strand and sister chromatid, normally ensured by cohesive proteins that maintain sequences in place to facilitate DSB repair, thereby limiting the use of any other homologous sequence, whether intrachromosomal or interchromosomal. This lack of cohesion would explain the appearance of CNV in cells whose mechanisms are designed to avoid the accumulation of repeats (Kobayashi and Ganley 2005). Note that if homologous recombination is a key repair mechanism, it can also induce CNV-type structural changes.

2.5.2. *Non-homologous recombination mechanisms*

Non-homologous recombination refers to DNA repair systems that use no – or only very little – homology between the DNA sequences concerned. The mechanisms are classified into two categories, replicative and non-replicative.

---

2 Crossover denotes the overlapping of homologous chromosomes during meiosis, prior to the mutual exchange of genetic material.

For non-replicative mechanisms, there are two pathways for DSB repair that require no homology or only a little, involving only a few bases: the NHEJ (non-homologous end joining) mechanism, which can lead to minor deletions or insertions, and the MMEJ (microhomology-mediated end joining) mechanism, which can lead to deletions.

For replicative mechanisms, the formation of microhomologous junctions would be linked to DNA replication. Among the potential replicative non-homologous recombination mechanisms at the origin of CNVs, there are (1) the sliding mechanism during replication, which would act a priori inside the replication fork and would have only a limited impact on human CNVs, whose sequences are separated by several thousand, or even several million bases; (2) FoSTeS (fork stalling and template switching), the erroneous positioning of homologous sequences between replication forks and not just within a single replication fork, and lastly, (3) the MMBIR (microhomology mediated break-induced replication) model, where the BIR mechanism could be induced by microhomologies and would be responsible for non-recurrent changes in CNVs. In stressful conditions (hypoxia, for example), certain enzymes (RAD51) would be repressed, resulting in a decrease in the efficiency of the repair systems, passing from a high-fidelity HR to more error-prone DSB repair systems (such as NHEJ) inducing an MMBIR mechanism (Hastings et al. 2009).

2.6. Regions rich in SDs/LCRs favor the creation of CNVs: insertions/duplications, deletions and inversions

Rearrangements involving SDs will affect large regions that may include one or several functional genes, which can strongly impact an individual's phenotype and cause certain pathologies (Sharp et al. 2006). It is the orientation of the different copies that will determine the type of structural variation generated, namely an insertion/duplication, a deletion or an inversion.

2.6.1. *Insertions/duplications and deletions*

Insertions (sometimes incorrectly noted as duplications) and deletions represent, respectively, the gain or loss of genetic material (see Figure 2.6). Carrying one or several coding genes or regulatory elements, their impact can be minor or significant. It is commonly reported that deletions are far less frequent than duplications, and thus would be negatively selected; nevertheless, it is difficult to conclude on this point, owing to the difficulty in accurately tracing the chronology of events.

The presence of one or several genes in the duplicated fragment can lead to an imbalance in gene expression by modifying not only the quantity of proteins produced, but also the phenotype and the associated evolutionary fitness, although some authors claim that genomes are particularly resilient to dose effects (Sharp et al. 2006). While this is indeed the case for certain categories of genes, those coding for proteins involved in macromolecular complexes, whose duplication modifies stoichiometry within these complexes, are highly sensitive to dose effects and more subject to purifying selection, as shown by the underrepresentation of SDs in many organisms such as thale cress, rice, poplar and mammals (Maere et al. 2005, Makino and McLysaght 2010, Davis and Petrov 2005, Hakes et al. 2007, Coate et al. 2011, Freeling et al. 2008, Rodgers-Melnick et al. 2012). The CNV study of genes strongly connected within interactomes generally exhibit underrepresentation of the polymorphism of CNV (Dopman and Hartl 2007, Korbel et al. 2008, Ionita-Laza et al. 2009). These observations thus illustrate a variable sensitivity of the genes to dose effects, and indeed to the process of structural variation, depending on their role and their connectivities.

Some of the most famous insertions/deletions associated with certain phenotypes include rhesus factor, loss of color vision, but also modifications to the metabolic system, change in the number of copies of the cytochrome P450 enzyme, CYP2D6, being associated with variability in the metabolism of tricyclic antidepressants and antipsychotic drugs (Buckland 2003).

CNVs pair in both germ and somatic cells. Their reduced quantity in newborns compared to adults points to their accumulation over the course of a lifetime (Flores et al. 2007). The measured frequency of CNV varies according to several factors: investigation methods, loci, tissues and species. Locus-specific, population-based and bioinformatic estimation methods yield highly variable levels.

Locus-specific approaches

Duplication rates vary between organisms, but also between loci for the same organism: for *Drosophila melanogaster*, this level varies from 1.7×10^{-4} for the Rosy locus (Shapira and Finnerty 1986) to 4.0×10^{-7} for eye and body color locus (Watanabe et al. 2009); in *Homo sapiens*, from 2.6×10^{-5} for the locus of α-globin (Lam and Jeffreys 2007) to 1.0×10^{-8} for the DMD locus (van Ommen 2005). The parallel analysis of semen and blood (Lam and Jeffreys 2007) showed a large variation in deletions generated by NAHR in blood 6.7 to 6.8×10^{-6} per haploid genome (close level between individuals) and a much higher and variable level between individuals in semen, from 16 to 68×10^{-6}.

Population-based approaches

Use the frequency of duplications observed in a population under the hypothesis of a mutation/selection balance. Extrapolating the mutation rates observed for DMD, the duplication rate on the genome scale would be 0.02 duplication per genome, per generation. The same measurement via the PMD22 gene estimates a spontaneous duplication rate of $1.7 - 2.6 \times 10^{-5}$ duplication per locus, per generation (Lupski 2007) – which illustrates the heterogeneity of estimations based on a particular locus.

Bioinformatic methods

Estimate the duplication rate over the entire genome, approaching the age distribution of duplicated genes through divergence of synonym sites between paralogous genes: the proposed rates vary from 3.2×10^{-3} by 1% silent-site divergence (*Arabidopsis thaliana*), 1.1×10^{-3} by 1% silent-site divergence (*Drosophila melanogaster*), and 3.0×10^{-3} per 1% silent-site divergence (*Mus musculus*), to 4.9×10^{-3} per 1% silent-site divergence in humans (Lynch and Conery 2003, Lynch 2007) and up to 1.1×10^{-9} gene per year (Cotton and Page 2005). But these rates overlook the actions of natural selection and gene conversion. These are taken into account, the estimated rates for *S. cerevisiae*, *D. melanogaster* and *C. elegans* rise to 0.028, 0.0014 and 0.024 duplication per gene per million years (Gu et al. 2002).

Variations in the proportion of CNVs observed in humans, particularly with respect to geographical areas, would be linked to the contribution of certain archaic populations in the latter's genome: the high proportion of CNVs detected in the Oceanian populations would appear to be due to this same incidence found and transmitted by the Denisovians (Bergström et al. 2020).

The overall results show that estimated rates vary widely between species, populations and individuals, but also between methods, location on the genome and tissues.

Box 2.2. *Estimation of CNVs: methods and levels (extracted from Katju and Bergthorson (2013); the duplication rates are usually given as duplication/gene/generation)*

Insertion/deletion structural rearrangements would affect 400 sites in the human genome, cumulating a length of 100 Mbp. Deletions would be more disadvantageous than insertions, the impact of a gene deletion being less easily offset by the cell than that of a duplication (Petrov 2001). The relative proportions and forces in action on these structural variants remain an open topic of discussion.

2.6.2. *Inversions*

Inversions correspond to a change in the orientation of a DNA fragment, without being necessarily associated with a gain or loss of biological material. Several studies have shown that more than 75% of inversions were associated with SDs, and were accompanied by a loss or gain of material (Newman et al. 2005, Tuzun et al. 2005). In humans, among the most commonly discussed inversions, the inversion polymorphism of 900 kbp on chromosome 17, confers higher fertility to women carrying it, and is likely to be under positive selection (Stefansson et al. 2005).

But inversions are rarely studied, since the molecular approaches used until now have prevented their fine analysis; so their incidence is certainly underestimated.

While inversions can theoretically affect gene expression, the inversions reported so far do not seem to drastically alter the gene copy number and expression in the genomes (Kleinjan and Van Heyningen 1998). Inversions can predispose a region to serial rearrangements, as observed for olfactory receptors (OR) on chromosomes 4 and 8 (Giglio et al. 2001). This is also true for the Y chromosome, where an inversion of 4 Mbp to Yp11.2 involved in sex-reversal cases through a translocation between PRKX and PRKY can be observed in one-third of healthy individuals (Schiebel et al. 1997, Rosser et al. 2009). Inversions have played a decisive role in the differentiation of autosomes in sex chromosomes: following a cessation of recombination further to the acquisition of a beneficial allele of the SRY gene on the Y chromosome, serial inversions would have gradually ceased recombination along the X and Y protochromosomes, leading to an almost-complete cessation of recombination between the sex chromosomes. This phenomenon has been observed in mammals (Lahn and Page 1999), birds (Zhou et al. 2014) and snakes (Singchat et al. 2020). Inversions could play (1) an organizational role in the genome by suppressing the recombination locally, (2) a functional role by inducing gene expression changes, and (3) a dynamic role by creating areas of instability potentially involved in dose effects.

2.7. From SDs to CNVs in humans and primates

SDs and regions exhibiting high intraspecific structural polymorphism (CNV) are not randomly distributed in genomes. This distribution could be explained, on the one hand, by the regional mutational characteristics conducive to a lesser or greater extent to the appearance of SDs and CNV and, on the other hand, by the differential impact of natural selection according to the composition of these segments (gene presence or absence) and their potential impact on the organism.

2.7.1. *General overview*

SDs are concentrated primarily in the subtelomeric and centromeric heterochromatin regions, and, to a lesser extent, interspersed in other regions of the genome. The high density of these regions in SDs provides a favorable substrate for homologous or nonhomologous recombination, as well as for the appearance of CNV explaining that SD and CNV occur in substantially the same places, as this has been demonstrated in mice (Graubert et al. 2007). In addition to this physical concentration, the presence of SD is correlated with other factors such as local gene density, repeated sequence density, recombination rates, and GC content, the most impactful factor being the local gene density (Zhang et al. 2005).

2.7.2. *Delineating regions of interest*

In order to evaluate the representation of SDs in the centromeric and subtelomeric regions, several methods can be used. The quantity of SDs can be established on the basis of (1) a physical delineation of the estimated region in a pair of bases from the telomere or the centromere, (2) a repeat of the content of a given area characterizing it such as α-satellites, β-satellites, CER-satellites, γ-satellites and CAGGG-repeats. In humans, SDs are primarily located in pericentromeric and subtelomeric regions, where they are on average 10 times more concentrated than in the rest of the genome (Bailey et al. 2001), making these regions veritable hot spots for structural polymorphism.

2.7.3. *Heterogeneity in the distribution of intra- and interchromosomal SDs*

Pericentromeric regions would be 5–12 times richer in SDs than the rest of the genome, depending on the approach used; this heterogeneity could be explained either by unreliable delineation of these regions – real mosaics with discontinuous properties that make them very difficult to delineate – or by a global problem assigning SDs to certain chromosomes when assembling genomes (about 23% of duplications are not attributed to humans in the 2003 sequencing). Subtelomeric regions, using a physical approach analyzing proximal parts located 500 kbp from the end of the chromosomes, exhibit a 2.7-fold enrichment in interchromosomal duplications, but only 0.76 in intrachromosome duplications.

These enrichments are heterogeneous and vary according to the chromosomes and regions concerned: the SD concentration varies from 0% (chromosome 19) to 7% (chromosome 7) in pericentromeric regions, from 0% (chromosome 17, Y chromosome) to 14.8% (chromosome 9) in subtelomeric regions, and from 0.2% (chromosome 18) to 5.4% (Y chromosome) elsewhere in the genome (Zhang et al. 2005). In addition, inter- and intrachromosomal SDs are not homogeneously distributed between these three major regions: the pericentromeric and subtelomeric regions are both enriched with interchromosomal duplications, but only the pericentromeric regions and the long arm of the Y chromosome are enriched with intrachromosomal duplications (Skaletsky et al. 2003, Bailey et al. 2001, Zhang et al. 2005). These heterogeneities could attest to either repair mechanisms, or a purifying selection, which are highly variable depending on the location of SDs and/or their intra- or interchromosomal natures.

2.7.4. *Intrachromosomal and interchromosomal SDs: what do they teach us about the evolutionary history and origin of SDs?*

In humans, the average size of SDs is 18,564 bp for intrachromosomal SDs, longer than interchromosal SDs, which measure in at an average of 14,759 bp (Zhang et al. 2005). The similarity rates between duplicated sequences are higher for intrachromosomal SDs (94.8%) than for interchromosomal SDs (93.7%). This result would demonstrate either a more recent intrachromosomal accumulation of SDs than that of interchromosomal SDs, or a greater homogenization rate of intrachromosomal SDs.

2.7.4.1. *Interchromosomal SDs*

A large number of interchromosomal SDs exist in the human genome (occupying 1.7% of the genome), which demonstrates an active duplicative transposition process at the interchromosomal scale. Similarity rates varying from 90% to 99% also suggest that these rearrangements occurred several times in the course of the evolution of primates (Bailey et al. 2002). Interchromosomal duplicative transposition activity appears to have peaked after the divergence of ancient-world monkeys from the hominoidea lineage around -25MYA: these interchromosomal SDs emerged through a series of translocations from the euchromatin to the pericentromeric and subtelomeric heterochromatin regions, today offering the complex mosaic that we know concerning these two centers of structural polymorphism (She et al. 2006). These mechanisms would have involved systems for ectopic DNA repair by NHEJ (Linardopoulou et al. 2005). The emergence of SDs at subtelomeres will lead to an accumulation of homologous sequence transfers between chromosomes, ensuring interchromosomal mixing and more-intense trafficking of genetic material. These key regions in the genome constitute areas of preferential transitions between the chromosome-specific parts and those comprising telomeric repeats covering the end of each chromosome, which, from this point of view, are pivotal. Among the known interchromosomal duplications of human subtelomeric regions, we can cite the polymorphism of the DUSP22 gene, which is transposed from chromosome 16 to chromosome 6, but also SDs comprising the gene coding for olfactory receptors, copies of which exist in the subtelomeric regions of half of the genome and the copy number of which varies according to the ethnic groups concerned.

2.7.4.2. *Intrachromosomal SDs*

Intrachromosomal SDs, which are more numerous in primate genomes than interchromosomal duplications (3% vs. 1.7%) experienced a completely different dynamic. Over the period during which interchromosomal SDs appeared to explode (i.e. the last 25 million years), SDs spread across the hominid ancestor's genome, accumulating more specifically on 11 chromosomes and 72 gene-rich regions (Marques-Bonet et al. 2009b). Over the past two million years, the number of intrachromosomal SDs has begun to decrease in the human lineage (She et al. 2006). The rate of similarity between intrachromosomal SDs decreases as the distance between the copies increases (Zhang et al. 2005); this could suggest different repair mechanisms involved according to the distancing of copies, advocating the intervention of homologous recombination and more specifically of a fairly conservative mechanism when copies are close and long; and conversely, the intervention of less-conservative mechanisms when copies are more distant.

2.7.4.3. *Palindromic intrachromosomal SDs: the particular case of the Y chromosome*

The Y chromosome, a haploid DNA fragment transmitted solely from father to son, is a unique piece in the human genome. Because of its haploid state, it enables a phylogenetic tree to be built on the basis of stable markers (typically SNPs) and thus the processes affecting each of the phylogenetic branches to be studied (Jobling and Tyler-Smith 2003).

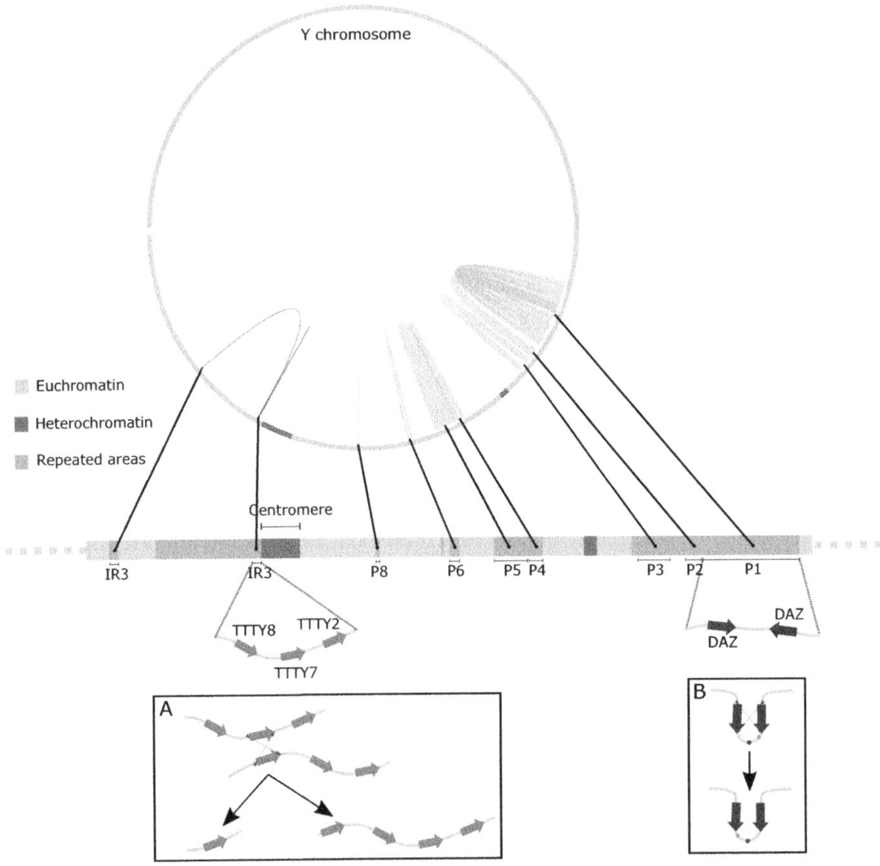

Figure 2.2. *The human Y chromosome, its major duplications, and the alterations that they cause: (A) insertion/deletion in TTTY tandems, (B) inversion in the DAZ area*

This chromosome has a very particular architecture since one third of the chromosome consists of mainly-intrachromosomal SDs (Skaletsky et al. 2003). Of these SDs, about 50% are palindromic, made of two mutually complementary copies, positioned head-to-tail. A total of eight palindromes 30 to 2,902 kb long occupy the long arm, presenting levels of identity varying from 99.95% (P1) to 99.997% (P8) and harboring numerous multicopy genes involved in human reproductive functions (DAZ, VCY and RBMY). It has been proposed that non-allelic homologous recombination (NAHR) or gene conversion be involved in maintaining homology between the Y-palindrome arms, with a rate of 2.2×10^{-4} conversion/duplicated nucleotide events per generation (Rozen et al. 2003). We have shown that strong differences in genomic architectures, due to SDs, exist between Y chromosomal lineages and impact the copy number of the genes involved in fertility (DAZ, RBMY, CDY, VCY). Our results showed that depending on their phylogenetic origin, certain Y chromosomes are prone to rearrangements, while others are much more stable (Balaresque et al. 2008), with a mutation rate of 5×10^{-5} to 5×10^{-6} per generation. The variable potential of lineages to do structural polymorphism should be taken into account in fertility research and especially to explain the variable impact of pesticides, heavy metals, and other pollutants. Depending on their initial structure, certain Y chromosomes may be more prone to being subjected to mechanisms of generations of CNV. This lineage-specific propensity for CNVs is also found on the scale of the whole genome (Bergström et al. 2020). All of these results suggest the existence of genomes that are more unstable than others, due to their genomic architecture's predisposition to structural variations, which could be exacerbated under the influence of intrinsic or extrinsic factors (Yona et al. 2012).

To go further in our understanding of the dynamics of intrachromosomal SDs, we chose to study one of the smallest palindromes, P6. We characterized the existing differences between the two arms, or PSV (Pseudo-Variants), and studied 378 men belonging to distinct phylogenetic branches of the tree of the Y chromosome (Jobling and Tyler-Smith 2003), then estimated a rate of gene conversion in the order of 2.9×10^{-4} to 8.46×10^{-4} events per locus and per generation (Hallast et al. 2013); this rate varies widely between phylogenetic branches of the Y chromosome, and the gene conversion appears significantly GC-biased, presenting preferential binding of the G or C sites. By sequencing 220 additional individuals (plus chimpanzees and gorillas), we showed that the gene conversion affects long fragments of 2,068 bp on average, reaching up to 9 kbp – and not 2,500 bp, then described as the length of the smallest convertible segment, all organisms combined (Mansai et al. 2011). There also exists a trend toward conversion to the ancestral state (i.e. return to the present state in chimpanzees and gorillas) suggesting the possible selection of a basic architecture by the action of a conservative mechanism. These results reveal a fast, GC-biased gene conversion, acting on large segments of the Y chromosome. From a fundamental point of view, this key result suggests that this force, while helping to conserve the original shape of a region, could also reduce the adaptive capacity of the region by limiting the plasticity of the Y chromosome and potentially the individual's resilience. The variation in the conversion rate between phylogenetic branches and the high concentration of genes involved in reproduction present on palindromes make this result all the more interesting to explore in fertility biology. These results suggest a more stabilizing than diversifying mechanism, but also variable between phylogenetic lineages; they raise the question of the role of fine chromosome architecture on sequence evolution and the individual adaptive potential through the arrangement of SDs and fertility genes. The copy variability motif is incompatible with a neutral evolution, where selection would have preserved a stable structure in the genome (Teitz et al. 2018). The observed replacement of the human lineages of Denisova (700kYA) and Neanderthal (370kYA) may have been precipitated by reduced adaptive potentials leading to rapid replacement of male lineages, with the hypothesis of a biological cause remaining open (Petr et al. 2020).

2.7.5. *Intra- and interchromosomal SDs: the specific case of sex chromosomes*

Lastly, complex regions exist comprising of not only intrachromosomal, but also interchromosomal duplications due to a complex evolutionary history. The sex chromosomes are a prime example of this. The X and Y intra- and inter-chromosomal glsSDs are derived from a pair of ancestral autosomes, and additional successive duplications and inversions (Lahn and Page 2000). This explains the very complex profile of certain SDs linked to sex chromosomes (see Box 2.3 for an example).

2.7.6. *SDs: an association with specific sequences?*

SDs appear to be associated with specific sequences such as replication origins, termination sequences, signal recognition particle RNAs (srpRNA), satellite DNAs, long terminal repeats (LTRs) and, especially, highly repetitive, interspersed sequences, such as Alu (SINEs) elements that are found to be enriched at the borders of SDs (Bailey et al. 2001, 2003, Cheung et al. 2003); nevertheless, repeat density does not seem to have a strong influence on the duplication rate (Zhang et al. 2005). These rearrangements resulted in the development of the fragile breakage model in mammals (Bailey et al. 2004, Pevzner and Tesler 2003): not only does a strong association exist between chromosome breakpoints and SDs (51%) in most mammalian genomes, but there is also a significant reuse of these breakpoints associated with NAHR events in mammals (20%). The link that exists between the fragility of some chromosomal regions and their density in SD remains to be investigated.

2.8. SDs in little-studied species: general genomic profiles

In the literature, the study of structural polymorphism and SDs is detailed primarily in humans, primates and some model species for which the genomes are most advanced. But with information on SDs of other genomes not yet available, we are unable to determine exactly what proportion of the total genome these sequences occupy, nor even whether they exhibit characteristics similar to those of the organisms already studied suggesting similar mechanisms at work. For example, the question of a physical constraint on the accumulation of large duplications continues to be asked.

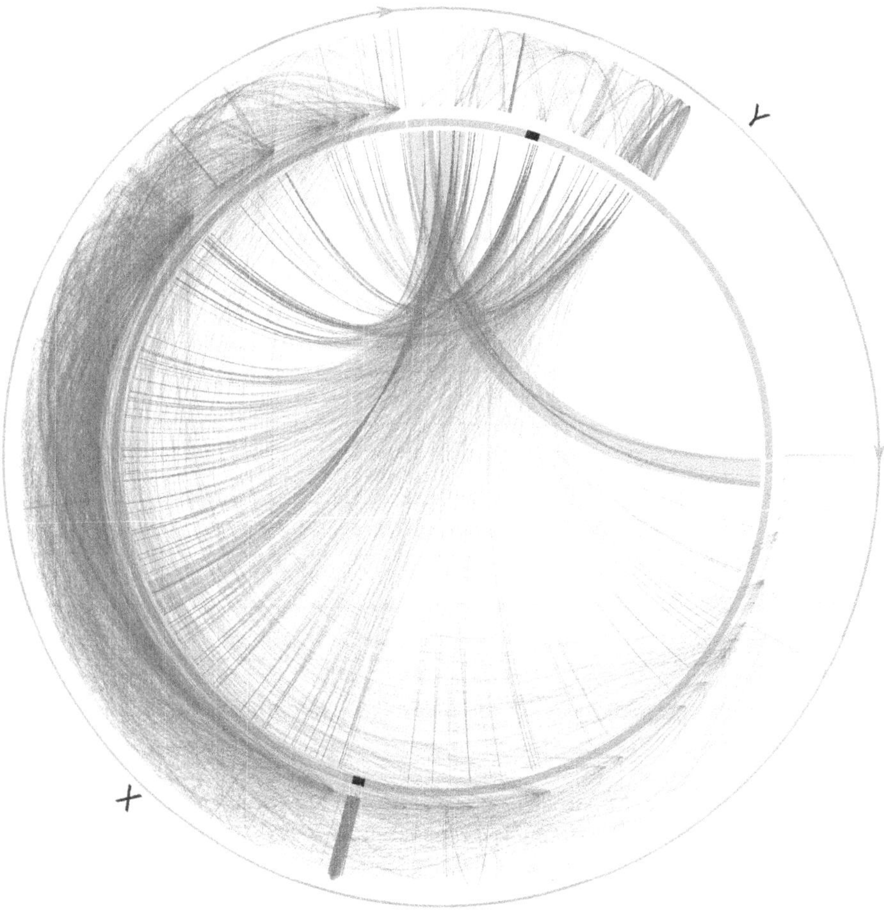

Figure 2.3. *Representation of interchromosomal (center) and intrachromosomal (outer edge) SDs of human X and Y chromosomes. Direct SDs are represented in orange, palindromic in ultramarine*

2.8.1. *Twelve genomes under study*

Within the framework of this review, we propose a succinct analysis of 12 genomes selected on the quality of their sequencing and representing the species discussed here (see Table 2.2) as much as possible.

Species	Sequencing	SD, % genome	% SD subtel.	% SD > 5 kbp	% SD > 20 kbp	Intra-/interchromosome ratio
Bos taurus	PacBio & Illumina, 80–	3.08	20.10	1.45	0.45	1.56
Homo sapiens	Sanger	6.20	8.10	4.80	3.70	
Ovis aries	Oxford Nanopore, PacBio RSII & Illumina, 55–	2.74	14.88	1.25	0.50	1.96
Gallus gallus	PacBio & Illumina, 102–	3.12	23.16	2.53	2.08	3.14
Canis familiaris	PacBio & Illumina, 57–	2.68	17.15	1.81	1.07	3.06
Sus scrofa	PacBio, 65–	2.99	20.32	1.85	0.91	1.53
Rattus norvegicus	Sanger, PacBio & BAC, 10–	9.33	12.82	6.57	3.59	2.75
Pan troglodytes	124–	3.83	18.06	2.42	1.45	1.66
Macaca fascicularis	Oxford Nanopore, 54–	3.79	16.88	2.71	1.49	3.19
Pongo abelii (PPYG2)	Sanger, 6–	6.10	10.71	0.97	0.08	3.80
Pongo abelii (PABv2)	PacBio & Illumina, 101–	2.60	20.31	1.28	0.56	1.87
Arabidopsis thaliana	Illumina & Sanger	7.30	1.85	4.10	2.30	N/A

Table 2.2. *Information on the SDs extracted from 12 genomes*

Our results show a variation in the quantity of SD in genomes from 2.6% to 9.33%, based on the most commonly accepted definition (>1 kb and 90% identity) and an identical process used to identify SD for all species (Delehelle et al. 2018). This result can be interpreted from several angles: (1) sequencing and assembly are not sufficiently accurate and a large number of SDs are not placed or are not placed correctly, which would explain the variability between organisms; (2) there is very high inter-individual variability, and we do not have access to an estimation of this variation through the analysis of one genome per individual and per species, or (3) the influence of forces limiting the accumulation of SDs beyond a certain threshold over the entire genome (Petrov 2001). However, the latter are multiple both in nature (biochemical, evolutionary and mechanical) and in scope (individual, population and species), and may enter into conflict, making the problem virtually impossible to study with today's data and technologies. Strict comparison of genomes acquired with the same sequencing, assembly and post-assembly technologies is the only option for an exact comparison.

Information on the percentage or gross quantity of duplications within a genome is ultimately only slightly, or not at all, correlated with the structural and functional interest of the latter (see Figure 2.4). This apparent inconsistency can be explained in many ways. Firstly, the term "segmental duplication" covers a broad spectrum of duplications that, while they all satisfy the definition, respond to multiple scenarios of formation, dynamics and function in the genome: there is ultimately little in common between, for example, a retrotransposon traveling through the genome and a tandem of genes. Secondly, the duplicated regions are notoriously complicated to sequence, and usually require several complementary sequencing techniques and tools. The effect will be all the more pronounced when the same comparison is made between different genomes of more and less distant species.

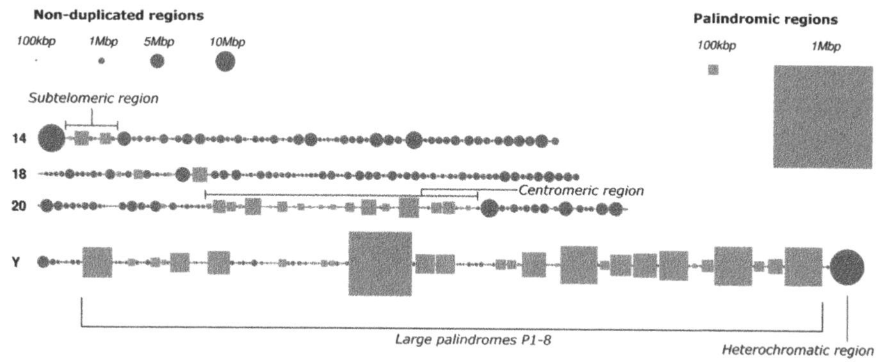

Figure 2.4. *Distribution of duplications on human Y chromosomes compared to a sample of autosomes. Palindromic duplications are shown in turquoise and non-duplicated regions in gray. The size of the palindromic regions is proportional to the square of their length, that of the non-palindromic regions to the logarithm of their length – this differential scale makes it possible to better grasp the relative distribution of the palindromic regions. The very large palindromic regions of the Y chromosome are exceptions, which, although they do not significantly affect the total length of duplications present in the human genome, nevertheless display a unique structure from which their very particular role arises. In contrast, palindromic duplications of autosomes are mostly composed of fairly common repeats charged with little coding material, the most massive being located in the largely repeated areas of telomeres (chromosomes 14) and centromeres (chromosome 20), those more modest in length breaking down into smaller elements over the entire chromosome (chromosome 18)*

Lastly, genome sequencing, even ideal, only ever represents a single individual. While continuing work is able to provide a more complete view of the polymorphic wealth of the human genome through the prism of the sequencing of several hundred genetically distinct individuals (Siva 2008), such a venture is not yet being pursued with respect to all the other species cited here; the results presented here are, so far, only reflecting partial information on these species.

As an illustration, we can note that two different versions of the orangutan genome, PPYG2 and PABv2, sequenced from samples from two different individuals more than 10 years apart, with biochemical and bioinformatic technologies equally separated, yield dramatically different results in terms of the quantity of duplications present.

2.8.2. *Distribution and characteristics of SDs in genomes*

The relationships observed between intra- and interchromosomal duplication content, although highly uncertain – due to the technical limitations detailed above – and covering a broad subset of the mammalian clade, remain included within a range from 1.5 to 4, which places humans (1.8) within this range. If we focus on the distribution of SDs on chromosomes, we again observe a data distribution close to humans. Since telomeres themselves are areas almost exclusively composed of high-frequency duplications, they are particularly difficult to sequence and delineate. Thus, by accepting any SD found in the first or last 5% of a chromosome as sub-telomeric, we observe a distribution for the species concerned, in a number of bases, ranging from 8% (for *H. sapiens*) to 24% (for *G. gallus*) of subtelomeric duplications. This notable enrichment, shared by all of the species represented, probably reflects a structural role of subtelomeric SDs, marred by imperfections and bias arising from the difficult assembly of areas composed primarily of duplications of all scales.

In the end, beyond the gross quantity of duplication per species, it is mainly their distribution, their content and their own dynamics that are of interest. For example, while the human genome does not stand out significantly from that of other primates in terms of the number of duplicated bases, the activity of a few large duplications enriched in genes of vital importance on the Y chromosome, which itself is endowed with a unique dynamic, creates a genomic area with activity that is unique, both in terms of its mechanisms and in terms of its evolutionary consequences. Thus, a macroscopic understanding of the duplications of a species is impossible. It is only by combining several types of information (sequence, genes, regulator network) that new scenarios can be proposed and the dynamics of each duplication family understood.

2.9. SD content: impact of a duplicated environment on sequences that make up the SDs

SDs have their own mutational dynamics. As a result, the sequences that make them up possess mutational dynamics that are all the more unique because they are located in a duplicated environment. Polymorphisms of conventional sequences/markers (SNP, SINE, microsatellites, minisatellites, etc.), when they are located on SDs, respond to theoretical models that differ greatly from what is expected in a single-copy region. These are subject not only to their own mutational mechanisms, but also to additional mutational mechanisms that apply to SDs. Unfortunately, the SNP, microsatellites and other commonly-used marker sets rarely take this factor into account.

2.9.1. *SDs and non-coding sequences: the case of microsatellites*

Microsatellites, discussed in Chapter 7, are tandem repeat sequences with a high level of polymorphism between individuals. Extensively studied in the context of individual identification – particularly in forensic medicine and population genetics, microsatellites have been explored in a wide range of theoretical and applied studies (Goldstein et al. 1999). The size polymorphism of these sequences has been the subject of the development of several mutation models (IAM, SMM, KAM) based on the hypotheses that (1) these sequences are neutral and the observed polymorphism results from a balance between mutation and drift, and (2) no additional evolutionary force influences the theoretical spectrum of the expected alleles.

But the broader regional context in which microsatellites sequences are located (single- or multicopy, near or in a gene or a regulatory element) is rarely taken into account, nor their polymorphism, which could be affected (impact of non-allelic homologous recombination, selection, etc.). In some regions, such as sex chromosomes, and more specifically the Y chromosome, microsatellites are subject to a structural polymorphism and are themselves found to be CNV. Yet the impact of duplication on the polymorphism of these markers has been studied very little (Balaresque et al. 2003); even for the microsatellites markers, most anciently used in forensic science to establish male genetic profiles, such as DYS19, DYS389I/II and DYS385 (Jobling 2008).

Microsatellites, subject to CNV presenting several undifferentiable copies, should not be used in the context of applied studies; because of the new alleles generated by gene conversion, their mutation spectrum is beyond any known theoretical framework and cannot be interpreted under classical mutation-drift hypotheses (Balaresque et al. 2014). For example, microsatellites DYS385ab presents the highest diversity index of all microsatellites linked to the Y chromosome (Ballantyne et al. 2010) induced by the regular appearance of new allelic associations generated by gene conversion (Balaresque et al. 2014).

In theory, any study involving microsatellites should begin with a preliminary analysis of the local genomic context and the expected allelic spectrum in order to verify whether the theoretical models used are compatible with the observed allelic spectra: this approach would exclude inadequate hypotheses concerning the expected polymorphism and derived diversity indices (heterozygosity), such as in a duplicated region or in the vicinity of genes or regulatory elements. It has been shown that the polymorphism of microsatellites was directly related to the surrounding genomic context, which can be modeled either by sex-specific selection (Balaresque et al. 2003) or by gene conversion (Balaresque et al. 2014) when they are located on SDs and subject to CNV (Balaresque et al. 2008). Conversion rates would vary strongly between phylogenetic lineages, attesting to mechanisms and lineage-specific rates, induced by the local structure (Hallast et al. 2013).

With this in mind, it is important to select the microsatellite markers as accurately as possible based on their genomic location, evolutionary forces modeling them and expected mutation rates – making it possible to estimate and interpret diversity metrics as best as possible within the framework of population genetic analysis. It is also important to analyze lineages with very different phylogenetic origins to maximize the diversity of observed characteristics. In the case of the Y chromosome, where many microsatellites present CNV profiles, using a set of unique microsatellites chosen and covering a broad mutation spectrum allows the estimation of genetic distances between individuals to be optimized. We have also shown that it is possible to estimate the phylogeographic lineage of an individual using microsatellites only, as an alternative to using SNP, when the set is well chosen (excluding SD-CNV like loci) and coupled with ad hoc learning methods, guaranteeing the prediction of the Y phylogenetic lineage without turning to SNPs with an accuracy of 97% on average, reaching 100% in the most favorable cases (Bouakaze et al. 2020).

2.9.2. *SDs and coding genes: the fate of genes in SDs*

2.9.2.1. *General overview*

The duplication process, as mentioned above, is one of the main mechanisms involved in the expansion of multigene families that can produce new copies of these genes, responsible for the creation of 50% of prokaryotic genes (Brenner et al. 1995, Teichmann et al. 1998) and 90% of eukaryotic genes (Gough et al. 2001). In the human genome, about 800 to 1,800 genes present a variation in the number of their copies (CNV), through different types of insertion/deletion, inversion or tandem rearrangements (Conrad et al. 2006, Lupski and Stankiewicz 2005, McCarroll et al. 2006, Sebat et al. 2004). This estimation is constantly evolving owing to increasingly reliable sequencing and assembly techniques that allow ever more precise detection and quantification of the variation in the number of copies.

Gene duplication can occur in three main ways: retroposition, unequal homologous recombination (ectopic recombination or NAHR) and non-homologous recombination. Once the gene is duplicated, its fate can be multiple; with duplication constituting a type of redundancy, the direct phenotypic consequences can be difficult to assess.

If the duplicated region is entirely devoid of genes, structural variations outside coding regions are more likely to make no change; or, if there is a change, this would be done by altering one or several regulatory elements (Deeb 2005) or by modifying the chromatin structure locally (Kleinjan and Van Heyningen 1998).

If the duplicated region contains only one gene fragment, protein variants of different types (exon shuffling[3] and alternative splicing[4]) can appear, altering or canceling the original functionality of the protein.

If the region contains one or several functional genes, then this is the only scenario where a high-dose effect can occur. Population genetics shows that the fixation of a new allele would be a rare event, evaluated at 1 per 100 duplicated genes and fixed per million years (Lynch and Conery 2000), although this rate varies greatly over time. It is certainly underestimated at present because of the technical limitations on detection and analysis of duplicated sequences discussed above.

Once fixed, the duplicated gene undergoes a period of time during which the selection pressure is released and can result in four destinies (see Figure 2.5):

1) The copy of the duplicated gene(s) will conserve the same function as the single-copy ancestral gene, producing a dose effect and a new phenotype.

2) Before being copied, a gene can have several distinct functions, and a partitioning of these functions can occur: one copy ensures some of the original functions and the other copies the rest (subfunctionalization), solving, at once, both functional and evolutionary conflicts that may exist when one gene has several functions.

This partitioning may be spatial (different tissues) or temporal, with several copies presenting distinct expressions at different stages of development. An example of this is the cluster of β-globin, whose three genes are expressed at different stages of development under the effect of regulators activated at the embryonic, fetal and newborn stages as a function of the circulating oxygen (Makova and Li 2003). These modified and specialized genes tend to present specific profiles of certain tissues, or play a very precise role in a more integrative function. The fate of the duplicated genes and the modification of their functions in relation to the unique ancestral gene have been the subject of numerous experimental and/or theoretical studies (Lynch and Conery 2000, Conant and Wolfe 2008), and have led to the development of the DDC

3 Process through which a novel gene appears from the exon mixture of two donor genes.

4 A new protein is created by an unprecedented assembly of messenger RNAs.

model[5] and to changes in regulation occurring following a duplication (Force et al. 1999).

3) By the accumulation of deleterious mutation, the gene will transform into non-functional pseudogene – it has been estimated that the time between fixation and degradation would be four million years on average (Lynch and Conery 2000).

4) A new allele emerges in one of the two copies and acquires a new function (neofunctionalization): the accelerated rate of nucleotide mutation after duplication in one of the copies reflects this intermittent release of selective pressure, which will then retighten (Zhang 2003, Zhang et al. 1998).

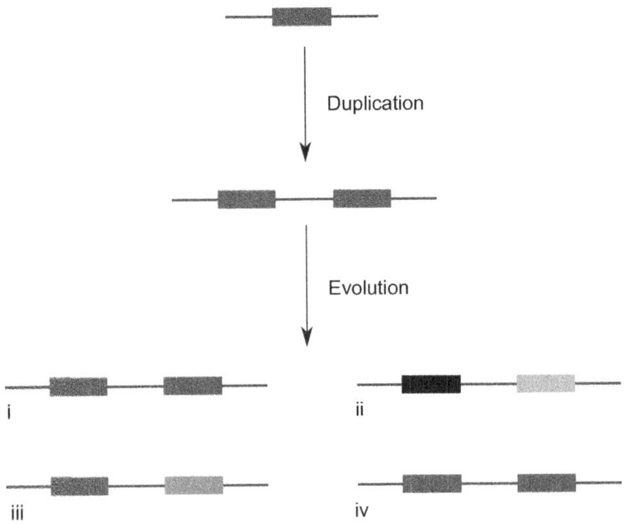

Figure 2.5. *Usual fates of a duplicated gene: (i) conservation; (ii) subfunctionalization; (iii) pseudogenization; (iv) neofunctionalization*

2.9.2.2. *The singularity of SDs containing genes*

SDs containing genes exhibit two singular characteristics compared to those devoid of them: (1) these SDs are greater in number and longer: if we consider a null model of gene accumulation in SDs, the frequency of SDs containing genes is much higher than expected, suggesting that most duplications containing genes would hold a selective advantage and be maintained by selection following duplication (Zhang et al. 2005); (2) it is also observed that the higher the proportion of SDs containing genes, the higher the similarity between sequences, unlike SDs devoid of genes.

5 Duplication–degeneration–complementation.

These results on SDs containing genes can be explained either (1) by a stronger action of the purifying selection and/or (2) by a more efficient homogenization process between copies through gene conversion due to their higher level of identity compared to non-coding regions.

2.9.2.3. *Duplication of genes involved in networks: a specific case within interactomes*

The duplication of a gene will not only modify the stoichiometry of genes in the genomes, but also the potential network this CNV gene is involved in. In baker's yeast *Saccharomyces cerevisiae*, it has been shown that duplicated copies of genes presenting multiple connections are frequently deleted, suggesting that natural selection acts in yeast against duplication of multiconnected groups. Nevertheless, because networks are labile over time, even if both copies of such genes persist, one of the duplicates may lose its many connections, while the other copy may conserve them, thereby restoring some balance (Hughes and Friedman 2005).

Beyond genes, the duplication-divergence phases also affect the regulatory elements of organisms, as they will indirectly modify protein/protein interactions: when a transcription factor is duplicated, the new factor will first recognize the same site and regulate the same target gene as the initial transcription factor and thus modulates the expression of this gene. It is during the divergence phase that it will either continue to regulate the same target gene, or respond to a different signal, recognize a new site and regulate other genes (Teichmann and Babu 2004). Duplication is therefore particularly studied in protein–protein interaction networks: more than one-third of known regulatory interactions are inherited from transcription factors or genes resulting from duplications. Studies of *Escherichia coli* and *Saccharomyces cerevisiae* show that more than half of the regulatory interactions are acquired during the divergence phase after duplication, making duplication a key factor in network evolution (Teichmann and Babu 2004).

2.10. SDs and epigenetic modifications

Given the multiplicity and complexity of the epigenetic modifications cataloged in the plant and animal kingdom, we have chosen to give only a few examples of the link existing between the process of duplication and epigenetic modifications. Epigenetic modifications can lead to changes in the DNA sequence through a number of mechanisms, with DNA methylation being one of the most studied. The modification of histones or non-coding RNAs directly affects chromatin, itself involved in the DNA-protein complex that makes up chromosomes. Chromatin modifications open the door to new mutational processes that can alter the DNA sequence, directly and indirectly, inducing a new type of evolutionary change.

The molecular mechanisms by which duplications are maintained in genomes remain poorly understood and constitute a real mystery for this disciplinary field. It has been proposed that epigenetic modifications, including DNA methylation, could play a key role in the evolution of duplicated genes. They would be involved in the preservation of duplicated copies and in the functional diversification of proteins from duplicated genes in mammals and plants, the rates of neofunctionalization and subfunctionalization increasing when the silencing of duplicated copies is possible (Rodin and Riggs 2003).

This has been observed in the genome of plants where duplicated genes are silenced by DNA methylation or chromatin alteration (Flavell 1994). It is also the case in insects, with the duplicated gene *obp11* in bees, which belong to a multigene family coding for odorant binding proteins (OBPs): this gene expressed only in females, unlike the ancestral gene *obp10*, expressed in males and females, is an example of the role played by methylation in subfunctionalizing proteins from duplicated genes by generating a different but a priori complementary expression pattern to the pattern of the *obp10* ancestral gene (Kucharski et al. 2016).

A study of DNA methylation signatures in humans and mice showed that duplicated genes would be more likely to be highly methylated than other genes (Chang and Liao 2012). Lastly, a study of 10 different tissues in humans demonstrated that while the duplicated genes were strongly methylated initially, they lost methylation with age; tissue-specific methylation would also be correlated with tissue-specific expression (observed in the promoter but not in the gene body). Lastly, for the majority of duplicated gene pairs, one copy is hyper-methylated compared to the other, which would demonstrate the key role of methylation in the evolution of the new duplications (Keller and Yi 2014).

The impact of epigenetic modifications will affect the genes involved in cellular machinery housekeeping genes (HKG) that are constitutively expressed in all tissues – and genes enriched only in certain tissues (TEGs). While many studies have already highlighted different levels of conservation, and the impact of natural selection on these two gene categories, the parallel study of these two gene categories showed that the functions and expression models of HKGs were conserved and did not present SD and CNV. TEGs, meanwhile, are both enriched in terms of SDs and they vary in their copy number. The high expression of HKGs in different tissues is associated with a decrease in nucleosome occupancy at the transcription start site (increased hypersensitivity to DNase) and not with an increase in copy number and dose effect. An enrichment of the CpG islands is observed at the transcription start sites for the HKGs, and conversely, a depletion of these islands is noticeable in TEGs. Histone methylation patterns differ significantly between HKG and TEGs, suggesting that methylation also contributes to differential expression patterns (She et al. 2009).

The study of the duplication and emergence of novel genes is usually carried out by comparing genomes of similar species and therefore concerns genes that are fixed over long time scales. Yet in the emergence of a novel gene, the creation of a new transcript is an essential early step that is worth exploring. The study of transcriptome from several tissues in nine geographically distinct populations of three-spined stickleback showed that novel genes typically start out as transcripts with low expression and high tissue specificity. Early expression regulation appears to be modulated by gene-body methylation. Although many novel, narrowly expressed genes are often lost, those that survive and subsequently spread through populations tend to gain broader and higher expression levels, while the properties of the associated proteins hardly change (e.g. aggregation). "Young" novel genes are not specially subject to positive selection, whereas the older novel genes are more so. Taken together, expression of the surviving novel genes is rapidly regulated, probably via epigenetic mechanisms, while structural properties of encoded proteins may only change much later (Schmitz et al. 2020). This observation raises the question of the impact of epigenetic processes in the selection of novel genes in the short term and long term.

Many multigene families presenting monoallelic expression are grouped in areas presenting excess SD (Gimelbrant and Chess 2006), areas prone to rapid evolutionary changes (e.g. olfactory receptor genes (Young et al. 2002)). While asynchronous DNA replication could affect the duplication rate in return, several chromatin states of different replicated regions could lower the recombination rate in these regions. Thus, a subtle balance would be struck between expanding gene families and maintaining a certain genomic stability. By allowing species to modify certain phenotypes rapidly in response to a changing environment, but while keeping a certain physical constraint on the genome, these mechanisms could constitute a real potentiometer for maintaining areas of high instability.

While whole-genome duplication (WGD) does not change the stoichiometry of genes within the genome, SDs can significantly unbalance the expression of the genome by duplicating only certain genes and/or regulatory elements. By triggering genome-wide trans-effects, these could modulate the expression of duplicated genes by more-complex cascading effects (Shi et al. 2020). The deleterious effects of large duplications could be potentially silenced by endogenous cellular machinery involving epigenetic repressive marks or RNA interferences (for a discussion, see Vaschetto and Ortiz (2019)).

Lastly, it should be noted that the impact of epigenetic modifications on gene expression and that of structural polymorphism are not always easy to distinguish. It has been shown that the overexpression of the FGF4 gene (a potential therapeutic target) initially attributed to methylation of an FGF regulator was due to the particular structure of the duplicated region closely correlated with the expression of FGF4 (proximity of a superactivator ANO1) (Urbini et al. 2020). This example shows the need to progress on the fine structural characterization of genomes in order to conclude as accurately as possible as to the mechanisms involved in gene evolution and expression, and the potential impact of epigenetic changes.

2.11. The adaptive potential of SDs: between the benefit of innovation and the cost of pathology

Mechanisms generating structural polymorphism are considered within the scientific community from two distinct angles: (1) an evolutionary mechanism that essentially generates redundant, unhelpful information that would not increase individuals' fitness and whose variants are often eliminated; or (2) an effective evolutionary mechanism to create novelty from the existing at a lower cost: by copying a matrix that is already present, the duplication can be seen as a recycling mechanism, reusing a proven system. The variants would be subject to drift and natural selection, but these variants nevertheless remain an opportunity given to the organism to acquire a new function or to make it more complex, allowing it to diversify, specialize, and access a new phenotype better adapted to its environment. Ninety percent of eukaryotic genes are the result of a duplication process, so the highly effective evolutionary mechanism is the preferred hypothesis. Compared to other sequence categories, SDs can induce many rearrangements leading to pathologies by non-allelic recombination, in different ways: the effects will be different depending on whether the SDs will directly include genes or only frame them, and whether they will be placed in direct or inverted orientation (see Figure 2.6 and previous discussions). The high variability in SD lengths (1,000–200,000bp) and the distances separating them (0–5 Mbp) are also responsible for a wide variety of rearrangements generated (Lupski 1998): these can induce dose effects on one or several genes, creating highly penetrating, alleles, as in neurodegenerative diseases (Marques-Bonet et al. 2009b).

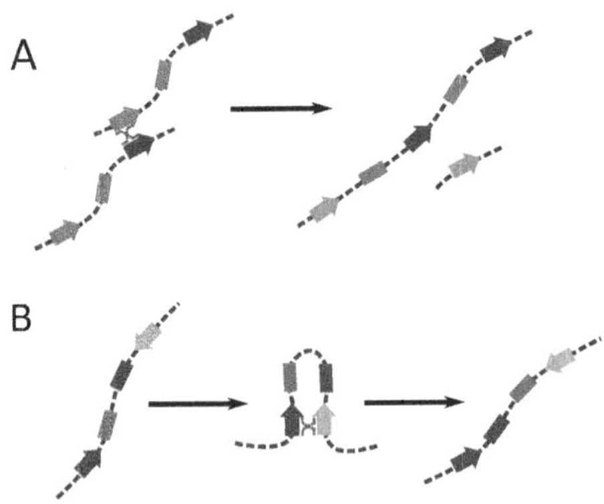

Figure 2.6. *Rearrangements involving SDs that are directly (duplication/deletion, A) or indirectly (inversion, B) orientated*

The first sequencings of mammals (human, mouse, dog, opossum, etc.) showed that the duplicated genes belonged to well-defined functional categories (Church et al. 2009, Plassais et al. 2019). To be found are phenotypes related to immune defense and host defense (T lymphocytes and immunoglobulins), sensory perception of the environment (vision, olfaction and taste), degradation of toxins (cytochrome P450 family), reproduction (placental growth factors), food assimilation and phenotypes more specifically linked to the sex chromosomes (cognition and reproduction).

2.11.1. *The organism's defense: immune system*

One of the first pathologies to be described involving rearrangement between SDs is α-thalassemia (Higgs et al. 1980). This disease is due to the deletion of the α-globin gene following an unequal crossing-over between two 4 kbp blocks located within the alpha-globin gene. Exchanges between these two blocks lead to individuals possessing one or three alpha-globin genes. It has been shown that the number of alpha-globin genes and the quantity of alpha-globin mRNA are correlated, and individuals carrying only one alpha-globin gene will be affected by this disease: homozygous deletion leads to disease, whereas heterozygous deletion will, conversely, protect carriers of this disease, an example of balanced selection.

The CCL3L1 (CC motif chemokine ligand 3-like 1) gene is one of many genes that encode cytokines, proteins involved in inflammatory and immunoregulatory processes. It exhibits structural polymorphism, and the higher the number of copies, the lower the risk of HIV infection and AIDS progression in humans (Liu et al. 2010). This effect is also found to be visible in macaques, with Indian-origin macaques having few CCL3L1 copies demonstrating far lower survival rates than Chinese-origin macaques, which have a high copy number (Degenhardt et al. 2009). Note that a high copy number of the CCL3L1 gene could also be responsible for the appearance of systemic lupus erythematosus (Mamtani et al. 2008) involving modulation of the maximum copy number. This would explain the contrasting results concerning the structural polymorphism of this gene in the literature (Perry 2008), suggesting the existence of a trade-off on the optimal copy number of CCL3L1.

Also, variation in the copy number of the FCGR3B gene (Fc receptor for immunoglobulins γ (IgG), low-affinity III) would determine the susceptibility to developing immune-induced glomerulonephritis (Aitman et al. 2006); a low copy number of the FCGR3B gene would predispose patients with lupus erythematosus to develop this kidney inflammation.

Lastly, genes coding for beta-defensins, which play an important role in the innate immune system, would also be subject to structural variation: a decrease in the copy number would result in a predisposition to Crohn's disease (inflammation of the colon) (Fellermann et al. 2006), whereas an increase in the copy number of beta-defensins would be associated with a high risk of psoriasis (Hollox et al. 2008).

2.11.2. *Nutrient/food assimilation*

The gene coding for the digestive enzyme amylase also exhibits structural polymorphism. Mammals produce amylase with their pancreas, but primates, rodents and lagomorphs also present salivary expression of this enzyme. The human genome contains two pancreatic genes (AMY2A and AMY2B) and three salivary amylase genes (AMY1A, AMY1B and AMY1C) with two clusters positioned side-by-side and head-to-tail on chromosome 1 (Perry et al. 2007). While we possess six copies on average, a very high CNV variability of amylase genes exists, ranging from 2 to 18 copies depending on the individual (Carpenter et al. 2017).

2.11.3. *Sensory perception of the environment*

Duplications are overrepresented in the genes involved in perception of the environment, mainly olfactory, gustatory, visual and auditory perception; this observation is consistent with the evolutionary plasticity that they offer – an asset for the adaptation of individuals to their environment.

The evolution of trichromatic vision, for example, is a recent sensory innovation in primates due to the duplication of the opsin (Deeb 2005) gene just after the divergence of the lineage of monkeys of the ancient world with those of the new world (Dominy and Lucas 2001, Jacobs et al. 1996). This is a prime example of the rapid evolution of a new function that occurs through tandem duplication (Samonte and Eichler 2002). This trichromatic vision, which consists of distinguishing red, blue and green, requires the synthesis of three photopigments. After the gene's duplication on the X chromosome, a mutation enabled detection of visible light on a wider spectrum, which would have been of key importance to the ancestors of these primates, who were then able to distinguish ripe (yellow and orange) from green fruit. Some authors suggest that the acquisition of this trichromatic vision may have led to the decline of the olfaction genes observed in primates by limiting the pressure of positive selection on them (Rouquier et al. 2000, Gilad et al. 2004).

Odor recognition is assured by the olfactory receptors (OR): receptors are coupled to G7-transmembrane proteins that are expressed in the nasal olfactory epithelium (Glusman et al. 2001, Niimura and Nei 2007). ORs constitute a curiosity in the genome and landscape of SDs, because they are distributed over half of the subtelomeric and pericentromeric regions, constituting one of the most impressive examples of interchromosomal polymorphism (Trask et al. 1998). With 400 apparently functional members (compared with 1,200 in mice), ORs are one of the largest multigene families in the mammalian genome (Zhang and Firestein 2002, Young et al. 2002). The proportion of functional OR families differs between species (from 80% in mice and dogs to 50% in humans). The human genome contains around 100 blocks of OR, which each contain between 1 and 105 genes. A variation also exists in the copy number of some of these genes in humans, not only between individuals of the same ethnic group, but also between ethnic groups (Young et al. 2008). It has been shown that both the mechanisms of NAHR and of NHEJ were involved in the rearrangements and CNV of functional OR CNV genes. It is difficult to exclude the action of drift to explain the very large variations in the copy number of olfactory receptors observed between two individuals in humans (Nozawa et al. 2007), which does not allow conclusions to be drawn on the evolutionary forces driving the OR family.

Taste perception begins at the papillae located on the surface of the tongue and the palate. Each contains one or several taste buds, which themselves are composed of hundreds of cells containing specialized receptors (Chaudhari and Roper 2010). These receptors will translate the signal and transmit it to the brain where it will be transformed into taste (Kinnamon 2013). Two major families of receptors are involved in the perception of three sensations – umami, sweet and bitter (salty and acidic passing through ion channels): the TAS1R and TAS2R families are coupled to G-chemoreceptors (GPCRs). In humans, the TAS1R family consists of three genes, TAS1R1, TAS1R2 and TAS1R3 (Zhao et al. 2003). The TAS2R family has 25 functional genes that code for bitter taste receptors in humans (Conte et al. 2003, Dobon et al. 2019). The TAS2R43 and TAS2R45 genes exhibit structural variation, with the copy number ranging from 0 to 2, in 946 individuals and 52 populations (Roudnitzky et al. 2016).

Lastly, hearing sensitivity is also affected by structural variation. This trait has been the subject of few evolutionary studies and the available information mainly concerns non-syndromic hearing loss (NSHL). Of the many genes potentially involved in hearing sensitivity (Angeli et al. 2012), deletion of an SD from chromosome 15 including the STRC gene causes an autosomal recessive NSHL (ARNSHL) (Francey et al. 2012), but also deafness-infertility syndrome (DIS) when the neighboring CATSPER2 gene is involved (Zhang et al. 2007). Of the 89 genes involved in NSHL, structural variations would affect 18 genes (Shearer et al. 2014). These CNVs would be caused mainly by the mechanism of NAHR and would mainly affect seven genes: TSPEAR, OTOA, STRC, MYO3A, ESPN, OTOGL and CACNA1D. The MYO3A gene is part of the multigene myosine superfamily, which consists of 35 classes of molecular motors based on actin (Odronitz and Kollmar 2007), several classes of which would play a key role in inner-ear integrity, the formation of stereocilia and the hearing process (Sellers 2000). Stereocilia are structures where the amplification of the received mechanical auditory signal occurs (Petit and Richardson 2009): the MYO3A gene (Class III) is crucial during the development of stereocilia, because it could function as a barrier to prevent their overelongation (Lelli et al. 2016, Cirilo Jr et al. 2021). Mutations in MYO6 appear to be associated with age-related progressive hearing loss (presbycusis (Oonk et al. 2013)), and MYO7A mutations are associated with both vision loss and profound deafness (Calabro et al. 2019), including the extensively-studied Usher syndrome type B (USH1B) (Kremer et al. 2006); mutations in MYO15A, involving the absence of expression of MYO15A, would lead to a defect in the elongation of the stereocilia and also in their morphology (Probst et al. 1998), and would be responsible for NSHL associated with the mutations in MYO15A (Zhang et al. 2019). The impact of CNV on myosine genes has been demonstrated in the stickleback model (Hirase et al. 2014), but remains little described in other species.

2.11.4. *Neurological processes*

Several pathologies, involving SDs, affect the cognitive and neurological sphere. The protein aggregation effect (Wong et al. 2005) is notably responsible for Parkinson's and Alzheimer's diseases, which are prime examples of pathological dosage effects induced by protein aggregation (Singleton and Gwinn-Hardy 2004). Duplication and triplication of the alpha-synuclein (SNCA) gene leads to increased expression of α-synuclein, a small protein involved in the transport of synaptic vesicles; the pathology and severity of the disease induced by this protein depend on its expression level (Singleton 2005).

Similarly, the accumulation of Aβ42, derived from the cleavage of the amyloid precursor protein β (APP), is responsible for the parenchymal and vascular deposition involved a priori in Alzheimer's disease. The pathology is caused by different mechanisms that lead to increased aggregation of APP proteins: among them we can note the duplication of the APP gene on chromosome 21 (Rovelet-Lecrux et al. 2006) and Down syndrome (trisomy 21), which provides three copies of the APP gene (38 of Conrad). Other neurodegenerative diseases involve similar mechanisms of protein deposition (Hardy 2006).

SDs and the associated duplications are responsible for strict dose effects causing different pathologies of the nervous system: Charcot-Marie-Tooth disease (CMT) (Lupski et al. 1992), which affects the peripheral nervous system, involves a tandem duplication of 1.4 Mb containing about 30 genes: it is found in about 70% of autosomal-dominant patients (MCT1A). This fragment contains the PMP22 gene that codes for the main myelin protein. It is the variation in the copy number of PMP22 and the associated dose effect of myelin production that is responsible for associated diseases: a PMP22 trisomy (duplication) leads to a nerve conduction deficit due to demyelination (CMT1A), whereas PMP22 monosomy (deletion) leads to a nerve conduction blockage. Other pathologies involving demyelination processes suggested that myelin formation was susceptible to a dose effect (Padiath et al. 2006). The dose effect, on the MECP2 gene, is involved in Rett syndrome, a progressive neurodegenerative disease that affects only women, but also men when there is duplication of the MECP2 gene (Van Esch et al. 2005).

2.11.5. *Reproduction and the X and Y chromosomes: true SD concentrates*

The sex chromosomes contain 20% of all SDs in the human genome (Zhang et al. 2005) (Y chromosome 15%, X chromosome 5%). As well as being responsible for determining sex, they are hotspots of genes involved in cognition and reproduction (Wang et al. 2001), with strong antagonistic effects between males and females (Gibson et al. 2002). The evolution of these chromosomes is unique: for many species, the sex chromosomes (X/Y, Z/W) evolved from an ancestral pair of autosomes (Lahn and Page 1999). In the case of the XY system, the acquisition of an allele advantageous for males on the future Y chromosome would have caused a sexual conflict leading to an interruption of recombination locally. This would have resulted in a series of inversions resulting in the cessation of recombination on all the X and Y protochromosomes. This scenario is not the only one that explains the evolution of these heterochromosomes, but they would be produced in different taxa, particularly in therian mammals, birds, snakes, fish and silene (Furman et al. 2020). Of the hundreds of homologous gene pairs that existed on the mammalian XY protochromosomes, only 29 XY homologous pairs would have survived the structural degradation of chromosomes and the process of pseudogenization.

Works on the X (Ross et al. 2005) and Y (Skaletsky et al. 2003) chromosomes have shown intra- and interchromosomal SD enrichment on the latter. Maintaining an interchromosomal sequence identity favored homologous (NAHR) and non-homologous recombination, and subsequent rearrangements intensified structural variations, affecting genes primarily involved in spermatogenesis, cognitive processes, but also CT tumor antigens (whose expression is limited to germ cells *cancer-testis*) (Scanlan et al. 2002). Interchromosomal exchanges are documented for many KALX/Y, VCX/Y, SRY/SOX3, ARSE/ARSEP, ZFY/ZFX, AMELY/X, STS/STSP and PRKX/PRKY genes: while these sequences with high levels of identity are not the result of duplicative transposition, their exchanges have favored local instability and intensified CNV.

911The VCXY[6] genes are located on SDs of about 7.5 kb in Xp22.3/Yq11 on an ancient homology area between these chromosomes. Originally described as CRI-S232 LCRs (Ballabio et al. 1990) and in more detail by other authors (Lahn and Page 2000, Fukami et al. 2000, Bailey et al. 2002, Balaresque et al. 2003), these SDs present between four and six copies on the X chromosome (VCX2r, 8r, 6r, 10r) and two copies on the Y chromosome (VCY, on palindrome 8).

Each SD is composed of several types of repeats (Alu, microsatellites, minisatellites) and each carries a copy of the VCX/Y: these are small, composed of a single intron

6 *Variable Charge XY.*

911(192 bp) and two exons: the first exon of 103 bp is stable, and the second exon is composed of 210 bp joined to one to 13 repeated units of 30 bp, of minisatellite type. The corresponding proteins therefore vary in terms of structure and charges: the more the gene possesses repeats containing arginine and lysine residues, the higher the negative charge of its protein. Because of their high charge, small size and hypothetical expression exclusive in male germ cells, the hypothesis was put forth (Fukami et al. 2000, Lahn and Page 2000) that VCXY proteins could interact with DNA, such as histones, and play a role in spermatogenesis – a hypothesis confirmed (Zou et al. 2003) later, the VCXY family regulating ribosome assembly during spermatogenesis.

An association between a high VCX gene copy number and the risk of non-obstructive azoospermia has also been demonstrated: an increasing number of VCX copies could overregulate gene expression, triggering cell proliferation followed by the process of apoptosis during spermatogenesis (Ji et al. 2016).

In addition, the absence of VCX8 (or VCX-A) causes intellectual disability (Fukami et al. 2000).

Lastly, VCX3A is expressed at the protein level in approximately 20% of pulmonary adenocarcinomas and 35% of squamous cell carcinomas, but does not occur in normal lung outcomes, making it a new antigen associated with CT *Cancer-testis*-type tumors, joining the XAGE1, IL13RA2 and SYCE1 already described (Taguchi et al. 2014).

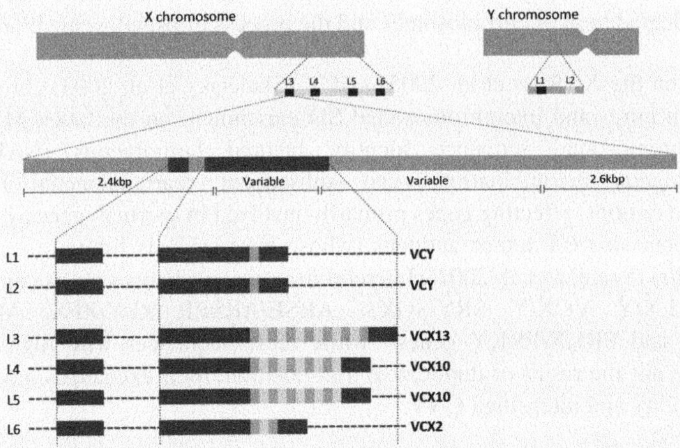

Box 2.4. *VCXY – At the height of structural complexity*

2.11.5.1. *X chromosome and intellectual disability*

The most unstable regions of the X chromosome are the strata 4 and 5, located in the Xp22 subtelomeric region (Lahn and Page 1999, Ross et al. 2005); these regions are the last to have ceased recombining with the Y chromosome. In fact, the X and Y homologous genes found there present higher levels of identity than elsewhere on these chromosomes, and a greater propensity for interchromosomal NAHR and CNVs.

The KAL gene, located on the X chromosome at Xp22.31 in stratum 4, involved in Kallmann syndrome (hypogonadotropic hypogonadism syndrome, IHH) is subject to CNVs. The latter, generated by non-homologous recombination, is induced by sequences around the KAL gene (Ballabio et al. 1990), notably the SDs CRI-S232 (Meitinger et al. 1990) including many sulfatase genes. Among the sulfatase genes (STS, ARSD, ARSE and ARSF), deletion of steroid sulfate (STS) is responsible for ichthyosis syndrome linked to the X chromosome (XLI), a steroid sulfatase deficiency (Kent et al. 2008). It is one example among others of the dose effect for certain genes involved in X-chromosome-related intellectual-disability syndromes, as well as those affecting the UTX, KDM5C, and NLGN4X genes (Bellott et al. 2014, Lindgren et al. 2013, Lawson-Yuen et al. 2008). Deletions of this region are associated with severe cognitive difficulties including autism. About 10% of patients showing an X-linked intellectual disability (XLID) carry CNV of more than 100 kbp on the X chromosome (Muthusamy et al. 2017). Among the structural polymorphisms linked to the X chromosome, large duplications affecting the Xq13.3–q21.1 region appear to be responsible for the dose increase of the ATRX gene involved in the ATRX intellectual disability syndrome. The duplication of the SOX3 gene, developmental transcription factor (homologue and precursor of the SRY gene), in turn would lead to a large deficit in the production of pituitary hormones (Woods et al. 2005).

2.11.5.2. *Y chromosome and SDs: impact on spermatogenesis and beyond*

The Y chromosome, responsible for the sex determinism of males, carries the SRY gene involved in the production of semen, and more generally in male fertility. But because of the presence/absence of the Y chromosome in an individual, a multitude of traits could be modified, including certain diseases (Ferguson-Smith 1965). The Y chromosome could be one of the factors explaining the large variation in the occurence of certain diseases between men and women (Ober et al. 2008). Of all its characteristics, the Y chromosome (or its other haploid homologue, the W chromosome) is the chromosome that has been the most affected structurally by rearrangements during its evolution. It is characterized by a structure made of sequences of different origins: some are inherited from its ancient homology with the X chromosome (X-degenerate) and others have been more recently acquired by transposition (X-transposed) or duplicated (X-ampliconic, large palindromes) (Skaletsky et al. 2003). This chromosome presents a very large duplicated fraction (also referred to as ampliconic), with 30% of the chromosome being made of recently amplified sequences, of which 50% exhibit a palindromic structure, with levels of identity ranging from 90 to 99.997%. The majority of genes, located in these ampliconic regions (palindromes P1 to P8), are therefore duplicated (Rozen et al. 2012). In fact, 60 of the 78 genes of the human Y chromosome are naturally present in several copies. The RBMY, TSPY, and HSFY genes, which had a homologue on the X chromosome, were preserved by becoming ampliconic: today they are multicopics, expressed only in the testicles, and play a critical role in spermatogenesis. Among the regions described as hosting the genes involved in spermatogenesis (AZFa, AZFb and AZFc – *Azoospermia factor* a, b, and c) (Reijo et al. 1995, Nakahori et al. 1996, Simoni et al. 1997), the AZFc region has proved to be particularly involved in spermatogenesis failures. This region includes three large palindromes, three other duplicated and inverted sequences, and three direct repeats, a substrate extremely favorable to rearrangements and to NAHR (Kuroda-Kawaguchi et al. 2001). The publication of this structure has led to a better understanding of the main classes of deletions that had been described as associated with spermatogenesis failures. But the indices of new rearrangements are multiplying (Balaresque et al. 2008, Hallast et al. 2021), showing the great diversity of rearrangements in the palindromic regions: it has been estimated that the prevalence of men exhibiting a CNV associated with a spermatogenesis failure is one in 27 (Rozen et al. 2012). The comparison of molecular and selective processes affecting the ampliconic (multicopy) region with the degenerated and transposed (single-copy) region will be highly informative, because while studies conducted on single regions have so far reinforced the idea that natural selection did not act on the Y, which instead evolved under the action of drift, in line with neutralist hypotheses, the hypotheses on the duplicated region are very different, since it has been shown that a minimal set of certain genes would have been strictly maintained during evolution, suggesting a strong action of natural selection (Teitz et al. 2018).

2.12. SDs and associated CNVs: their roles in species adaptation to changes in environments

2.12.1. *SDs: a link between genomic architecture, adaptive potential and environmental changes?*

In many organisms, SDs and the associated CNV profiles can be related to individuals' adaptation to environmental changes (Kondrashov 2012). Among the most discussed adaptations, we can cite those to temperature, nutrient composition of the soil, toxicity (heavy metals, pesticides, antibiotics, antimalaria treatments) or the use of species by humans (domestication).

2.12.2. *Adaptation to global environmental stress*

Studies on the associations between SV-CNV and stress have recently been multiplying: many organisms show an association of their structural profile, or CNV, in response to an environmental stress such as for the papaya (Zhenyang et al. 2021), pine (Julien et al. 2017), *Arabidopsis halleri* (Suryawanshi et al. 2016) or that of the potato (Hardigan et al. 2016). A detailed study of the structural polymorphism of the latter revealed that more than 30% of the CNV genes were involved in the adaptation of this species to its environment. As previously discussed for others plants, CNVs

Gene/loci	Description	Size	Modification	Consequences	Mechanism	Reference
RDH	Blood group rhesus, D antigen	60kbp	Reduced CNV (0-2 copies)	Rhesus sensitivity	Loss of gene function by dose effect	(Wagner and Flegel 2000)
OPN1LW, OPN1MW	Opsin 1 (cone pigment), medium spectrum sensitivity	13-38kbp	Reduced CNV (0-4, resp. 0-7 copies)	Altered color vision	Loss of gene function by dose effect	(Neitz and Neitz 1995)
Testis-specific genes	Spermatogenesis genes and testicular cancer antigens	Several kbp to 1.6Mbp	Insertion, deletion, CNV	Altered spermatogenesis	Loss of gene function by dose effect	(Repping et al. 2003)
β-defensin (DEFB4)	Defensin $\beta 4$	240-310kbp	Reduced or increased CNV	Psoriasis, Crohn's disease	Loss of gene function by dose effect	(Hollox et al. 2003)
SOX3	Developmental regulator	685kbp	Duplication of SOX3	Hypopituitarism (XLHP) and infundibular hypoplasia	Loss of gene function by dose effect	(Wagner and Flegel 2000, Woods et al. 2005)
PMP22	Myelin protein	From 186kbp to 1.5Mbp	Duplication of PMP22	Chacot-Marie-Tooth 1A, peripheral myelin neuropathy	Loss of gene function by dose effect	(Moog et al. 1996, Lupski and Stankiewicz 2005, Weterman et al. 2010)
SNCA	α-synuclein (molecular chaperone)	Unknown	Duplication	Parkison's disease	Protein aggregation and mutation of the underlying gene	(Shiu et al. 2006)
APP	Amyloid precursor protein	Variable	APP duplication	Alzheimer's disease (parenchyma or vascular amyloid deposition)	Protein aggregation and mutation of the underlying gene	(Gu et al. 2004, Potocki et al. 2000, Hunter and Brayne 2018)
CYP2D6	Cytochrome P450 subfamily II D, polypeptide 6 (P450 isoenzyme)	5kbp	Increased CNV	Altered drug metabolism	Altered dosage	(Caraco 2004, Padiath et al. 2006, Turner et al. 2021)
CCL3L1	Chemokine ligand 3-like1 precursor	64kbp	CNV decrease	Increased HIV-AIDS susceptibility and Kawasaki	Altered dosage	(Gimelbrant and Chess 2006)

Table 2.3. *SDs in humans – Gene, modification, phenotype-disease, mechanism and reference*

usually affect genes involved in pathogen resistance, UV-Bs, resistance to glyphosate-type herbicides (Gaines et al. 2010), nematodes (Cook et al. 2012), frost (Knox et al. 2010), submersion (Xu et al. 2006) and aluminum and boron toxicity (Maron et al. 2013).

2.12.3. *Adaptation to nutrient-poor surroundings*

Many sources of stress exist, including nutrient deficiency within the surroundings. To respond to this, the most commonly observed adaptation involves an increase in the transport of this nutrient from the surroundings to the organism. In the yeast *Saccharomyces cerevisiae*, the recent duplication of the HXT6 and HXT7 genes is associated with the increased expression level of the hexose transporter and the level of glucose imported into the cell (Brown et al. 1998). Lineages possessing duplicated genes are more efficient, compared to parental lineages, at importing glucose and will be favored in a glucose-poor stressful environment. Similary it has also been shown that the yeasts adapting best to sulfate-poor environments present duplications of the genes coding for sulfide transporters (Gresham et al. 2008).

2.12.4. *Adaptation to low and high temperatures*

In *Escherichia coli*, adaptation to high temperatures requires the duplication of several stress-related genes (Christ and Chin 2008). The duplication of the *evgA* gene, a major transcriptional regulator that is part of the signal transduction system, allows this bacterium to withstand temperatures above 50°C.

Antarctic cod (*Gadus morhua*) has adapted to cold waters through duplications. Compared to the genome of its living counterparts in warmer waters, the Antarctic cod genome has 10 times more duplications, including the AntiFreeze GlycoProtein (AFGP) involved in resistance to very low temperatures (Chen et al. 1997).

The expression profiles of the two species have shown that 177 genes are overexpressed in Antarctic cod, including 118 genes derived from multiple duplications. Of these overregulated genes, some have very different functions and include the AFGP, the FBP32II genes, and the F-type lectin, a protein close to the type II antifreeze protein playing a key role in this species adapting to the cold (Chen et al. 2008).

In ectothermic species, temperature is a key factor in life history strategies, since organisms living in warm environments show an accelerated growth, a reduced longevity and early aging. The SNP and CNV patterns acquired over 14 years on *Rana luteiventris* frogs showed that 39 regions of the genome presented CNV, linked to TE, associated with the temperature increase over the period studied. This illustrates the key role that structural variation could play in response to global warming (Cayuela et al. 2021a).

A study conducted on more than a thousand American lobsters (*Homarus americanus*), subject to the highest annual thermal variance of Canadian coastal marine waters, located at 21 sites south of the Gulf of St. Lawrence noted that populations exhibit extreme differentiations based on 48 regions presenting CNV profiles, which are not detectable on the basis of SNPs alone. These variations in CNV are significantly associated with the annual variance in sea surface temperature, resulting in a grouping of populations based on temperature, not geographical separation (Dorant et al. 2020). A similar study conducted on marine fish (*Mallotus villosus*) in the North Atlantic Ocean, showed that 105 regions affected by CNVs are associated with water temperature, suggesting an adaptation, particularly in very cold water, through a gene amplification; additionally, other CNV regions could be associated with potential gene candidates for reproductive isolation, explaining the divergence of three glacial parametric lineages (Cayuela et al. 2021b)[7].

2.12.5. *Heavy-metal adaptation*

Adaptation to high concentrations of heavy metals appears to be related to the ability of different species to export cations out of the cell. Soil bacteria *Ralstonia pickettii* are adapted to high copper concentrations as found in the sediments of contaminated lakes. These bacteria accumulate copper in their outer membranes and exhibit duplications of regions involved in resistance to metals and cation transporters (Yang et al. 2010). Similarly, the fungal pathogen *Cryptococcus neoformans var. grubii* (subclade VNI A5) contains a tandem duplication of the arsenic outward transporter gene that confers resistance to arsenic linked to the number of copies present. An analysis of the CNV pattern in *Arabidopsis halleri*, a species capable of growing on highly toxic soils and able to accumulate extreme concentrations of zinc and cadium, has revealed a link between this species' adaptation to heavy metal and its CNV profile; the genes involved in homeostasis functions exhibit twice as much CNV as the other gene categories and have higher copy numbers. Similarly, genes involved in biotic stress functions – including disease resistance – are twice as affected by CNV as other gene categories with very small copy counts. These results indicate that *Arabidopsis halleri* could offset heavy-metal hyperaccumulation by lowering its threshold of defense to disease, an example of an evolutionary trade-off (Suryawanshi et al. 2016).

---

7 With adjacent but disjointed geographical distributions.

2.12.6. *Antibiotics and drugs*

The routine use of antibiotics and different molecules in anti-cancer or antimalarial treatments induces a response from bacteria, somatic cells, and other organisms; this adaptive response usually occurs via gene duplications. In *Plasmodium falciparum*, duplication of the pfmdr1 gene, coding for an ABC protein involved in transporting chloroquine out of the cell, confers resistance to mefloquine and other antimalarial molecules (Duraisingh and Cowman 2005, Chavchich et al. 2010). Duplication of the pfmdr1 gene has occurred independently multiple times, and an increase in the resistance of plasmodium to these different molecules has been observed in Africa, Asia and South America (Gadalla et al. 2011). The amplification of the gene pfmdr1 appears to be due to a positive selection action (Nair et al. 2007). Similarly, the bacterium *Escherichia coli* also shows an amplification of several genes when it is found in toxic environments (i.e. in the presence of antibiotics), giving it an increase in its fitness (Soo et al. 2011). Lastly, in *Candida*, the duplication and expansion of the mediator subunit[8] has occurred on three occasions, suggesting an important role of the mediator in the evolution of these pathogens (Moran et al. 2019).

2.12.7. *Pesticide resistance*

The resistance of several mosquito species to pesticides, used to limit the spread of malaria, has been studied for more than 40 years. But for the past 10 years, the key role played by duplications and many other variations in genome structure in this resistance phenomenon has become clear. Sequencing of the genome of *Anopheles gambiae* has enabled the identification of 250 CNV genes, 28 of them organized into multigenerational families related to metabolic resistance to insecticides (Lucas et al. 2019).

The adaptation of the mosquito *Culex pipiens* to pesticides was achieved through the duplication of two non-specific esterases (Est-2, Est-3) and the Ace1 gene coding for acetylcholinesterase, one of the targets of organophosphate insecticides; acetylcholinesterase playing a key role in cell apoptosis in insects (Knorr et al. 2020). These duplications would have been maintained by the rapid action of positive selection as soon as the duplication occurred (Lenormand et al. 1998). Duplication of the Ace1 gene in pesticide-exposed populations has been observed independently worldwide (Labbé et al. 2007a): this duplication is associated with a high fitness cost, which would explain the maintenance of duplications in heterozygous state. Mutations in the regulator of the Ace1 gene coupled with an increase in the copy number (causing overregulation of the Ace1 gene) also plays a key role in the

8 An important regulator of gene expression for eukaryotes.

resistance of the domestic fly to Propoxur (You et al. 2020). These observations would suggest the same genomic response through the mutation/duplication mechanism of the Ace1 gene and of these regulatory sequences in several insect species when exposed to different insecticides targeting acetylcholinesterase (Labbé et al. 2007b) in all of these populations.

Structural variation is one of the most effective mechanisms of resistance to glyphosate herbicide: this resistance involves the duplication of the EPSPS gene (5-enolpyruvylshikimate-3-phosphate synthase), whose copy number can vary from 3 to 150. Among these plants, note that Palmer amaranth (*Amaranthus palmeri*) – or quelite – contains duplicated fragments of more than 300 kbp including EPSPS and other genes, which are inserted at a position distinct from that of the ancestral EPSPS gene in the genome. Mobile elements, helitrons, contained in these duplicated fragments, suggest the potential amplification of the EPSPS gene by an extra-chromosomal pathway; whereas the 3 to 10 copies of the EPSPS gene located in tandem repeats, near the ancestral gene in the summer cypress (*Kochia scoparia*), suggest a more classical mechanism for acquiring these copies. This evolutionary convergence, reflected by the accumulation of CNV of the EPSPS gene by different multiplication pathways, reveals the importance of this adaptation mechanism in plants (Patterson et al. 2018). The direct relationship existing between CNVs and resistance to fungicides places these plants in the face of an important evolutionary trade-off, because the duplication rate of the EPSPS gene and the associated enzyme induces a change in the plant's catalytic activity: the mutations that confer the highest resistance to glyphosate are also those that decrease the catalytic activity. Understanding the relationship between the overexpression of the EPSPS gene and the biochemical and physiological consequences will help to evaluate the cost of glyphosate resistance (Vila-Aiub et al. 2019).

CNVs are also involved in resistance to fungicides, as illustrated by the duplication of the EnCYP51 gene in the powdery mildew of grapes (*Erysiphe necator*). This filamentous fungus exhibits an increase in the copy number of the EnCYP51 gene after it has acquired a specific allele making it tolerant to fungicides: a correlation between the increased expression of the EnCYP51 gene due to a higher copy number and fungal growth in the presence of DMI (demethylase inhibitor) fungicides has been confirmed (Jones et al. 2014).

2.12.8. *Domestication and post-domestication of plant and animal species*

Domestication is the evolutionary process by which a species will control the reproduction and dispersal of another species for its own benefit (Lye and Purugganan 2019). Domestication began in human societies in the Paleolithic, continuing to the present day. Humans started modifying hundreds of species of plants and animals for their needs. Domestication is made up of two distinct phases: (i) first the initial domestication phase, which aims to control the reproduction of a species at a given time and place, then (ii) the diversification and improvement phase, which consists of improving certain characteristics of a species so that it corresponds even more to the expectations of humans (Lye and Purugganan 2019).

Species	Locus	Type	Phenotype	Description	Phase	Ref.
Wheat	Pbd-B1	mCNV 25kbp	Early flowering	Pseudo-response regulator (Ppd-B1)	Diversification	(Diaz et al. 2012)
Wheat	Pbd-D1a	Deletion 2kbp	Photoperiod insensitivity	Pseudo-response regulator (Ppd-D1a)	Diversification	(Beales et al. 2007)
Rice	GL7	Duplication 17.1kbp	Grain length	Uncharacterized function (homologous ti LONGIFOLIA *A. thaliana*	Domestication, diversification	(Wang et al. 2015)
Sorghum	Sh1	Deletion 2.2kbp	Shattering	YABBY-like transcription factor	Domestication	(Lin et al. 2012)
Sheep	ASIP, AHCY	Duplication 190kbp	White coat	ASIP, AHCY	Diversification	(Norris and Whan 2008)
Dog	AMYB2	mCNV	Starch diet	AMYB2 pancreatic amylase	Diversification	(Axelsson et al. 2013)
Dog	Intergenic region	Duplication 98kbp	Blue eyes	Intergenic region adjacent to ALX4, plays a role in eye development	Diversification	(Deane-Coe et al. 2018)

Table 2.4. *CNVs and examples of traits affected during domestication*

In order to study the impact of domestication on species, the number of studies and conclusions based on SNPs has seen an increase in recent years; *a contrario*, studies of structural variations have emerged with difficulty due to the diversity of CNV sequencing and estimation methods, but also to the diversity of varieties studied, making direct comparisons difficult. Very recently, comparative studies have offered up their initial conclusions: some tend to show that CNVs are primarily deleterious, and subject to purifying selection affecting coding sequences less frequently than non-coding sequences (Redon et al. 2006, Katju and Bergthorsson 2013). Other studies prefer to highlight the key role of CNVs in rapid adaptation to environmental change, a process more subject to processes of positive selection adaptation (Kondrashov 2012).

CNV-type structural variations are widely found in domesticated plant and animal species. The diversity of CNVs in domesticated species is less important than in their wild counterparts, an expected result similar to that observed for SNP (Swanson-Wagner et al. 2010, Dolatabadian et al. 2017, Muñoz-Amatriain et al. 2013). However, exceptions exist, including the amylase gene, a digestive enzyme, which exhibits structural polymorphism in dogs. The number of copies of the AMY2B gene has been shown to be seven times higher than those of wolves (Axelsson et al. 2013). This change is correlated with the level of gene expression and enzyme activity, indicating that alpha-amylase duplications conferred a selective advantage in the early ancestors of the modern dog, improving management and assimilation for better growth and survival. The increased enzyme activity allowed them to switch to a diet rich in starch compared to that of wild carnivores such as wolves, which would represent a crucial step in the domestication of dogs (Axelsson et al. 2013, Innan and Kondrashov 2010).

Currently, we know that in plants, CNVs primarily affect transcription factors associated with the photoperiod, development, stress tolerance or resistance (R genes). In animals, CNVs mainly affect genes coding for growth factors and developmental receptors. Consequently, structural variation seems to affect the whole domestication process in plants (initial domestication and diversification), although in animals CNVs mainly affect the diversification and improvement phases (Lye and Purugganan 2019). The types of duplications also differ; they are mainly driven by insertions/deletions in plants and tandem duplications in animals; they are likely to reflect distinct duplication history, noticeably driven in plants by one or multiple whole genome duplications.

2.12.9. *Competition and evolutionary success: invasive species and hybridization*

In the case of invasive species, genomic characteristics and the propensity to make CNV could play a decisive role in the evolutionary success of certain species. The Mediterranean mussel (*Mytilus galloprovincialis*), whose invasiveness and resilience to biotic and abiotic stress factors are well known, is involved in the disappearance of many bivalves. The long-range sequencing of its genome has revealed an entirely unexpected proportion of duplicated genes: more than 20,000 recently duplicated genes, enriched in survival functions, have been proposed to explain the resilience and invasiveness of this species (Gerdol et al. 2020). The rapid adaptation of invasive species to new habitats, in the presence of reduced genetic variability and intense stress, constitutes one of the key questions on these species today; among the possible fields of investigation, beyond certain shared ecological traits, it is the genomic and molecular mechanisms – in the presence of biotic and abiotic stress – that seem fundamental to investigate. It has notably been hypothesized that stress would induce changes in the activity of TE, inducing cascading modifications in structural dynamics and notably of CNVs and the associated gene expression; this could facilitate species' rapid adaptation to new environments (Stapley et al. 2015). The hybrid species of invasive fish, Cottus, exhibits a very different structural polymorphism compared to its parental lineages *Cottus rhenanus* and *Cottus perifretum*. An increase in the copy number of the genes involved in vision, immune system and muscles was observed, accompanied by a dose effect; these are believed to be the result of accumulation over several generations. A bias also appears to exist in the origin of CNVs observed with more than 54% found in *C. perifretum*, compared to only 1.54% in *C. rhenanus* (Dennenmoser et al. 2017).

2.13. Conclusion

Structural polymorphism corresponds to architectural changes in genomes. The high proportion of genes and repeated elements generated by the duplication process illustrates the key role of structural modifications in species evolution. These changes are capable of generating new genetic architectures, modifying the expression of genes or networks of genes either by their direct expressions or with the help of epigenetic modifications involved notably in modifying the expression of novel duplicated genes. If the fate of new duplications depends on the intensity of the selection and the effective population size of the species, it seems important to emphasize that the rate of duplications and deletions is sufficiently high to constitute an excellent source of functional innovation; it is therefore necessary to seek to clearly define the processes at work that are responsible for these new architectures, but also the whole of the mechanisms at work involved in maintaining the integrity of the genome, while allowing the latter some flexibility. If the constraints to which each species is subject are highly likely to influence the fixation of new architectures, it would be interesting to evaluate the extent to which external stresses induce these changes and advance our knowledge of the mechanisms involved. While epigenetic modifications can affect the fate of duplicated copies by changing their expression level in particular, structural modifications can also act on epigenetic mechanisms in return by modifying the spread of epigenetic signals through modifications in the sequence composition of the genome. Structural modifications are therefore capable of modifying phenotypes through several pathways of action, being potentially very impactful. Sequencing techniques based on long fragments are becoming more commonplace, revealing the Pandora's box of this disciplinary field – for humans, but also for other species. As our knowledge of the SD and CNV composition of genomes continues to grow, all models of the relative impact of selection and drift can also be refined and with them of the potential impact of the duplication process in species evolution. The science of redundancy is emerging.

2.14. Glossary of terms

CNV (*Copy Number Variant*)**:** a genomic variant or polymorphism involving a segment of DNA at least one thousand bases long at a specific location in a chromosomal region that has a different copy number than the reference genome. Classes of CNVs include insertions, deletions and duplications.

Duplicon: a duplication, or a portion thereof, of a genomic sequence that shows a high degree of homology with another region of the reference genome.

Gene amplification: the increase in the number of copies of a particular gene.

Gene conversion: a recombination process that homogenizes gene sequences in a non-reciprocal way.

HKG: *housekeeping genes* are a set of highly conserved genes shared by large parts of life. They are linked to the fundamental functions of the cell.

Microsatellite: a sequence of a few base pairs repeated from a few times to a few dozen times.

Minisatellite: a sequence of a few dozen base pairs repeated from a few times to a few dozen times.

NAHR (*Non-allelic homologous recombination*)**:** non-allelic recombination between paralogous segmental duplications; a key mechanism responsible for the appearance of deletions, duplications and inversions, and thus complex structural polymorphism and rearrangements in the human genome.

Penetrating: the intensity with which a certain genotype manifests itself in a given phenotype. The penetrance of certain genotypes for certain diseases is linked to age, complicating real penetrance being determined.

Positive selection: natural selection promoting the fixation of advantageous alleles.

Pseudogene: a DNA sequence derived from a functional gene, later made non-functional by mutation.

Purifying selection: natural selection preventing the fixation of deleterious alleles.

SD: *segmental duplications*, or *low-copy repeats* (LCR), are DNA segments more than a thousand bases long that occur in two or more copies per haploid genome and whose *duplicons* share more than 90% sequence identity. They often vary in copy number between individuals and are therefore also sometimes grouped with CNVs.

SINE: *short interspersed nuclear elements* are non-condensing retrotransposons of a few hundred base pairs in length, abundant in mammalian genomes.

SNP: *Single Nucleotide Polymorphism*, inter-individual variation on a single nucleotide either by mutation, insertion or deletion.

Structural variants: structural variations are defined as microscopic or submicroscopic genomic alterations. This term does not imply any information about size, gene content, location, or phenotypic consequence. Here, structural variations, or structural variants, will refer to variations and polymorphisms involving changes in the copy number of a DNA segment such as insertions or deletions.

Translocation: the change in position of a chromosomal segment with in the genome, which does not involve a change in total DNA content. Translocations can be intra- or inter-chromosomal.

2.15. References

Aitman, T.J., Dong, R., Vyse, T.J., Norsworthy, P.J., Johnson, M.D., Smith, J., Mangion, J., Roberton-Lowe, C., Marshall, A.J., Petretto, E. et al. (2006). Copy number polymorphism in Fcgr3 predisposes to glomerulonephritis in rats and humans. *Nature*, 439(7078), 851–855.

Angeli, S., Lin, X., Liu, X.Z. (2012). Genetics of hearing and deafness. *The Anatomical Record: Advances in Integrative Anatomy and Evolutionary Biology*, 295(11), 1812–1829.

Audemard, E., Schiex, T., Faraut, T. (2012). Detecting long tandem duplications in genomic sequences. *BMC Bioinformatics*, 13(1), 1–13.

Axelsson, E., Ratnakumar, A., Arendt, M.-L., Maqbool, K., Webster, M.T., Perloski, M., Liberg, O., Arnemo, J.M., Hedhammar, Å., Lindblad-Toh, K. (2013). The genomic signature of dog domestication reveals adaptation to a starch-rich diet. *Nature*, 495(7441), 360–364.

Bailey, J.A., Yavor, A.M., Massa, H.F., Trask, B.J., Eichler, E.E. (2001). Segmental duplications: Organization and impact within the current human genome project assembly. *Genome Research*, 11(6), 1005–1017.

Bailey, J.A., Gu, Z., Clark, R.A., Reinert, K., Samonte, R.V., Schwartz, S., Adams, M.D., Myers, E.W., Li, P.W., Eichler, E.E. (2002). Recent segmental duplications in the human genome. *Science*, 297(5583), 1003–1007.

Bailey, J.A., Liu, G., Eichler, E.E. (2003). An alu transposition model for the origin and expansion of human segmental duplications. *The American Journal of Human Genetics*, 73(4), 823–834.

Bailey, J.A., Church, D.M., Ventura, M., Rocchi, M., Eichler, E.E. (2004). Analysis of segmental duplications and genome assembly in the mouse. *Genome Research*, 14(5), 789–801.

Balaresque, P., Toupance, B., Heyer, E., Crouau-Roy, B. (2003). Evolutionary dynamics of duplicated microsatellites shared by sex chromosomes. *Journal of Molecular Evolution*, 57(1), S128–S137.

Balaresque, P., Bowden, G.R., Parkin, E.J., Omran, G.A., Heyer, E., Quintana-Murci, L., Roewer, L., Stoneking, M., Nasidze, I., Carvalho-Silva, D.R. et al. (2008). Dynamic nature of the proximal AZFc region of the human Y chromosome: Multiple independent deletion and duplication events revealed by microsatellite analysis. *Human Mutation*, 29(10), 1171–1180.

Balaresque, P., King, T.E., Parkin, E.J., Heyer, E., Carvalho-Silva, D., Kraaijenbrink, T., de Knijff, P., Tyler-Smith, C., Jobling, M.A. (2014). Gene conversion violates the stepwise mutation model for microsatellites in Y-chromosomal palindromic repeats. *Human Mutation*, 35(5), 609–617.

Ballabio, A., Bardoni, B., Guioli, S., Basler, E., Camerino, G. (1990). Two families of low-copy-number repeats are interspersed on Xp22.3: Implications for the high frequency of deletions in this region. *Genomics*, 8(2), 263–270.

Ballantyne, K.N., Goedbloed, M., Fang, R., Schaap, O., Lao, O., Wollstein, A., Choi, Y., van Duijn, K., Vermeulen, M., Brauer, S. et al. (2010). Mutability of Y-chromosomal microsatellites: Rates, characteristics, molecular bases, and forensic implications. *The American Journal of Human Genetics*, 87(3), 341–353.

Bannister, A.J. and Kouzarides, T. (2011). Regulation of chromatin by histone modifications. *Cell Research*, 21(3), 381–395.

Bao, Z. and Eddy, S.R. (2002). Automated de novo identification of repeat sequence families in sequenced genomes. *Genome Research*, 12(8), 1269–1276.

Beales, J., Turner, A., Griffiths, S., Snape, J.W., Laurie, D.A. (2007). A pseudo-response regulator is misexpressed in the photoperiod insensitive Ppd-D1a mutant of wheat (*Triticum aestivum* L.). *Theoretical and Applied Genetics*, 115(5), 721–733.

Bellott, D.W., Hughes, J.F., Skaletsky, H., Brown, L.G., Pyntikova, T., Cho, T.-J., Koutseva, N., Zaghlul, S., Graves, T., Rock, S. et al. (2014). Mammalian Y chromosomes retain widely expressed dosage-sensitive regulators. *Nature*, 508(7497), 494–499.

Bergström, A., McCarthy, S.A., Hui, R., Almarri, M.A., Ayub, Q., Danecek, P., Chen, Y., Felkel, S., Hallast, P., Kamm, J. et al. (2020). Insights into human genetic variation and population history from 929 diverse genomes. *Science*, 367(6484), eaay5012.

Bernstein, E., Goldberg, A.D., Allis, C.D. (2007). Epigenetics: A landscape takes shape. *Cell*, 128(4), 635–638.

Bouakaze, C., Delehelle, F., Sáenz-Oyhéréguy, N., Moreira, A., Schiavinato, S., Croze, M., Delon, S., Fortes-Lima, C., Gibert, M., Bujan, L. et al. (2020). Predicting haplogroups using a versatile machine learning program (PredYMaLe) on a new mutationally balanced 32 Y-STR multiplex (combyplex): Unlocking the full potential of the human STR mutation rate spectrum to estimate forensic parameters. *Forensic Science International: Genetics*, 48, 102342.

Boža, V., Brejová, B., Vinař, T. (2017). Deepnano: Deep recurrent neural networks for base calling in minion nanopore reads. *PLoS One*, 12(6), e0178751.

Brenner, S.E., Hubbard, T., Murzin, A., Chothia, C. (1995). Gene duplications in *H. influenzae*. *Nature*, 378(6553), 140.

Bridges, C.B. (1936). The bar "gene" a duplication. *Science*, 83(2148), 210–211.

Brown, C.J., Todd, K.M., Rosenzweig, R.F. (1998). Multiple duplications of yeast hexose transport genes in response to selection in a glucose-limited environment. *Molecular Biology and Evolution*, 15(8), 931–942.

Buckland, P.R. (2003). Polymorphically duplicated genes: Their relevance to phenotypic variation in humans. *Annals of Medicine*, 35(5), 308–315.

Calabro, K.R., Boye, S.L., Choudhury, S., Fajardo, D., Peterson, J.J., Li, W., Crosson, S.M., Kim, M.-J., Ding, D., Salvi, R. et al. (2019). A novel mouse model of MYO7A USH1B reveals auditory and visual system haploinsufficiencies. *Frontiers in Neuroscience*, 13, 1255.

Caraco Y. (2004). Genes and the response to drugs. *N. Engl. J. Med.*, 351(27), 2867–2869.

Carpenter, D., Mitchell, L.M., Armour, J.A. (2017). Copy number variation of human AMY1 is a minor contributor to variation in salivary amylase expression and activity. *Human Genomics*, 11(1), 1–6.

Cayuela, H., Dorant, Y., Forester, B.R., Jeffries, D.L., Mccaffery, R.M., Eby, L.A., Hossack, B.R., Gippet, J.M.W., Pilliod, D.S., Chris, F.W. (2021a). Genomic signatures of thermal adaptation are associated with clinal shifts of life history in a broadly distributed frog. *J. Anim. Ecol.*, 91(6), 1222–1238.

Cayuela, H., Dorant, Y., Mérot, C., Laporte, M., Normandeau, E., Gagnon-Harvey, S., Clément, M., Sirois, P., Bernatchez, L. (2021b). Thermal adaptation rather than demographic history drives genetic structure inferred by copy number variants in a marine fish. *Mol. Ecol.*, 30(7), 1624–1641.

Chang, A.Y.-F. and Liao, B.-Y. (2012). DNA methylation rebalances gene dosage after mammalian gene duplications. *Mol. Biol. Evol.*, 29(1), 133–144.

Chaudhari, N. and Roper, S.D. (2010). The cell biology of taste. *Journal of Cell Biology*, 190(3), 285–296.

Chavchich, M., Gerena, L., Peters, J., Chen, N., Cheng, Q., Kyle, D.E. (2010). Role of pfmdr1 amplification and expression in induction of resistance to artemisinin derivatives in *Plasmodium falciparum*. *Antimicrobial Agents and Chemotherapy*, 54(6), 2455–2464.

Chen, L., DeVries, A.L., Cheng, C.-H.C. (1997). Evolution of antifreeze glycoprotein gene from a trypsinogen gene in antarctic notothenioid fish. *Proceedings of the National Academy of Sciences*, 94(8), 3811–3816.

Chen, Z., Cheng, C.-H.C., Zhang, J., Cao, L., Chen, L., Zhou, L., Jin, Y., Ye, H., Deng, C., Dai, Z. et al. (2008). Transcriptomic and genomic evolution under constant cold in Antarctic notothenioid fish. *Proceedings of the National Academy of Sciences*, 105(35), 12944–12949.

Cheung, J., Estivill, X., Khaja, R., MacDonald, J.R., Lau, K., Tsui, L.-C., Scherer, S.W. (2003). Genome-wide detection of segmental duplications and potential assembly errors in the human genome sequence. *Genome Biology*, 4(4), 1–10.

Christ, D. and Chin, J.W. (2008). Engineering *Escherichia coli* heat-resistance by synthetic gene amplification. *Protein Engineering, Design & Selection*, 21(2), 121–125.

Church, D.M., Goodstadt, L., Hillier, L.W., Zody, M.C., Goldstein, S., She, X., Bult, C.J., Agarwala, R., Cherry, J.L., DiCuccio, M. et al. (2009). Lineage-specific biology revealed by a finished genome assembly of the mouse. *PLoS Biology*, 7(5), e1000112.

Cirilo Jr., J.A., Gunther, L.K., Yengo, C.M. (2021). Functional role of class iii myosins in hair cells. *Frontiers in Cell and Developmental Biology*, 9, 285.

Coate, J.E., Schlueter, J.A., Whaley, A.M., Doyle, J.J. (2011). Comparative evolution of photosynthetic genes in response to polyploid and nonpolyploid duplication. *Plant Physiology*, 155(4), 2081–2095.

Conant, G.C. and Wolfe, K.H. (2008). Turning a hobby into a job: How duplicated genes find new functions. *Nature Reviews Genetics*, 9(12), 938–950.

Conrad, D.F., Andrews, T.D., Carter, N.P., Hurles, M.E., Pritchard, J.K. (2006). A high-resolution survey of deletion polymorphism in the human genome. *Nature Genetics*, 38(1), 75–81.

Conte, C., Ebeling, M., Marcuz, A., Nef, P., Andres-Barquin, P.J. (2003). Evolutionary relationships of the Tas2r receptor gene families in mouse and human. *Physiological Genomics*, 14(1), 73–82.

Cook, D.E., Lee, T.G., Guo, X., Melito, S., Wang, K., Bayless, A.M., Wang, J., Hughes, T.J., Willis, D.K., Clemente, T.E. et al. (2012). Copy number variation of multiple genes at rhg1 mediates nematode resistance in soybean. *Science*, 338(6111), 1206–1209.

Cotton, J.A. and Page, R.D.M. (2005). Rates and patterns of gene duplication and loss in the human genome. *Proc. Biol. Sci.*, 272(1560), 277–283.

Davis, J.C. and Petrov, D.A. (2005). Do disparate mechanisms of duplication add similar genes to the genome? *Trends in Genetics*, 21(10), 548–551.

De Coster, W., Weissensteiner, M.H., Sedlazeck, F.J. (2021). Towards population-scale long-read sequencing. *Nat. Rev. Genet.*, 22(9), 572–587.

Deane-Coe, P.E., Chu, E.T., Slavney, A., Boyko, A.R., Sams, A.J. (2018). Direct-to-consumer DNA testing of 6,000 dogs reveals 98.6-kb duplication associated with blue eyes and heterochromia in Siberian huskies. *PLoS Genetics*, 14(10), e1007648.

Deeb, S. (2005). The molecular basis of variation in human color vision. *Clinical Genetics*, 67(5), 369–377.

Degenhardt, J.D., de Candia, P., Chabot, A., Schwartz, S., Henderson, L., Ling, B., Hunter, M., Jiang, Z., Palermo, R.E., Katze, M. et al. (2009). Copy number variation of CCL3-like genes affects rate of progression to simian-AIDS in rhesus macaques (*Macaca mulatta*). *PLoS Genetics*, 5(1), e1000346.

Delehelle, F., Cussat-Blanc, S., Alliot, J.-M., Luga, H., Balaresque, P. (2018). ASGART: Fast and parallel genome scale segmental duplications mapping. *Bioinformatics*, 34(16), 2708–2714.

Dennenmoser, S., Sedlazeck, F.J., Iwaszkiewicz, E., Li, X.-Y., Altmüller, J., Nolte, A.W. (2017). Copy number increases of transposable elements and protein-coding genes in an invasive fish of hybrid origin. *Molecular Ecology*, 26(18), 4712–4724.

Derti, A., Roth, F.P., Church, G.M., Wu, C.-T. (2006). Mammalian ultraconserved elements are strongly depleted among segmental duplications and copy number variants. *Nature Genetics*, 38(10), 1216–1220.

Diaz, A., Zikhali, M., Turner, A.S., Isaac, P., Laurie, D.A. (2012). Copy number variation affecting the photoperiod-B1 and vernalization-A1 genes is associated with altered flowering time in wheat (Triticum aestivum). *PLoS One*, 7(3), e33234.

Djedatin, G., Monat, C., Engelen, S., Sabot, F. (2017). DuplicationDetector, a light weight tool for duplication detection using NGS data. *Current Plant Biology*, 9, 23–28.

Dobon, B., Rossell, C., Walsh, S., Bertranpetit, J. (2019). Is there adaptation in the human genome for taste perception and phase I biotransformation? *BMC Evolutionary Biology*, 19(1), 1–10.

Dolatabadian, A., Patel, D.A., Edwards, D., Batley, J. (2017). Copy number variation and disease resistance in plants. *Theoretical and Applied Genetics*, 130(12), 2479–2490.

Dominy, N.J. and Lucas, P.W. (2001). Ecological importance of trichromatic vision to primates. *Nature*, 410(6826), 363–366.

Dopman, E.B. and Hartl, D.L. (2007). A portrait of copy-number polymorphism in *Drosophila melanogaster. Proceedings of the National Academy of Sciences of the United States of America*, 104(50), 19920–19925.

Dorant, Y., Cayuela, H., Wellband, K., Laporte, M., Rougemont, Q., Mérot, C., Normandeau, E., Rochette, R., Bernatchez, L. (2020). Copy number variants outperform SNPs to reveal genotype-temperature association in a marine species. *Mol. Ecol.*, 29(24), 4765–4782.

Duraisingh, M.T. and Cowman, A.F. (2005). Contribution of the pfmdr1 gene to antimalarial drug-resistance. *Acta Tropica*, 94(3), 181–190.

Fellermann, K., Stange, D.E., Schaeffeler, E., Schmalzl, H., Wehkamp, J., Bevins, C.L., Reinisch, W., Teml, A., Schwab, M., Lichter, P. et al. (2006). A chromosome 8 gene-cluster polymorphism with low human beta-defensin 2 gene copy number predisposes to Crohn disease of the colon. *The American Journal of Human Genetics*, 79(3), 439–448.

Ferguson-Smith, M.A. (1965). Karyotype-phenotype correlations in gonadal dysgenesis and their bearing on the pathogenesis of malformations. *Journal of Medical Genetics*, 2(2), 142.

Flavell, R.B. (1994). Inactivation of gene expression in plants as a consequence of specific sequence duplication. *Proc. Natl. Acad. Sci. USA*, 91(9), 3490–3496.

Flores, M., Morales, L., Gonzaga-Jauregui, C., Domínguez-Vidaña, R., Zepeda, C., Yañez, O., Gutiérrez, M., Lemus, T., Valle, D., Avila, M.C. et al. (2007). Recurrent DNA inversion rearrangements in the human genome. *Proceedings of the National Academy of Sciences*, 104(15), 6099–6106.

Force, A., Lynch, M., Pickett, F.B., Amores, A., Yan, Y.-l., Postlethwait, J. (1999). Preservation of duplicate genes by complementary, degenerative mutations. *Genetics*, 151(4), 1531–1545.

Francey, L.J., Conlin, L.K., Kadesch, H.E., Clark, D., Berrodin, D., Sun, Y., Glessner, J., Hakonarson, H., Jalas, C., Landau, C. et al. (2012). Genome-wide SNP genotyping identifies the stereocilin (STRC) gene as a major contributor to pediatric bilateral sensorineural hearing impairment. *American Journal of Medical Genetics Part A*, 158(2), 298–308.

Freeling, M., Lyons, E., Pedersen, B., Alam, M., Ming, R., Lisch, D. (2008). Many or most genes in arabidopsis transposed after the origin of the order brassicales. *Genome Research*, 18(12), 1924–1937.

Fukami, M., Kirsch, S., Schiller, S., Richter, A., Benes, V., Franco, B., Muroya, K., Rao, E., Merker, S., Niesler, B. et al. (2000). A member of a gene family on Xp22. 3, VCX-A, is deleted in patients with X-linked nonspecific mental retardation. *The American Journal of Human Genetics*, 67(3), 563–573.

Furman, B.L., Metzger, D.C., Darolti, I., Wright, A.E., Sandkam, B.A., Almeida, P., Shu, J.J., Mank, J.E. (2020). Sex chromosome evolution: So many exceptions to the rules. *Genome Biology and Evolution*, 12(6), 750–763.

Gadalla, N.B., Adam, I., Elzaki, S.-E., Bashir, S., Mukhtar, I., Oguike, M., Gadalla, A., Mansour, F., Warhurst, D., El-Sayed, B.B. et al. (2011). Increased pfmdr1 copy number and sequence polymorphisms in plasmodium falciparum isolates from Sudanese malaria patients treated with artemether-lumefantrine. *Antimicrobial Agents and Chemotherapy*, 55(11), 5408–5411.

Gaines, T.A., Zhang, W., Wang, D., Bukun, B., Chisholm, S.T., Shaner, D.L., Nissen, S.J., Patzoldt, W.L., Tranel, P.J., Culpepper, A.S. et al. (2010). Gene amplification confers glyphosate resistance in *Amaranthus palmeri*. *Proceedings of the National Academy of Sciences of the United States of America*, 107(3), 1029–1034.

Gerdol, M., Moreira, R., Cruz, F., Gómez-Garrido, J., Vlasova, A., Rosani, U., Venier, P., Naranjo-Ortiz, M.A., Murgarella, M., Greco, S. et al. (2020). Massive gene presence-absence variation shapes an open pan-genome in the Mediterranean mussel. *Genome Biol.*, 21(1), 275.

Gibbs, R.A., Rogers, J., Katze, M.G., Bumgarner, R., Weinstock, G.M., Mardis, E.R., Remington, K.A., Strausberg, R.L., Venter, J.C., Wilson, R.K. et al. (2007). Evolutionary and biomedical insights from the rhesus macaque genome. *Science*, 316(5822), 222–234.

Gibson, J.R., Chippindale, A.K., Rice, W.R. (2002). The X chromosome is a hot spot for sexually antagonistic fitness variation. *Proceedings of the Royal Society of London. Series B: Biological Sciences*, 269(1490), 499–505.

Giglio, S., Broman, K.W., Matsumoto, N., Calvari, V., Gimelli, G., Neumann, T., Ohashi, H., Voullaire, L., Larizza, D., Giorda, R. et al. (2001). Olfactory receptor–gene clusters, genomic-inversion polymorphisms, and common chromosome rearrangements. *The American Journal of Human Genetics*, 68(4), 874–883.

Gilad, Y., Wiebe, V., Przeworski, M., Lancet, D., Pääbo, S., Hillis, D. (2004). Loss of olfactory receptor genes coincides with the acquisition of full trichromatic vision in primates. *PLoS Biology*, 2(1), e5.

Gimelbrant, A.A. and Chess, A. (2006). An epigenetic state associated with areas of gene duplication. *Genome Research*, 16(6), 723–729.

Girgis, H.Z. (2015). Red: An intelligent, rapid, accurate tool for detecting repeats de-novo on the genomic scale. *BMC Bioinformatics*, 16(1), 1–19.

Glusman, G., Yanai, I., Rubin, I., Lancet, D. (2001). The complete human olfactory subgenome. *Genome Research*, 11(5), 685–702.

Goldstein, D.B., Roemer, G.W., Smith, D.A., Reich, D.E., Bergman, A., Wayne, R.K. (1999). The use of microsatellite variation to infer population structure and demographic history in a natural model system. *Genetics*, 151(2), 797–801.

Goodwin, S., McPherson, J.D., McCombie, W.R. (2016). Coming of age: Ten years of next-generation sequencing technologies. *Nature Reviews Genetics*, 17(6), 333–351.

Gough, J., Karplus, K., Hughey, R., Chothia, C. (2001). Assignment of homology to genome sequences using a library of hidden Markov models that represent all proteins of known structure. *Journal of Molecular Biology*, 313(4), 903–919.

Graubert, T.A., Cahan, P., Edwin, D., Selzer, R.R., Richmond, T.A., Eis, P.S., Shannon, W.D., Li, X., McLeod, H.L., Cheverud, J.M. et al. (2007). A high-resolution map of segmental DNA copy number variation in the mouse genome. *PLoS Genetics*, 3(1), e3.

Gresham, D., Desai, M.M., Tucker, C.M., Jenq, H.T., Pai, D.A., Ward, A., DeSevo, C.G., Botstein, D., Dunham, M.J. (2008). The repertoire and dynamics of evolutionary adaptations to controlled nutrient-limited environments in yeast. *PLoS Genetics*, 4(12), e1000303.

Gu, Z., Cavalcanti, A., Chen, F.-C., Bouman, P., Li, W.-H. (2002). Extent of gene duplication in the genomes of drosophila, nematode, and yeast. *Mol. Biol. Evol.*, 19(3), 256–262.

Gu, Z., Rifkin, S.A., White, K.P., Li, W.-H. (2004). Duplicate genes increase gene expression diversity within and between species. *Nature Genetics*, 36(6), 577–579.

Haas, B.J., Delcher, A.L., Wortman, J.R., Salzberg, S.L. (2004). Dagchainer: A tool for mining segmental genome duplications and synteny. *Bioinformatics*, 20(18), 3643–3646.

Hakes, L., Pinney, J.W., Lovell, S.C., Oliver, S.G., Robertson, D.L. (2007). All duplicates are not equal: The difference between small-scale and genome duplication. *Genome Biology*, 8(10), R209.

Haldane, J. (1933). The part played by recurrent mutation in evolution. *The American Naturalist*, 67(708), 5–19.

Hallast, P., Balaresque, P., Bowden, G.R., Ballereau, S., Jobling, M.A. (2013). Recombination dynamics of a human Y-chromosomal palindrome: Rapid GC-biased gene conversion, multi-kilobase conversion tracts, and rare inversions. *PLoS Genetics*, 9(7), e1003666.

Hallast, P., Kibena, L., Punab, M., Arciero, E., Rootsi, S., Grigorova, M., Flores, R., Jobling, M.A., Poolamets, O., Pomm, K. et al. (2021). A common 1.6 mb Y-chromosomal inversion predisposes to subsequent deletions and severe spermatogenic failure in humans. *Elife*, 10, e65420.

Hardigan, M.A., Emily, C., Hamilton, J.P., Jeongwoon, K., Parker, L., Leisner, C.P., Manrique-Carpintero, N.C., Linsey, N., Pham, G.M., Brieanne, V. et al. (2016). Genome reduction uncovers a large dispensable genome and adaptive role for copy number variation in asexually propagated *Solanum tuberosum*. *The Plant Cell*, 28(2), 388–405.

Hardy, J. (2006). Amyloid double trouble. *Nature Genetics*, 38(1), 11–12.

Hastings, P.J., Lupski, J.R., Rosenberg, S.M., Ira, G. (2009). Mechanisms of change in gene copy number. *Nature Reviews Genetics*, 10(8), 551–564.

Higgs, D., Old, J., Pressley, L., Clegg, J., Weatherall, D. (1980). A novel α-globin gene arrangement in man. *Nature*, 284(5757), 632–635.

Hirase, S., Ozaki, H., Iwasaki, W. (2014). Parallel selection on gene copy number variations through evolution of three-spined stickleback genomes. *BMC Genomics*, 15, 735.

Hollox, E.J., Armour, J.A., Barber, J.C. (2003). Extensive normal copy number variation of a β-defensin antimicrobial-gene cluster. *The American Journal of Human Genetics*, 73(3), 591–600.

Hollox, E.J., Huffmeier, U., Zeeuwen, P.L., Palla, R., Lascorz, J., Rodijk-Olthuis, D., Van De Kerkhof, P.C., Traupe, H., De Jongh, G., Den Heijer, M. et al. (2008). Psoriasis is associated with increased β-defensin genomic copy number. *Nature Genetics*, 40(1), 23–25.

Hughes, A.L. and Friedman, R. (2005). Gene duplication and the properties of biological networks. *Journal of Molecular Evolution*, 61(6), 758–764.

Hunter, S. and Brayne, C. (2018). Understanding the roles of mutations in the amyloid precursor protein in Alzheimer disease. *Molecular Psychiatry*, 23(1), 81–93.

Huxley, J. (1944). Evolution: The modern synthesis. *Philosophy*, 19(73), 166–170.

Innan, H. and Kondrashov, F. (2010). The evolution of gene duplications: Classifying and distinguishing between models. *Nature Reviews Genetics*, 11(2), 97–108.

Ionita-Laza, I., Rogers, A.J., Lange, C., Raby, B.A., Lee, C. (2009). Genetic association analysis of copy-number variation (CNV) in human disease pathogenesis. *Genomics*, 93(1), 22–26.

Jacobs, G.H., Neitz, M., Deegan, J.F., Neitz, J. (1996). Trichromatic colour vision in new world monkeys. *Nature*, 382(6587), 156–158.

Ji, J., Qin, Y., Wang, R., Huang, Z., Zhang, Y., Zhou, R., Song, L., Ling, X., Hu, Z., Miao, D. et al. (2016). Copy number gain of VCX, X-linked multi-copy gene, leads to cell proliferation and apoptosis during spermatogenesis. *Oncotarget*, 7(48), 78532.

Jobling, M.A. (2008). Copy number variation on the human Y chromosome. *Cytogenetic and Genome Research*, 123(1–4), 253–262.

Jobling, M.A. and Tyler-Smith, C. (2003). The human Y chromosome: An evolutionary marker comes of age. *Nature Reviews Genetics*, 4(8), 598–612.

Johnson, M.E., Cheng, Z., Morrison, V.A., Scherer, S., Ventura, M., Gibbs, R.A., Green, E.D., Eichler, E.E. (2006). Recurrent duplication-driven transposition of DNA during hominoid evolution. *Proceedings of the National Academy of Sciences*, 103(47), 17626–17631.

Jones, L., Riaz, S., Morales-Cruz, A., Amrine, K.C.H., McGuire, B., Gubler, W.D., Walker, M.A., Cantu, D. (2014). Adaptive genomic structural variation in the grape powdery mildew pathogen, *Erysiphe necator*. *BMC Genomics*, 15(1), 1081.

Julien, P., Sébastien, C., John, M. (2017). CNVs into the wild: Screening the genomes of conifer trees (Picea spp.) reveals fewer gene copy number variations in hybrids and links to adaptation. *BMC Genomics*, 18(1), 97.

Katju, V. and Bergthorsson, U. (2013). Copy-number changes in evolution: Rates, fitness effects and adaptive significance. *Frontiers in Genetics*, 4, 273.

Keller, T.E. and Yi, S.V. (2014). DNA methylation and evolution of duplicate genes. *Proc. Natl. Acad. Sci. USA*, 111(16), 5932–5937.

Kent, L., Emerton, J., Bhadravathi, V., Weisblatt, E., Pasco, G., Willatt, L.R., McMahon, R., Yates, J.R. (2008). X-linked ichthyosis (steroid sulfatase deficiency) is associated with increased risk of attention deficit hyperactivity disorder, autism and social communication deficits. *Journal of Medical Genetics*, 45(8), 519–524.

Kinnamon, S.C. (2013). Neurosensory transmission without a synapse: New perspectives on taste signaling. *BMC Biology*, 11(1), 1–4.

Kleinjan, D.-J. and Van Heyningen, V. (1998). Position effect in human genetic disease. *Human Molecular Genetics*, 7(10), 1611–1618.

Knorr, D.Y., Georges, N.S., Pauls, S., Heinrich, R. (2020). Acetylcholinesterase promotes apoptosis in insect neurons. *Apoptosis*, 25(9–10), 730–746.

Knox, A.K., Dhillon, T., Cheng, H., Tondelli, A., Pecchioni, N., Stockinger, E.J. (2010). CBF gene copy number variation at frost resistance-2 is associated with levels of freezing tolerance in temperate-climate cereals. *Theor. Appl. Genet.*, 121(1), 21–35.

Kobayashi, T. and Ganley, A.R. (2005). Recombination regulation by transcription-induced cohesin dissociation in rDNA repeats. *Science*, 309(5740), 1581–1584.

Kondrashov, F.A. (2012). Gene duplication as a mechanism of genomic adaptation to a changing environment. *Proceedings of the Royal Society B: Biological Sciences*, 279(1749), 5048–5057.

Korbel, J.O., Kim, P.M., Chen, X., Urban, A.E., Weissman, S., Snyder, M., Gerstein, M.B. (2008). The current excitement about copy-number variation: How it relates to gene duplications and protein families. *Current Opinion in Structural Biology*, 18(3), 366–374.

Kremer, H., van Wijk, E., Märker, T., Wolfrum, U., Roepman, R. (2006). Usher syndrome: Molecular links of pathogenesis, proteins and pathways. *Human Molecular Genetics*, 15(Suppl. 2), R262–R270.

Kucharski, R., Maleszka, J., Maleszka, R. (2016). A possible role of DNA methylation in functional divergence of a fast evolving duplicate gene encoding odorant binding protein 11 in the honeybee. *Proceedings of the Royal Society B: Biological Sciences*, 283(1833), 20160558.

Kuroda-Kawaguchi, T., Skaletsky, H., Brown, L.G., Minx, P.J., Cordum, H.S., Waterston, R.H., Wilson, R.K., Silber, S., Oates, R., Rozen, S. et al. (2001). The AZFc region of the Y chromosome features massive palindromes and uniform recurrent deletions in infertile men. *Nature Genetics*, 29(3), 279–286.

Labbé, P., Berthomieu, A., Berticat, C., Alout, H., Raymond, M., Lenormand, T., Weill, M. (2007a). Independent duplications of the acetylcholinesterase gene conferring insecticide resistance in the mosquito *Culex pipiens*. *Mol. Biol. Evol.*, 24(4), 1056–1067.

Labbé, P., Berticat, C., Berthomieu, A., Unal, S., Bernard, C., Weill, M., Lenormand, T. (2007b). Forty years of erratic insecticide resistance evolution in the mosquito *Culex pipiens*. *PLoS Genetics*, 3(11), e205.

Lahn, B.T. and Page, D.C. (1999). Four evolutionary strata on the human X chromosome. *Science*, 286(5441), 964–967.

Lahn, B.T. and Page, D.C. (2000). A human sex-chromosomal gene family expressed in male germ cells and encoding variably charged proteins. *Human Molecular Genetics*, 9(2), 311–319.

Lam, K.-W.G. and Jeffreys, A.J. (2007). Processes of de novo duplication of human alpha-globin genes. *Proc. Natl. Acad. Sci. USA*, 104(26), 10950–10955.

Lawson-Yuen, A., Saldivar, J.-S., Sommer, S., Picker, J. (2008). Familial deletion within NLGN4 associated with autism and Tourette syndrome. *European Journal of Human Genetics*, 16(5), 614–618.

Lelli, A., Michel, V., Boutet de Monvel, J., Cortese, M., Bosch-Grau, M., Aghaie, A., Perfettini, I., Dupont, T., Avan, P., El-Amraoui, A. et al. (2016). Class III myosins shape the auditory hair bundles by limiting microvilli and stereocilia growth. *Journal of Cell Biology*, 212(2), 231–244.

Lenormand, T., Guillemaud, T., Bourguet, D., Raymond, M. (1998). Appearance and sweep of a gene duplication: Adaptive response and potential for new functions in the mosquito *Culex pipiens*. *Evolution*, 52(6), 1705–1712.

Li, E. and Zhang, Y. (2014). DNA methylation in mammals. *Cold Spring Harbor Perspectives in Biology*, 6(5), a019133.

Lin, Z., Li, X., Shannon, L.M., Yeh, C.-T., Wang, M.L., Bai, G., Peng, Z., Li, J., Trick, H.N., Clemente, T.E. et al. (2012). Parallel domestication of the shattering1 genes in cereals. *Nature Genetics*, 44(6), 720–724.

Linardopoulou, E.V., Williams, E.M., Fan, Y., Friedman, C., Young, J.M., Trask, B.J. (2005). Human subtelomeres are hot spots of interchromosomal recombination and segmental duplication. *Nature*, 437(7055), 94–100.

Lindgren, A.M., Hoyos, T., Talkowski, M.E., Hanscom, C., Blumenthal, I., Chiang, C., Ernst, C., Pereira, S., Ordulu, Z., Clericuzio, C. et al. (2013). Haploinsufficiency of kdm6a is associated with severe psychomotor retardation, global growth restriction, seizures and cleft palate. *Human Genetics*, 132(5), 537–552.

Lipinski, K.J., Farslow, J.C., Fitzpatrick, K.A., Lynch, M., Katju, V., Bergthorsson, U. (2011). High spontaneous rate of gene duplication in *Caenorhabditis elegans*. *Current Biology*, 21(4), 306–310.

Liu, S., Yao, L., Ding, D., Zhu, H. (2010). Ccl3l1 copy number variation and susceptibility to HIV-1 infection: A meta-analysis. *PLoS One*, 5(12), e15778.

Lucas, E.R., Miles, A., Harding, N.J., Clarkson, C.S., Lawniczak, M.K., Kwiatkowski, D.P., Weetman, D., Donnelly, M.J., Anopheles gambiae 1000 Genomes Consortium. (2019). Whole-genome sequencing reveals high complexity of copy number variation at insecticide resistance loci in malaria mosquitoes. *Genome Research*, 29(8), 1250–1261.

Lupski, J.R. (1998). Genomic disorders: Structural features of the genome can lead to DNA rearrangements and human disease traits. *Trends in Genetics*, 14(10), 417–422.

Lupski, J.R. (2007). Genomic rearrangements and sporadic disease. *Nature Genetics*, 39(7), S43–S47.

Lupski, J.R. and Stankiewicz, P. (2005). Genomic disorders: Molecular mechanisms for rearrangements and conveyed phenotypes. *PLoS Genetics*, 1(6), e49.

Lupski, J.R., Wise, C.A., Kuwano, A., Pentao, L., Parke, J.T., Glaze, D.G., Ledbetter, D.H., Greenberg, F., Patel, P.I. (1992). Gene dosage is a mechanism for Charcot-Marie-Tooth disease type 1A. *Nature Genetics*, 1(1), 29–33.

Lye, Z.N. and Purugganan, M.D. (2019). Copy number variation in domestication. *Trends in Plant Science*, 24(4), 352–365.

Lynch, M. (2007). *The Origins of Genome Architecture.* Sinauer, Sunderland, MA.

Lynch, M. and Conery, J.S. (2000). The evolutionary fate and consequences of duplicate genes. *Science*, 290(5494), 1151–1155.

Lynch, M. and Conery, J.S. (2003). The evolutionary demography of duplicate genes. *J. Struct. Funct. Genomics.*, 3(1–4), 35–44.

Lynch, V.J. and Wagner, G.P. (2008). Resurrecting the role of transcription factor change in developmental evolution. *Evolution: International Journal of Organic Evolution.* 62(9), 2131–2154.

Maere, S., De Bodt, S., Raes, J., Casneuf, T., Van Montagu, M., Kuiper, M., Van de Peer, Y. (2005). Modeling gene and genome duplications in eukaryotes. *Proceedings of the National Academy of Sciences of the United States of America*, 102(15), 5454–5459.

Makino, T. and McLysaght, A. (2010). Ohnologs in the human genome are dosage balanced and frequently associated with disease. *Proceedings of the National Academy of Sciences of the United States of America*, 107(20), 9270–9274.

Makova, K.D. and Li, W.-H. (2003). Divergence in the spatial pattern of gene expression between human duplicate genes. *Genome Res.*, 13(7), 1638–1645.

Mamtani, M., Rovin, B., Brey, R., Camargo, J.F., Kulkarni, H., Herrera, M., Correa, P., Holliday, S., Anaya, J.-M., Ahuja, S.K. (2008). CCL3l1 gene-containing segmental duplications and polymorphisms in CCR5 affect risk of systemic lupus erythaematosus. *Annals of the Rheumatic Diseases*, 67(8), 1076–1083.

Mansai, S.P., Kado, T., Innan, H. (2011). The rate and tract length of gene conversion between duplicated genes. *Genes*, 2(2), 313–331.

Maron, L.G., Guimarães, C.T., Kirst, M., Albert, P.S., Birchler, J.A., Bradbury, P.J., Buckler, E.S., Coluccio, A.E., Danilova, T.V., Kudrna, D. et al. (2013). Aluminum tolerance in maize is associated with higher mate1 gene copy number. *Proceedings of the National Academy of Sciences of the United States of America*, 110(13), 5241–5246.

Marques-Bonet, T., Kidd, J.M., Ventura, M., Graves, T.A., Cheng, Z., Hillier, L.W., Jiang, Z., Baker, C., Malfavon-Borja, R., Fulton, L.A. et al. (2009a). A burst of segmental duplications in the genome of the African great ape ancestor. *Nature*, 457(7231), 877–881.

Marques-Bonet, T., Girirajan, S., Eichler, E.E. (2009b). The origins and impact of primate segmental duplications. *Trends in Genetics*, 25(10), 443–454.

McCarroll, S.A., Hadnott, T.N., Perry, G.H., Sabeti, P.C., Zody, M.C., Barrett, J.C., Dallaire, S., Gabriel, S.B., Lee, C., Daly, M.J. et al. (2006). Common deletion polymorphisms in the human genome. *Nature Genetics*, 38(1), 86–92.

Mefford, H.C. and Trask, B.J. (2002). The complex structure and dynamic evolution of human subtelomeres. *Nature Reviews Genetics*, 3(2), 91–102.

Meitinger, T., Heye, B., Petit, C., Levilliers, J., Golla, A., Moraine, C., Dalla Piccola, B., Sippell, W., Murken, J., Ballabio, A. (1990). Definitive localization of X-linked Kallman syndrome (hypogonadotropic hypogonadism and anosmia) to Xp22. 3: Close linkage to the hypervariable repeat sequence CRI-S232. *American Journal of Human Genetics*, 47(4), 664.

Miga, K.H., Koren, S., Rhie, A., Vollger, M.R., Gershman, A., Bzikadze, A., Brooks, S., Howe, E., Porubsky, D., Logsdon, G.A. et al. (2020). Telomere-to-telomere assembly of a complete human X chromosome. *Nature*, 585(7823), 79–84.

Moog, U., Engelen, J., Albrechts, J., Hoorntje, T., Hendrikse, F., Schrander-Stumpel, C. (1996). Alagille syndrome in a family with duplication 20p11. *Clinical Dysmorphology*, 5(4), 279–288.

Moran, G.P., Anderson, M.Z., Myers, L.C., Sullivan, D.J. (2019). Role of mediator in virulence and antifungal drug resistance in pathogenic fungi. *Curr. Genet.*, 65(3), 621–630.

Muller, H.J. and Prokofyeva, A. (1935). The individual gene in relation to the chromomere and the chromosome. *Proceedings of the National Academy of Sciences of the United States of America*, 21(1), 16.

Muñoz-Amatriain, M., Eichten, S.R., Wicker, T., Richmond, T.A., Mascher, M., Steuernagel, B., Scholz, U., Ariyadasa, R., Spannagl, M., Nussbaumer, T. et al. (2013). Distribution, functional impact, and origin mechanisms of copy number variation in the barley genome. *Genome Biology*, 14(6), 1–17.

Muthusamy, B., Selvan, L.D.N., Nguyen, T.T., Manoj, J., Stawiski, E.W., Jaiswal, B.S., Wang, W., Raja, R., Ramprasad, V.L., Gupta, R. et al. (2017). Next-generation sequencing reveals novel mutations in X-linked intellectual disability. *Omics: A Journal of Integrative Biology*, 21(5), 295–303.

Nair, S., Nash, D., Sudimack, D., Jaidee, A., Barends, M., Uhlemann, A.-C., Krishna, S., Nosten, F., Anderson, T.J. (2007). Recurrent gene amplification and soft selective sweeps during evolution of multidrug resistance in malaria parasites. *Molecular Biology and Evolution*, 24(2), 562–573.

Nakahori, Y., Kuroki, Y., Komaki, R., Kondoh, N., Namiki, M., Iwamoto, T., Toda, T., Kobayashi, K. (1996). The Y chromosome region essential for spermatogenesis. *Hormone Research in Paediatrics*, 46(Suppl. 1), 20–23.

Neitz, M. and Neitz, J. (1995). Numbers and ratios of visual pigment genes for normal red-green color vision. *Science*, 267(5200), 1013–1016.

Newman, T.L., Tuzun, E., Morrison, V.A., Hayden, K.E., Ventura, M., McGrath, S.D., Rocchi, M., Eichler, E.E. (2005). A genome-wide survey of structural variation between human and chimpanzee. *Genome Research*, 15(10), 1344–1356.

Niimura, Y. and Nei, M. (2007). Extensive gains and losses of olfactory receptor genes in mammalian evolution. *PLoS One*, 2(8), e708.

Norris, B.J. and Whan, V.A. (2008). A gene duplication affecting expression of the ovine asip gene is responsible for white and black sheep. *Genome Research*, 18(8), 1282–1293.

Nozawa, M., Kawahara, Y., Nei, M. (2007). Genomic drift and copy number variation of sensory receptor genes in humans. *Proceedings of the National Academy of Sciences*, 104(51), 20421–20426.

Numanagić, I., Gökkaya, A.S., Zhang, L., Berger, B., Alkan, C., Hach, F. (2018). Fast characterization of segmental duplications in genome assemblies. *Bioinformatics*, 34(17), i706–i714.

Ober, C., Loisel, D.A., Gilad, Y. (2008). Sex-specific genetic architecture of human disease. *Nature Reviews Genetics*, 9(12), 911–922.

Odronitz, F. and Kollmar, M. (2007). Drawing the tree of eukaryotic life based on the analysis of 2,269 manually annotated myosins from 328 species. *Genome Biology*, 8(9), 1–23.

Ohno, S. (1970). *Evolution by Gene Duplication*. Springer Berlin, Heidelberg.

van Ommen, G.-J.B. (2005). Frequency of new copy number variation in humans. *Nat. Genet.*, 37(4), 333–334.

Oonk, A., Leijendeckers, J., Lammers, E., Weegerink, N., Oostrik, J., Beynon, A., Huygen, P., Kunst, H., Kremer, H., Snik, A. et al. (2013). Progressive hereditary hearing impairment caused by a MYO6 mutation resembles presbyacusis. *Hearing Research*, 299, 88–98.

Padiath, Q.S., Saigoh, K., Schiffmann, R., Asahara, H., Yamada, T., Koeppen, A., Hogan, K., Ptáček, L.J., Fu, Y.-H. (2006). Lamin B1 duplications cause autosomal dominant leukodystrophy. *Nature Genetics*, 38(10), 1114–1123.

Patterson, E.L., Pettinga, D.J., Ravet, K., Neve, P., Gaines, T.A. (2018). Glyphosate resistance and EPSPS gene duplication: Convergent evolution in multiple plant species. *J. Hered.*, 109(2), 117–125.

Perry, G.H. (2008). The evolutionary significance of copy number variation in the human genome. *Cytogenetic and Genome Research*, 123(1–4), 283–287.

Perry, G.H., Dominy, N.J., Claw, K.G., Lee, A.S., Fiegler, H., Redon, R., Werner, J., Villanea, F.A., Mountain, J.L., Misra, R. et al. (2007). Diet and the evolution of human amylase gene copy number variation. *Nature Genetics*, 39(10), 1256–1260.

Petit, C. and Richardson, G.P. (2009). Linking genes underlying deafness to hair bundle development and function. *Nature Neuroscience*, 12(6), 703–710.

Petr, M., Hajdinjak, M., Fu, Q., Essel, E., Rougier, H., Crevecoeur, I., Semal, P., Golovanova, L.V., Doronichev, V.B., Lalueza-Fox, C. et al. (2020). The evolutionary history of Neanderthal and Denisovan Y chromosomes. *Science*, 369(6511), 1653–1656.

Petrov, D.A. (2001). Evolution of genome size: New approaches to an old problem. *Trends in Genetics*, 17(1), 23–28.

Pevzner, P. and Tesler, G. (2003). Human and mouse genomic sequences reveal extensive breakpoint reuse in mammalian evolution. *Proceedings of the National Academy of Sciences*, 100(13), 7672–7677.

Plassais, J., Kim, J., Davis, B.W., Karyadi, D.M., Hogan, A.N., Harris, A.C., Decker, B., Parker, H.G., Ostrander, E.A. (2019). Whole genome sequencing of canids reveals genomic regions under selection and variants influencing morphology. *Nature Communications*, 10(1), 1–14.

Potocki, L., Chen, K.-S., Park, S.-S., Osterholm, D.E., Withers, M.A., Kimonis, V., Summers, A.M., Meschino, W.S., Anyane-Yeboa, K., Kashork, C.D. et al. (2000). Molecular mechanism for duplication 17p11.2 – The homologous recombination reciprocal of the smith-magenis microdeletion. *Nature Genetics*, 24(1), 84–87.

Probst, F.J., Fridell, R.A., Raphael, Y., Saunders, T.L., Wang, A., Liang, Y., Morell, R.J., Touchman, J.W., Lyons, R.H., Noben-Trauth, K. et al. (1998). Correction of deafness in shaker-2 mice by an unconventional myosin in a BAC transgene. *Science*, 280(5368), 1444–1447.

Redon, R., Ishikawa, S., Fitch, K.R., Feuk, L., Perry, G.H., Andrews, T.D., Fiegler, H., Shapero, M.H., Carson, A.R., Chen, W. et al. (2006). Global variation in copy number in the human genome. *Nature*, 444(7118), 444–454.

Reijo, R., Lee, T.-Y., Salo, P., Alagappan, R., Brown, L.G., Rosenberg, M., Rozen, S., Jaffe, T., Straus, D., Hovatta, O. et al. (1995). Diverse spermatogenic defects in humans caused by Y chromosome deletions encompassing a novel RNA-binding protein gene. *Nature Genetics*, 10(4), 383–393.

Repping, S., Skaletsky, H., Brown, L., van Daalen, S.K., Korver, C.M., Pyntikova, T., Kuroda-Kawaguchi, T., de Vries, J.W., Oates, R.D., Silber, S. et al. (2003). Polymorphism for a 1.6-mb deletion of the human Y chromosome persists through balance between recurrent mutation and haploid selection. *Nature Genetics*, 35(3), 247–251.

Rinn, J.L. and Chang, H.Y. (2012). Genome regulation by long noncoding RNAs. *Annual Review of Biochemistry*, 81, 145–166.

Rodgers-Melnick, E., Mane, S.P., Dharmawardhana, P., Slavov, G.T., Crasta, O.R., Strauss, S.H., Brunner, A.M., Difazio, S.P. (2012). Contrasting patterns of evolution following whole genome versus tandem duplication events in populus. *Genome Research*, 22(1), 95–105.

Rodin, S.N. and Riggs, A.D. (2003). Epigenetic silencing may aid evolution by gene duplication. *J. Mol. Evol.*, 56(6), 718–729.

Rogers, J. and Gibbs, R.A. (2014). Comparative primate genomics: Emerging patterns of genome content and dynamics. *Nature Reviews Genetics*, 15(5), 347–359.

Ross, M.T., Grafham, D.V., Coffey, A.J., Scherer, S., McLay, K., Muzny, D., Platzer, M., Howell, G.R., Burrows, C., Bird, C.P. et al. (2005). The DNA sequence of the human X chromosome. *Nature*, 434(7031), 325–337.

Rosser, Z.H., Balaresque, P., Jobling, M.A. (2009). Gene conversion between the X chromosome and the male-specific region of the Y chromosome at a translocation hotspot. *The American Journal of Human Genetics*, 85(1), 130–134.

Roudnitzky, N., Risso, D., Drayna, D., Behrens, M., Meyerhof, W., Wooding, S.P. (2016). Copy number variation in TAS2R bitter taste receptor genes: Structure, origin, and population genetics. *Chemical Senses*, 41(8), 649–659.

Rouquier, S., Blancher, A., Giorgi, D. (2000). The olfactory receptor gene repertoire in primates and mouse: Evidence for reduction of the functional fraction in primates. *Proceedings of the National Academy of Sciences*, 97(6), 2870–2874.

Rovelet-Lecrux, A., Hannequin, D., Raux, G., Le Meur, N., Laquerrière, A., Vital, A., Dumanchin, C., Feuillette, S., Brice, A., Vercelletto, M. et al. (2006). App locus duplication causes autosomal dominant early-onset Alzheimer disease with cerebral amyloid angiopathy. *Nature Genetics*, 38(1), 24–26.

Rozen, S.G., Skaletsky, H., Marszalek, J.D., Minx, P.J., Cordum, H.S., Waterston, R.H., Wilson, R.K., Page, D.C. (2003). Abundant gene conversion between arms of palindromes in human and ape Y chromosomes. *Nature*, 423(6942), 873–876.

Rozen, S.G., Marszalek, J.D., Irenze, K., Skaletsky, H., Brown, L.G., Oates, R.D., Silber, S.J., Ardlie, K., Page, D.C. (2012). AZFc deletions and spermatogenic failure: A population-based survey of 20,000 Y chromosomes. *The American Journal of Human Genetics*, 91(5), 890–896.

Samonte, R.V. and Eichler, E.E. (2002). Segmental duplications and the evolution of the primate genome. *Nature Reviews Genetics*, 3(1), 65–72.

Scanlan, M.J., Gure, A.O., Jungbluth, A.A., Old, L.J., Chen, Y.-T. (2002). Cancer/testis antigens: An expanding family of targets for cancer immunotherapy. *Immunological Reviews*, 188(1), 22–32.

Schiebel, K., Winkelmann, M., Mertz, A., Xu, X., Page, D.C., Weil, D., Petit, C., Rappold, G.A. (1997). Abnormal XY interchange between a novel isolated protein kinase gene, PRKY, and its homologue, PRKX, accounts for one third of all (Y+) XX males and (Y-) XY females. *Human Molecular Genetics*, 6(11), 1985–1989.

Schmitz, J.F., Chain, F.J.J., Bornberg-Bauer, E. (2020). Evolution of novel genes in three-spined stickleback populations. *Heredity (Edinb)*, 125(1–2), 50–59.

Schrider, D.R., Navarro, F.C., Galante, P.A., Parmigiani, R.B., Camargo, A.A., Hahn, M.W., de Souza, S.J. (2013). Gene copy-number polymorphism caused by retrotransposition in humans. *PLoS Genetics*, 9(1), e1003242.

Schwartz, D.C., Li, X., Hernandez, L.I., Ramnarain, S.P., Huff, E.J., Wang, Y.-K. (1993). Ordered restriction maps of *Saccharomyces cerevisiae* chromosomes constructed by optical mapping. *Science*, 262(5130), 110–114.

Sebat, J., Lakshmi, B., Troge, J., Alexander, J., Young, J., Lundin, P., Månér, S., Massa, H., Walker, M., Chi, M. et al. (2004). Large-scale copy number polymorphism in the human genome. *Science*, 305(5683), 525–528.

Sellers, J.R. (2000). Myosins: A diverse superfamily. *Biochimica et Biophysica Acta (BBA)–Molecular Cell Research*, 1496(1), 3–22.

Shapira, S.K. and Finnerty, V.G. (1986). The use of genetic complementation in the study of eukaryotic macromolecular evolution: Rate of spontaneous gene duplication at two loci of *Drosophila melanogaster*. *J. Mol. Evol.*, 23(2), 159–167.

Sharma, D., Issac, B., Raghava, G., Ramaswamy, R. (2004). Spectral repeat finder (SRF): Identification of repetitive sequences using Fourier transformation. *Bioinformatics*, 20(9), 1405–1412.

Sharp, A.J., Cheng, Z., Eichler, E.E. (2006). Structural variation of the human genome. *Annu. Rev. Genomics Hum. Genet.*, 7, 407–442.

She, X., Horvath, J.E., Jiang, Z., Liu, G., Furey, T.S., Christ, L., Clark, R., Graves, T., Gulden, C.L., Alkan, C. et al. (2004). The structure and evolution of centromeric transition regions within the human genome. *Nature*, 430(7002), 857–864.

She, X., Liu, G., Ventura, M., Zhao, S., Misceo, D., Roberto, R., Cardone, M.F., Rocchi, M., Green, E.D., Archidiacano, N. et al. (2006). A preliminary comparative analysis of primate segmental duplications shows elevated substitution rates and a great-ape expansion of intrachromosomal duplications. *Genome Research*, 16(5), 576–583.

She, X., Rohl, C.A., Castle, J.C., Kulkarni, A.V., Johnson, J.M., Chen, R. (2009). Definition, conservation and epigenetics of housekeeping and tissue-enriched genes. *BMC Genomics*, 10, 269.

Shearer, A.E., Kolbe, D.L., Azaiez, H., Sloan, C.M., Frees, K.L., Weaver, A.E., Clark, E.T., Nishimura, C.J., Black-Ziegelbein, E.A., Smith, R.J. (2014). Copy number variants are a common cause of non-syndromic hearing loss. *Genome Medicine*, 6(5), 1–10.

Shi, X., Chen, C., Yang, H., Hou, J., Ji, T., Cheng, J., Veitia, R.A., Birchler, J.A. (2020). The gene balance hypothesis: Epigenetics and dosage effects in plants. *Methods Mol. Biol.*, 2093, 161–171.

Shiu, S., Byrnes, J., Pan, R., Zhang, P., Li, W. (2006). Role of positive selection in the retention of duplicate genes in mammalian genomes. *Proc. Natl. Acad. Sci. USA*, 103, 2232–2236.

Simoni, M., Gromoll, J., Dworniczak, B., Rolf, C., Abshagen, K., Kamischke, A., Carani, C., Meschede, D., Behre, H.M., Horst, J. et al. (1997). Screening for deletions of the Y chromosome involving the daz (deleted in azoospermia) gene in azoospermia and severe oligozoospermia. *Fertility and Sterility*, 67(3), 542–547.

Sinclair, A.H., Berta, P., Palmer, M.S., Hawkins, J.R., Griffiths, B.L., Smith, M.J., Foster, J.W., Frischauf, A.M., Lovell-Badge, R., Goodfellow, P.N. (1990). A gene from the human sex-determining region encodes a protein with homology to a conserved DNA-binding motif. *Nature*, 346(6281), 240–244.

Singchat, W., Ahmad, S.F., Laopichienpong, N., Suntronpong, A., Panthum, T., Griffin, D.K., Srikulnath, K. (2020). Snake W sex chromosome: The shadow of ancestral amniote super-sex chromosome. *Cells*, 9(11), 2386.

Singleton, A.B. (2005). Altered α-synuclein homeostasis causing Parkinson's disease: The potential roles of dardarin. *Trends in Neurosciences*, 28(8), 416–421.

Singleton, A.B. and Gwinn-Hardy, K. (2004). Parkinson's disease and dementia with lewy bodies: A difference in dose? *The Lancet*, 364(9440), 1105–1107.

Sinha, A.U. and Meller, J. (2007). Cinteny: Flexible analysis and visualization of synteny and genome rearrangements in multiple organisms. *BMC Bioinformatics*, 8(1), 1–9.

Siva, N. (2008). 1000 genomes project. *Nature Biotechnology*, 26(3), 256–257.

Skaletsky, H., Kuroda-Kawaguchi, T., Minx, P.J., Cordum, H.S., Hillier, L., Brown, L.G., Repping, S., Pyntikova, T., Ali, J., Bieri, T. et al. (2003). The male-specific region of the human Y chromosome is a mosaic of discrete sequence classes. *Nature*, 423(6942), 825–837.

Soo, V.W., Hanson-Manful, P., Patrick, W.M. (2011). Artificial gene amplification reveals an abundance of promiscuous resistance determinants in *Escherichia coli*. *Proceedings of the National Academy of Sciences*, 108(4), 1484–1489.

Spielmann, M., Lupiáñez, D.G., Mundlos, S. (2018). Structural variation in the 3d genome. *Nature Reviews Genetics*, 19(7), 453–467.

Stankiewicz, P. and Lupski, J.R. (2002). Genome architecture, rearrangements and genomic disorders. *TRENDS in Genetics*, 18(2), 74–82.

Stapley, J., Santure, A.W., Dennis, S.R. (2015). Transposable elements as agents of rapid adaptation may explain the genetic paradox of invasive species. *Mol. Ecol.*, 24(9), 2241–2252.

Stefansson, H., Helgason, A., Thorleifsson, G., Steinthorsdottir, V., Masson, G., Barnard, J., Baker, A., Jonasdottir, A., Ingason, A., Gudnadottir, V.G. et al. (2005). A common inversion under selection in Europeans. *Nature Genetics*, 37(2), 129–137.

Sturtevant, A.H. (1925). The effects of unequal crossing over at the bar locus in Drosophila. *Genetics*, 10(2), 117.

Suryawanshi, V., Talke, I.N., Weber, M., Eils, R., Brors, B., Clemens, S., Krämer, U. (2016). Between-species differences in gene copy number are enriched among functions critical for adaptive evolution in *Arabidopsis halleri*. *BMC Genomics*, 17(Suppl. 13), 1034.

Swanson-Wagner, R.A., Eichten, S.R., Kumari, S., Tiffin, P., Stein, J.C., Ware, D., Springer, N.M. (2010). Pervasive gene content variation and copy number variation in maize and its undomesticated progenitor. *Genome Research*, 20(12), 1689–1699.

Taguchi, A., Taylor, A.D., Rodriguez, J., Çeliktaş, M., Liu, H., Ma, X., Zhang, Q., Wong, C.-H., Chin, A., Girard, L. et al. (2014). A search for novel cancer/testis antigens in lung cancer identifies VCX/Y genes, expanding the repertoire of potential immunotherapeutic targets. *Cancer Research*, 74(17), 4694–4705.

Teichmann, S.A. and Babu, M.M. (2004). Gene regulatory network growth by duplication, *Nature Genetics*, 36(5), 492–496.

Teichmann, S.A., Park, J., Chothia, C. (1998). Structural assignments to the mycoplasma genitalium proteins show extensive gene duplications and domain rearrangements. *Proceedings of the National Academy of Sciences*, 95(25), 14658–14663.

Teitz, L.S., Pyntikova, T., Skaletsky, H., Page, D.C. (2018). Selection has countered high mutability to preserve the ancestral copy number of Y chromosome amplicons in diverse human lineages. *The American Journal of Human Genetics*, 103(2), 261–275.

Trask, B.J., Friedman, C., Martin-Gallardo, A., Rowen, L., Akinbami, C., Blankenship, J., Collins, C., Giorgi, D., Iadonato, S., Johnson, F. et al. (1998). Members of the olfactory receptor gene family are contained in large blocks of DNA duplicated polymorphically near the ends of human chromosomes. *Human Molecular Genetics*, 7(1), 13–26.

Turner, A.J., Aggarwal, P., Boone, E.C., Haidar, C.-E., Relling, M.V., Derezinski, A.D., Broeckel, U., Gaedigk, A. (2021). Identification of CYP2D6 haplotypes that interfere with commonly used assays for copy number variation characterization. *The Journal of Molecular Diagnostics*, 23(5), 577–588.

Tuzun, E., Sharp, A.J., Bailey, J.A., Kaul, R., Morrison, V.A., Pertz, L.M., Haugen, E., Hayden, H., Albertson, D., Pinkel, D. et al. (2005). Fine-scale structural variation of the human genome. *Nature Genetics*, 37(7), 727–732.

Urbini, M., Astolfi, A., Indio, V., Nannini, M., Schipani, A., Bacalini, M.G., Angelini, S., Ravegnini, G., Calice, G., Del Gaudio, M. et al. (2020). Gene duplication, rather than epigenetic changes, drives FGF4 overexpression in KIT/PDGFRA/SDH/RAS-P WT GIST. *Sci. Rep.*, 10(1), 19829.

Van Esch, H., Bauters, M., Ignatius, J., Jansen, M., Raynaud, M., Hollanders, K., Lugtenberg, D., Bienvenu, T., Jensen, L.R., Gécz, J. et al. (2005). Duplication of the MECP2 region is a frequent cause of severe mental retardation and progressive neurological symptoms in males. *The American Journal of Human Genetics*, 77(3), 442–453.

Vaschetto, L.M. and Ortiz, N. (2019). The role of sequence duplication in transcriptional regulation and genome evolution. *Curr. Genomics*, 20(6), 405–408.

Vila-Aiub, M.M., Yu, Q., Powles, S.B. (2019). Do plants pay a fitness cost to be resistant to glyphosate? *New. Phytol.*, 223(2), 532–547.

Wagner, F.F. and Flegel, W.A. (2000). RHD gene deletion occurred in the rhesus box. *Blood, The Journal of the American Society of Hematology*, 95(12), 3662–3668.

Wang, P.J., McCarrey, J.R., Yang, F., Page, D.C. (2001). An abundance of X-linked genes expressed in spermatogonia. *Nature Genetics*, 27(4), 422–426.

Wang, Y., Xiong, G., Hu, J., Jiang, L., Yu, H., Xu, J., Fang, Y., Zeng, L., Xu, E., Xu, J. et al. (2015). Copy number variation at the GL7 locus contributes to grain size diversity in rice. *Nature Genetics*, 47(8), 944–948.

Watanabe, Y., Takahashi, A., Itoh, M., Takano-Shimizu, T. (2009). Molecular spectrum of spontaneous de novo mutations in male and female germline cells of *Drosophila melanogaster*. *Genetics*, 181(3), 1035–1043.

Weiner, P. (1973). Linear pattern matching algorithms. *14th Annual Symposium on Switching and Automata Theory (SWAT 1973)*. IEEE, USA.

Wenger, A.M., Peluso, P., Rowell, W.J., Chang, P.-C., Hall, R.J., Concepcion, G.T., Ebler, J., Fungtammasan, A., Kolesnikov, A., Olson, N.D. et al. (2019). Accurate circular consensus long-read sequencing improves variant detection and assembly of a human genome. *Nature Biotechnology*, 37(10), 1155–1162.

Weterman, M.A., Van Ruissen, F., De Wissel, M., Bordewijk, L., Samijn, J.P., Van Der Pol, W.L., Meggouh, F., Baas, F. (2010). Copy number variation upstream of PMP22 in Charcot–Marie–Tooth disease. *European Journal of Human Genetics*, 18(4), 421–428.

Wong, P., Fritz, A., Frishman, D. (2005). Designability, aggregation propensity and duplication of disease-associated proteins. *Protein Engineering Design and Selection*, 18(10), 503–508.

Woods, K.S., Cundall, M., Turton, J., Rizotti, K., Mehta, A., Palmer, R., Wong, J., Chong, W., Al-Zyoud, M., El-Ali, M. et al. (2005). Over- and underdosage of SOX3 is associated with infundibular hypoplasia and hypopituitarism. *The American Journal of Human Genetics*, 76(5), 833–849.

Xu, K., Xu, X., Fukao, T., Canlas, P., Maghirang-Rodriguez, R., Heuer, S., Ismail, A.M., Bailey-Serres, J., Ronald, P.C., Mackill, D.J. (2006). Sub1A is an ethylene-response-factor-like gene that confers submergence tolerance to rice. *Nature*, 442(7103), 705–708.

Yang, F., Pecina, D.A., Kelly, S.D., Kim, S.-H., Kemner, K.M., Long, D.T., Marsh, T.L. (2010). Biosequestration via cooperative binding of copper by *Ralstonia pickettii*. *Environmental Technology*, 31(8–9), 1045–1060.

Yona, A.H., Manor, Y.S., Herbst, R.H., Romano, G.H., Mitchell, A., Kupiec, M., Pilpel, Y., Dahan, O. (2012). Chromosomal duplication is a transient evolutionary solution to stress. *Proceedings of the National Academy of Sciences*, 109(51), 21010–21015.

You, C., Shan, C., Xin, J., Li, J., Ma, Z., Zhang, Y., Zeng, X., Gao, X. (2020). Propoxur resistance associated with insensitivity of acetylcholinesterase (ACHE) in the housefly, *Musca domestica* (diptera: Muscidae). *Sci. Rep.*, 10(1), 8400.

Young, J.M., Friedman, C., Williams, E.M., Ross, J.A., Tonnes-Priddy, L., Trask, B.J. (2002). Different evolutionary processes shaped the mouse and human olfactory receptor gene families. *Human Molecular Genetics*, 11(5), 535–546.

Young, J.M., Endicott, R.M., Parghi, S.S., Walker, M., Kidd, J.M., Trask, B.J. (2008). Extensive copy-number variation of the human olfactory receptor gene family. *The American Journal of Human Genetics*, 83(2), 228–242.

Zhang, J. (2003). Evolution by gene duplication: An update. *Trends in Ecology & Evolution*, 18(6), 292–298.

Zhang, X. and Firestein, S. (2002). The olfactory receptor gene superfamily of the mouse. *Nature Neuroscience*, 5(2), 124–133.

Zhang, J., Rosenberg, H.F., Nei, M. (1998). Positive Darwinian selection after gene duplication in primate ribonuclease genes. *Proceedings of the National Academy of Sciences*, 95(7), 3708–3713.

Zhang, L., Lu, H.H., Chung, W.-Y., Yang, J., Li, W.-H. (2005). Patterns of segmental duplication in the human genome. *Molecular Biology and Evolution*, 22(1), 135–141.

Zhang, Y., Malekpour, M., Al-Madani, N., Kahrizi, K., Zanganeh, M., Mohseni, M., Mojahedi, F., Daneshi, A., Najmabadi, H., Smith, R.J. (2007). Sensorineural deafness and male infertility: A contiguous gene deletion syndrome. *Journal of Medical Genetics*, 44(4), 233–240.

Zhang, F., Gu, W., Hurles, M.E., Lupski, J.R. (2009). Copy number variation in human health, disease, and evolution. *Annual Review of Genomics and Human Genetics*, 10, 451–481.

Zhang, J., Guan, J., Wang, H., Yin, L., Wang, D., Zhao, L., Zhou, H., Wang, Q. (2019). Genotype-phenotype correlation analysis of MYO15A variants in autosomal recessive non-syndromic hearing loss. *BMC Medical Genetics*, 20(1), 1–16.

Zhao, G.Q., Zhang, Y., Hoon, M.A., Chandrashekar, J., Erlenbach, I., Ryba, N.J., Zuker, C.S. (2003). The receptors for mammalian sweet and umami taste. *Cell*, 115(3), 255–266.

Zhenyang, L., Xunxiao, Z., Shengcheng, Z., Zhicong, L., Xingtan, Z., Ray, M. (2021). Structural variations in papaya genomes. *BMC Genomics*, 22(1), 335.

Zhou, Q., Zhang, J., Bachtrog, D., An, N., Huang, Q., Jarvis, E.D., Gilbert, M.T.P., Zhang, G. (2014). Complex evolutionary trajectories of sex chromosomes across bird taxa. *Science*, 346(6215).

Zou, S.W., Zhang, J.C., Zhang, X.D., Miao, S.Y., Zong, S.D., Sheng, Q., Wang, L.F. (2003). Expression and localization of VCX/Y proteins and their possible involvement in regulation of ribosome assembly during spermatogenesis. *Cell Research*, 13(3), 171–177.

3

Transposable Elements: Parasites that Shape Genome Evolution

Amandine BONNET[1], Karine CASIER[2], Clément CARRÉ[2], Laure TEYSSET[2] and Pascale LESAGE[1]

[1]*Génomes, biologie cellulaire et thérapeutique, Université Paris Cité, CNRS UMR7212, Inserm U944, France*

[2]*Epigénétique transgénérationnelle et biologie des petits ARN, Institut de Biologie Paris Seine, Sorbonne Université, CNRS UMR7622, France*

3.1. Transposable elements in eukaryotic genomes

Transposable elements (TEs) are mobile repetitive DNA sequences that propagate within genomes by a mechanism called transposition. To ensure transposition, TEs are transcribed and encode their own proteins necessary for their mobility. TEs were first discovered in maize by Barbara McClintock in 1950 (McClintock 1950), who was awarded the Nobel Prize in Physiology or Medicine in 1983. TEs are present in all eukaryotic and prokaryotic organisms. This chapter focuses mainly on eukaryotic TEs.

For a color version of all figures in this chapter, see http://www.iste.co.uk/richard/function.zip.

Function and Evolution of Repeated DNA Sequences,
coordinated by Guy-Franck RICHARD.

3.1.1. *TEs: essential components of eukaryotic genomes*

In the 2000s, the development of high-throughput sequencing techniques revealed that TEs constitute a significant part of eukaryotic genomes (see Figure 3.1). For example, in *Homo sapiens*, the proportion of TE-derived sequences is estimated to be 45% and in maize (*Zea mays*) over 80% (Biémont 2010; Wells and Feschotte 2020), while about 2% of these genomes are coding. TEs can be complete and autonomous for their transposition, non-autonomous and dependent on autonomous elements for their transposition, or they may be present only in the form of truncated fragments incapable of transposing. TEs have long been considered selfish, parasitic and even useless. In this chapter, we will see that, while their mobility poses a threat to genome stability when uncontrolled, TEs also represent powerful drivers of genome evolution.

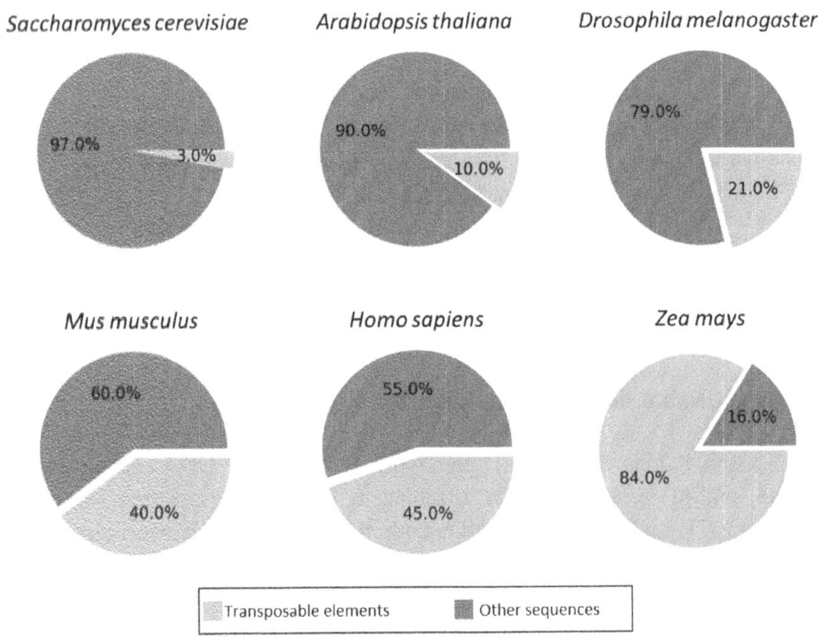

Figure 3.1. *Proportion of TE sequences in several genomes. While the proportion of the coding genome is stable (around 2%) regardless of the species apart from certain compact genomes such as Saccharomyces cerevisiae (70% of coding sequences), the proportion of TEs in the genomes varies from species to species. Some examples are presented here for different species: baker's yeast (Saccharomyces cerevisiae), thale cress (Arabidopsis thaliana), the fruit fly (Drosophila melanogaster), mice (Mus musculus), humans (Homo sapiens) and maize (Zea mays)*

3.1.2. *Acquisition of new TEs by horizontal transfer*

Horizontal transfer is a process in which an organism receives genetic material (gene, TE, plasmid, etc.) from a species different from its own. In prokaryotes, these transfers are relatively frequent as a result of transformation, conjugation, or transduction events (Garcia-Vallvé et al. 2000). In eukaryotes, sequencing data indicate that horizontal transfers also occur very frequently, the majority of them involving TEs. Many horizontal transfers of TEs have indeed been identified between insects and vertebrates (Peccoud et al. 2017; Zhang et al. 2020), between different plant species (El Baidouri et al. 2014), and even between insects and plants (Lin et al. 2016; Gao et al. 2018). It has been proposed that viruses, endosymbiotic bacteria or parasites would be the vectors for these exchanges of genetic material between species (Gilbert and Cordaux 2017; Loreto et al. 2008). A genome analysis of a pool of baculoviruses, double-stranded DNA viruses, from infected noctuid larvae (*Spodoptera exigua*) revealed the presence of numerous TE insertions, indicating that viruses could indeed be vectors of horizontal transfers of TEs from one species to another (Loiseau et al. 2020).

The best-documented horizontal transfer of TEs is the case of the *P* element absent from the *Drosophila melanogaster* species until the 1950s. The first strains of *D. melanogaster* containing *P* elements in their genome were identified in North America in 1954 and the last *P* element-free strain was identified in Ukraine in 1970 (Anxolabéhère et al. 1988). The rapid invasion of the *P* element in natural fly populations was probably made possible by intense trade in fruits, on which *D. melanogaster* feeds and grows. This allowed researchers to monitor the dynamics of the *P* element in real time and study the establishment of the mechanisms of its repression (see section 3.4.3). Molecular studies have determined that the horizontal transfer of the *P* element occurred from *D. willistoni* to *D. melanogaster* (Daniels et al. 1990), two species about 60 million years apart, incapable of interbreeding.

The invasion of a genome by a new TE can also occur as a result of a retrovirus infection. Retroviruses generally infect somatic cells, and are therefore not transmitted to offspring. In rare cases, germ cells may be infected. The retrovirus is then inherited by the next generations like any other gene (vertical transfer), and become an "endogenous retrovirus" (ERV), generally unable to infect other cells. The human genome consists of 8% of sequences derived from ancient retroviruses. For most of the known ERVs, endogenization occurred millions of years ago. Interestingly, a retrovirus began invading the germline genome of koalas in Australia "only" 50,000 years ago. As this germline infection is recent, not all koalas are infected, allowing the natural process of endogenization to be monitored (Löber et al. 2018).

3.2. Classification of TEs and transposition mechanisms

There is a great diversity in TE structures and transposition mechanisms (Bourque et al. 2018). The first TE classification, proposed in 1989 by Finnegan and his colleagues, distinguished two classes according to their transposition strategy (Finnegan 1989). Class I elements, or retrotransposons, transpose through an RNA intermediate via a "copy and paste" mechanism. Class II elements, or DNA transposons, transpose via a "cut and paste" mechanism. The discovery of new elements has revealed the need for a new classification based on the structural, mechanistic and enzymatic characteristics of TEs (Curcio and Derbyshire 2003; Wicker et al. 2007; Beauregard et al. 2008; Arkhipova 2017) (see Figure 3.2). Of note, this new classification focuses on eukaryotic TEs and therefore excludes certain transposition mechanisms such as group-I and group-II autocatalytic introns, and the inteins described in prokaryotes (Belfort 2017).

3.2.1. *Class I retrotransposons*

Because of their "copy and paste" transposition mechanism, the number of copies of class I retrotransposons, also called retroelements, increases mechanically. It has been shown that the increase in size of the human genome is mainly due to retrotransposition events (Liu et al. 2003). The retroelements are subdivided into four subclasses (see Figure 3.2).

The long terminal repeat (LTR) retrotransposon subclass comprises all TEs that have long repeated sequences flanking two open reading frames (ORFs) *GAG* and *POL* (Havecker et al. 2004). The *GAG* gene encodes a structural protein that has RNA-binding domains and encapsidates the LTR retrotransposon RNA in virus-like particles (VLP). The *POL* gene encodes the protease (PR), reverse transcriptase ribonuclease H (RT-RH) and integrase (IN) enzymatic activities required for retrotransposition. LTR retrotransposons are structurally and functionally related to retroviruses except that unlike the latter, they do not have an envelope gene and are not infectious. However, a third open reading frame is present in certain insect LTR retrotransposons (e.g. gypsy, ZAM) and appears to have a common origin with baculovirus envelope proteins (Malik et al. 2000; Rohrmann and Karplus 2001) and functional similarities with retrovirus envelope proteins (Kim et al. 1994; Song et al. 1994; Teysset et al. 1998). LTR retrotransposons are first transcribed in the nucleus into an RNA intermediate that is translated into Gag and Gag-Pol proteins in the cytoplasm and/or reverse transcribed into a linear double-stranded complementary DNA (cDNA) by the RT-RH enzyme in the cytoplasmic VLP (Hughes 2015). To initiate cDNA synthesis, RT-RH uses a transfer RNA (tRNA) that is packaged in the VLP and whose sequence is partially complementary to a region of the LTR retrotransposon transcript, called the primer binding site (PBS). The integrase binds the cDNA, which is imported into the nucleus and integrated into the host genome by the integrase.

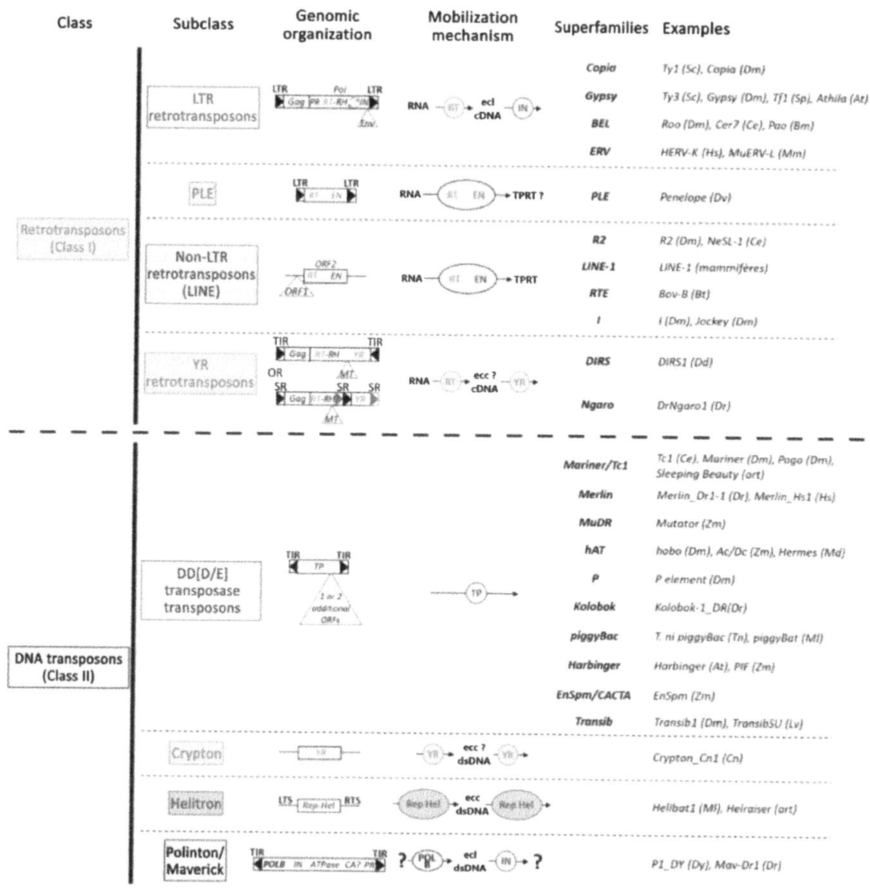

Figure 3.2. *Classification of autonomous TEs in eukaryotes. This classification is mainly based on the mechanisms and machineries involved in transposition events. There are two classes: class I retrotransposons, which encode for reverse transcriptase (RT) and transpose through an RNA intermediate, and class II DNA transposons, which are excised and move to a new location. Each class is subdivided into subclasses, based on the genome structure and/or transposition mechanisms. The figure illustrates the diversity of the main TE superfamilies of each subclass in main model organisms*

NOTES ON FIGURE 3.2.– *LTR, long terminal repeat; PLE, penelope-like element; PR, protease; RT-RH, reverse transcriptase-RNase H; IN, integrase with DD[D/E] catalytic domain; EN, apurinic/apyrimidinic or restriction-like endonuclease (purple) or with a GIY-YIG amino acid motif (blue); TIR, terminal inverted repeat; SR, split repeat; YR, tyrosine recombinase; MT, methyltransferase; TP, transposase with DD[D/E] catalytic domain; LTS, left terminal sequence; RTS, right terminal sequence containing a palindromic sequence; Rep, replication initiator HUH nuclease; Hel, DNA helicase; POLB, protein-primed type B DNA polymerase; CA, capsid protein; ecl, extrachromosomal linear; TPRT, target-primed reverse transcription; ecc, extrachromosomal circular; cDNA, complementary DNA; dsDNA, double-stranded DNA; ssDNA, single-stranded DNA; Sc, Saccharomyces cerevisiae; Dm, Drosophila melanogaster; Sp, Schizosaccharomyces pombe; At, Arabidopsis thaliana; Ce, Caenorhabditis elegans; Bm, Bombyx mori; Hs, Homo sapiens; Mm, Mus musculus; Dv, Drosophila virilis; Bt, Bos Taurus; Dd, Dictyostelium discoideum; Dr, Danio rerio; Zm, Zea mays; Md, Musca domestica; Tn, Trichoplusia ni; Ml, Myotis lucifugus; Lv, Lytechinus variegatus; Cn, Cryptococcus neoformans; Dy, Drosophila yakuba; art, artificially reconstructed.*

The subclass of non-LTR retrotransposons contains the LINE (long interspersed nuclear element), SINE (short interspersed nuclear element) and SVA (SINE-R/VNTR/Alu-like) elements. LINEs generally harbor a 5'UTR (untranslated region) and a poly(A) sequence or simple repeats at their 3' end. Their internal sequence contains a promoter and two ORFs. ORF1 encodes an RNA binding protein, ORF1p, which is absent from some non-LTR retrotransposons and whose function is poorly known. ORF2 encodes ORF2p, which has endonuclease (EN) and reverse transcriptase (RT) activities (Moran et al. 1996). SINEs do not possess coding sequences and are therefore not autonomous. Their structure consists of a sequence derived from the 5' end of a tRNA gene and a sequence derived from the 3' end of LINEs. They are transcribed by RNA polymerase III. In humans and primates, the most abundant SINEs are Alu elements. SVAs are composite SINEs, containing *Alu*-like sequences and a repeat region rich in C-T, and are also non-autonomous. SINEs and SVAs are mobilized by the LINE machinery. Reverse transcription of non-LTR retrotransposons begins with a single-stranded nick of the chromosomal DNA by the endonuclease, exposing a 3' hydroxyl end that serves as the primer for reverse transcription of the non-LTR retrotransposon transcript to DNA. This process, called target-primed reverse transcription (TPRT), leads to the integration of frequent 5'-truncated copies into the host genome.

The last two retrotransposon subclasses correspond to the elements *Penelope* (*PLE*) and *YR*. *PLE* has a structure similar to that of an LTR retrotransposon but is probably replicated by TPRT (Kojima 2020). *YR* retrotransposons carry a tyrosine recombinase instead of an integrase and can involve a circular cDNA in their replication cycle (Poulter and Butler 2015).

3.2.2. *Class II DNA transposons*

The class II elements or DNA transposons consist of four subclasses (Feschotte and Pritham 2007; Kojima 2020) (see Figure 3.2). The elements encoding a transposase characterized by a DD[D/E] catalytic motif are the most studied. The transposase recognizes repeated inverted sequences at the ends of the element (terminal inverted repeat [TIR]), excises the entire transposon sequence and inserts it in a new genomic site (Finnegan 1992). This "cut and paste" transposition mechanism is therefore conservative. The transposase of an autonomous element can also mobilize non-autonomous elements such as MITEs (miniature inverted repeats), which have typical DNA transposon ends but do not encode active transposase. The copy number of DNA transposons can increase in genomes either when their excision occurs after passing the replication fork and insertion in a new site before passing the replication fork, or when the double-strand break generated by excision is repaired by restoring a new copy of the element at the donor site, while the excised copy is inserted into a new location (for further details, see Majumdar and Rio (2015)).

The mobility of the Crypton subclass depends on a tyrosine recombinase (YR), while Helitron transposition involves the displacement of a single-stranded DNA intermediate (Arkhipova 2017). Helitrons generally code for a protein that includes two enzymatic domains, a DNA helicase (Hel) and a rolling-circle replication initiator. Lastly, Polintons/Mavericks can contain a dozen genes, including a DNA polymerase and an integrase sharing similarities with LTR retrotransposons. Their mode of transposition involves excision as single-stranded DNA, followed by double-stranded synthesis by DNA polymerase and integration into the genome (Wicker et al. 2007).

3.3. TE self-regulation

As genomic parasites, the fate of TEs is linked to their host. To limit their deleterious effects on the host, some TEs have developed self-regulating mechanisms (see Figure 3.3).

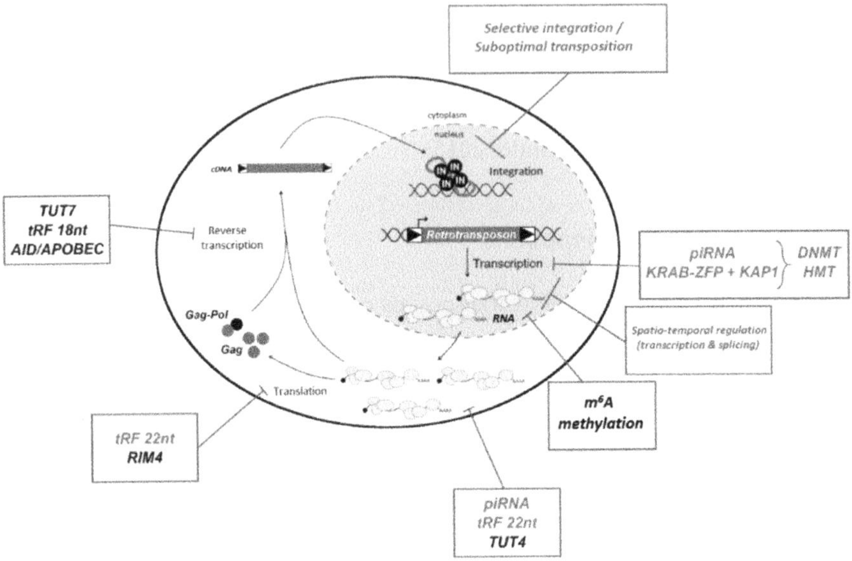

Figure 3.3. *Main mechanisms controlling the expression and mobility of TEs. Evolutionarily selected mechanisms of self-regulation of TE mobility (see sections 3.3.1 to 3.3.3) are framed in blue, while the restriction mechanisms put in place by the cell (see section 3.4) are framed in red. The mechanisms that target both retrotransposons and DNA transposons are written in green, while the restriction mechanisms that have been described only for retroelements are written in black. Note that the replication cycle of an LTR retrotransposon is depicted in order to position all the mechanisms discussed in sections 3.3.3 and 3.3.4*

3.3.1. *Spatio-temporal regulation of TE expression*

A recurrent strategy of TEs is to restrict their activity to certain cell types or developmental stages of host organisms. Transposition of TEs into somatic cells may affect host survival, but the new copies are not transmitted to offspring, unlike the newly integrated copies in germ cells that will be transmitted to offspring. Mammalian retrotransposons can benefit from epigenetic reprogramming that erases repressive chromatin marks at each generation before being *de novo* re-established, promoting their transcription during certain developmental stages (for review, see Wicker et al. 2007; Molaro and Malik 2016).

In other organisms, active mechanisms to induce TE expression in germ cells have occasionally been selected during evolution. In *Drosophila*, for example, the gene encoding the *P*-element transposase is interrupted by an intron that is only spliced in the germline (Laski et al. 1986). In a unicellular organism such as the yeast *Saccharomyces cerevisiae*, the retrotransposon Ty3 uses the meiotic transcription factor Ndt80 to activate its transcription specifically during meiosis (Laureau et al. 2021).

3.3.2. *Self-regulation of transposition efficiency*

The identification of hyperactive mutants of *Tc1/mariner* model DNA transposons indicates that these transposons encode suboptimal transposases (Lampe et al. 1999; Mátés et al. 2009; Liu and Chalmers 2014). This self-regulation is based on several strategies, including transposase aggregation over a certain concentration in the cells (Lohe and Hartl 1996) or the use of imperfect TIRs that decrease transposition efficiency (Augé-Gouillou et al. 2001). These strategies appear to be shared by many DNA transposons (for review, see Cosby et al. 2019). Another example is Ty1 retrotransposition in the yeast *S. cerevisiae*, which is subject to copy number control (CNC) enabling a high level of transposition when few Ty1 elements are present in the genome but limiting this activity when the copy number is high. Ty1 CNC involves an RNA transcribed from an internal Ty1 promoter, which produces a truncated Gag protein that interferes with the formation of virus-like particles (Saha et al. 2015; Cotte et al. 2021). Ty1 CNC in yeast can be considered as a protein equivalent of PIWI-interacting RNA (piRNA) regulation in animals, the abundance of which depends on the transcriptional activity of TE copies in genomes (see section 3.4.3).

3.3.3. *Selective integration to better protect the genome*

The impact of a new TE copy on host biology and genome integrity is strongly linked to its integration site. Whole-genome sequencing initiated in the 1990s has allowed TEs to be identified and their distribution in a large panel of genomes to be described. In the early stages, the difficulty of assembling repeated and/or heterochromatic regions of the genomes complicated these studies, and sophisticated bioinformatics tools had to be developed (Goerner-Potvin and Bourque 2018; Amselem et al. 2019).

A non-random distribution of TEs has been observed in many organisms, revealing their accumulation in specific regions of host genomes. This non-random distribution results in part from natural selection processes that eliminate deleterious insertions and maintain beneficial insertions. For some TEs, it is also the result of an active process that targets their integration into specific regions of the genome, where insertions will have little or no adverse effect on the physiology of the organism. The targeted regions often correspond to repeated and individually non-essential sequences, such as telomeric regions, ribosomal DNAs, upstream and downstream regions of tRNA genes, or regions surrounding open reading frames (see Figure 3.4 and, for review, see Sultana et al. 2017). The integration preference is not shared by all TEs and is of lesser or greater prominence. It depends on the enzyme responsible for the integration, as well as on the primary DNA sequence at the integration site, the DNA structure and its accessibility, or on so-called tethering protein factors, as illustrated in the examples below.

Integrases and many transposases have a DD[D/E] catalytic triad. They bind to the target DNA and make a staggered cut of the two strands by a transesterification reaction of phosphodiester bonds to integrate the TE double-stranded DNA. Integration sites of several DNA transposons, such as *Mariner/Tc1* or *piggyBack* families, or retroelements such as Ty1, change in human immunodeficiency retrovirus-1 (HIV-1) or murine leukemia retrovirus (MLV) present a weak nucleotide signature that includes a flexible pyrimidine-purine dinucleotide at the center of the integration site (Vigdal et al. 2002; Serrao et al. 2015). Because these sequences are very short, they are abundant in genomes and do not explain the genomic distribution of TEs. A notable exception is the integrase from the *Drosophila* ERV ZAM, which recognizes the CGCGCG consensus sequence as a specific integration site (Leblanc et al. 1999).

The integration of non-LTR retrotransposons that are replicated by TPRT at the integration site (see Figure 3.2) is initiated by cleavage of the target DNA by the "apurinic and apyrimidinic" (APE) or restriction enzyme-like (RLE) endonuclease activity of ORF2p. The APE domain contributes directly to the recognition of the target site sequence. For the RLE endonucleases, recognition of the integration site depends on one (or more) additional DNA binding domain(s) of ORF2p. Thus, most RLE elements, such as R2 retrotransposons in insects, insert in specific target sites, while most APE elements, including LINE-1 in humans, integrate less precisely into genomes (Fujiwara 2015). The ORF1p protein, which interacts with its own RNA, can also be involved in recognition of the integration site (Chung et al. 2007), as illustrated by the collaborative integration of non-LTR retrotransposons HeT-A, TART and TAHRE at telomeres in *D. melanogaster* (Rashkova et al. 2002; Fuller et al. 2010; Zhang et al. 2014). In this organism, HeT-A ORF1p forms spherical structures at the ends of the chromosomes and recruits ORF1p of TART and TAHRE. In return, HeT-A benefits from the RT activity of TART and TAHRE ORF2p to be reverse transcribed. The integration of these three retrotransposons by TPRT protects the end of the chromosomes and compensates for the absence of telomerase in *D. melanogaster*.

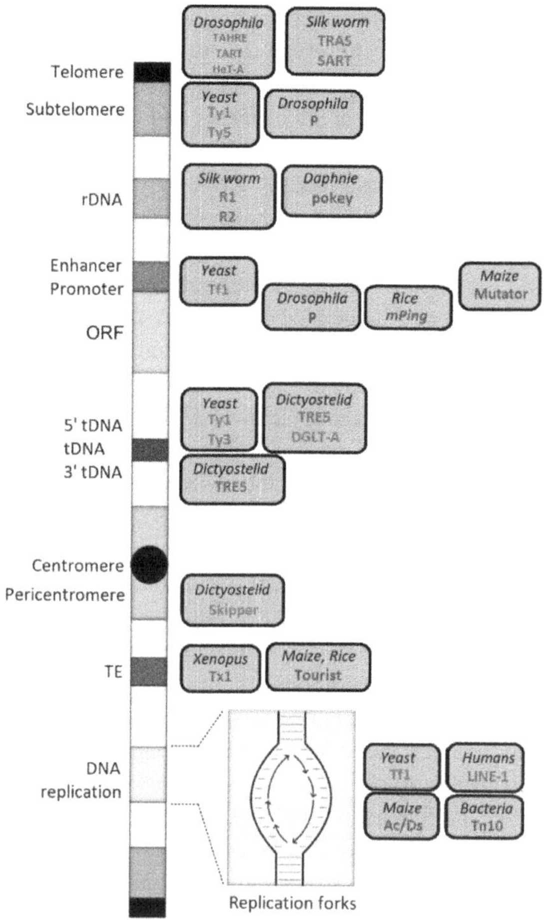

Figure 3.4. *Examples of integration sites targeted by TEs. Chromosome features and TE targeting these features are represented on the left and right of the chromosome, respectively. The DNA transposons are shown in red, LTR retrotransposons in green and retrotransposons without LTR in purple. References: TAHRE, TART HeT-A (Casacuberta 2017), TRAS and SART (Fujiwara et al. 2005), Ty1, Ty3 and Ty5 (Bonnet and Lesage 2021), P (Ronsseray et al. 1989; Spradling et al. 1995), R1 and R2 (Ye et al. 2005), Pockey (Penton and Crease 2004), Tf1 (Guo and Levin 2010; Hickey et al. 2015; Jacobs et al. 2015), mPing (Naito et al. 2009), mutator (Dietrich et al. 2002), TRE3, TRE5 and DGLT-A, Skipper (Spaller et al. 2016), Tx1 (Christensen et al. 2000), Tourist (Jiang and Wessler 2001), LINE-1 (Mita et al. 2018; Flasch et al. 2019; Sultana et al. 2019), Ac/Ds (Ros and Kunze 2001), Tn10 (Roberts et al. 1985)*

In the yeast *S. cerevisiae*, for which coding sequences represent 70% of the genome, evolution has selected a conserved mechanism to target integration of LTR retrotransposons, called *Ty*, into regions devoid of essential genes. The genome of *S. cerevisiae* contains five families of Ty elements and this mechanism was initially proposed to explain the selective integration of Ty3 and Ty5. Ty3 integrates several base pairs upstream of genes transcribed by RNA polymerase III (Pol III), mainly tRNAs (Kirchner et al. 1995) and Ty5 in heterochromatin of telomeres and silent cassettes *HML* and *HMR* (Gai and Voytas 1998; Xie et al. 2001). Their selective integration is based on a specific interaction of their integrase with tethering proteins, directly or indirectly associated with DNA or chromatin at preferential insertion sites. Ty3 integrase interacts with the Brf1 and TBP subunits of the TFIIIB complex involved in the recruitment of Pol III to the genome (Yieh et al. 2002; Qi and Sandmeyer 2012). The tethering factor of Ty5 is the heterochromatin protein Sir4 that interacts with a peptide of six amino acids at the C-terminus of Ty5 integrase (called the targeting domain). In the absence of interaction, Ty5 integrates randomly and becomes mutagenic (Gai and Voytas 1998). The interaction with Sir4 requires phosphorylation of a serine within the targeting domain of Ty5 integrase. This phosphorylation is abolished by nutrient starvation, suggesting a control of Ty5 integration profile by stress (Dai et al. 2007). Retrotransposons Ty1, Ty2 and Ty4 have the same integration preference for a 700 base pair window upstream of Pol III-transcribed genes and their integrases all interact with the AC40 subunit of Pol III (Bridier-Nahmias et al. 2015). The loss of interaction between Ty1 integrase and AC40 drastically reduces Ty1 integration upstream of Pol III-transcribed genes and induces a redistribution of integrations to subtelomeric regions (Bridier-Nahmias et al. 2015). The interaction with AC40 requires a short targeting domain in the C-terminus of Ty1 integrase that is conserved in Ty2 and Ty4 integrases (Asif-Laidin et al. 2020). When the Ty5 targeting domain is replaced by that of Ty1, the integration of Ty5 is redirected to Pol III-transcribed genes, revealing that a targeting domain can remain functional outside its initial context (Asif-Laidin et al. 2020). Tethering factors have also been described for the LTR retrotransposon Tf1 of *Schizosaccharomyces pombe* (Hickey et al. 2015), and also for retroviruses, whose integrases share common characteristics with those of LTR retrotransposons, such as HIV-1 (Cherepanov et al. 2003; Llano et al. 2006) and MLV (De Rijck et al. 2013; Gupta et al. 2013; Sharma et al. 2013).

The structure and three-dimensional organization of chromatin can also locally influence the access of the integration complex to the target DNA and thus the integration process. Certain TE families favor integration into open chromatin or into nucleosome-free regions, as described for Ty3 and Ty5 in yeast (Baller et al. 2011; Qi et al. 2012) and the *Mu* transposon in maize (Sanzhen Liu et al. 2009). Others favor their insertion in the nucleosomal DNA, such as yeast Ty1 (Baller et al.

2012; Mularoni et al. 2012) or display a tropism for particular chromatin marks, as described for *Ty1/copia* elements in *Arabidopsis thaliana* (Quadrana et al. 2019). During host DNA replication, the replication forks are also particular chromatin contexts that favor integration of many TEs (Zaratiegui 2017). In humans, the ORF2p protein of LINE-1 interacts with several proteins present at the replication forks and LINE-1 mainly retrotransposes during the S phase of the cell cycle (Mita et al. 2018; Flasch et al. 2019; Sultana et al. 2019). On the other hand, several studies suggest that the proximity of the integration site to the nuclear envelope, and particularly to the nuclear pore through which the cDNA is imported into the nucleus, could also influence retroelement integration site selection (Lelek et al. 2015; Marini et al. 2015; Lusic and Siliciano 2017; Manhas et al. 2018).

3.4. TE restriction by the host

To limit the cost and risks associated with TE activity in their cells, host organisms have developed a wide variety of defense mechanisms that can target most steps of their nuclear and cytoplasmic replication cycle (see Figure 3.3).

3.4.1. *Transcriptional repression of genomic copies*

The expression profiles of the different TE families and different copies of TEs within the same family are very often specific to the cell type or stage of development (Seifarth et al. 2005; Göke et al. 2015; Molaro and Malik 2016; Tokuyama et al. 2018). However, in a given cell, the expression of the majority of TEs is repressed at the transcription stage. As with other genes, this repression involves cytosine methylation on DNA and/or repressive marks on histone tails, including histone H3 lysines 9 and 27 methylation (H3K9me and H3K27me).

3.4.1.1. *Two complementary lines of defense: methylation of DNA and histones*

In mammals, DNA methylation is based on the action of three DNA methyltransferases (DNMTs). DNMT1 preferentially methylates hemi-methylated DNA and is therefore the main enzyme for the maintenance of DNA methylation during replication. DNMT3A and DNMT3B have the ability to methylate *de novo* the cytosines of the CpG dinucleotides, and are particularly important during the epigenetic reprogramming that takes place during development of the germline. DNMT1 is particularly important for the methylation of IAP (intracisternal A particle) LTR retrotransposons in mouse embryo somatic cells (Walsh et al. 1998), whereas DNMT3A and DNMT3B are involved in the methylation of LINE-1 retrotransposons (also called L1 in mice) and IAP in the male germ cells of mice (Kato et al. 2007). However, the absence of DNMT3A or DNMT3B individually has little effect on the expression of TEs, suggesting functional redundancy. Conversely, deletion of their non-enzymatic cofactor DNMT3L results in a strong decrease in TE methylation, associated with a significant increase in their expression (Kato et al. 2007). The rodent genome encodes another *de novo* DNMT derived from a duplication of DNMT3B. This DNMT, DNMT3C, specifically targets the promoters of "young" TEs in the genome of male fetal germ cells (Barau et al. 2016). Although all of these DNMTs have their specificities, they cooperate to repress most of the TEs in the genome and play a major role in the face of the threat that these elements may pose. As TE promoters are particularly rich in methylated CpGs, it has been suggested that cytosine methylation was selected during the course of evolution to defend the host genome against TE activity (Yoder et al. 1997).

Interestingly, it has been shown that retrotransposons present in two distant eukaryotic lineages, dinoflagellates and charophytes, acquired functional DNMT domains independently during evolution (de Mendoza et al. 2018). This DNMT domain could methylate LTR retrotransposon cDNA during reverse transcription to avoid being recognized as exogenous DNA (De Mendoza et al. 2018) but it could also limit their expression following their integration through a new self-regulation mechanism (see section 3.3.1).

As with DNA methylation, the different cellular histone methyltransferases (HMT) cooperate to deposit repressive chromatin marks on different TE classes. Studies conducted on mice have shown that different H3K9 HMTs are involved in TE repression depending on the cell type or stage of development (Kato et al. 2018) (for review, see Geis and Goff 2020). TE repression by H3K9 methylation is considered particularly important during the development of the germline, which undergoes two waves of DNA demethylation, during which the genome is particularly vulnerable to transposition. Thus, the repression of most ERVs is strongly dependent on the HMT SETDB1, whereas the expression of the ERV-L and LINE-1 retroelements is instead regulated by the HMTs G9a and Suv39h, respectively (Maksakova et al. 2013; Bulut-Karslioglu et al. 2014; Sheng Liu et al. 2014).

As DNMTs and HMTs are general factors used by the cell to repress gene expression, it is important to understand how TEs are recognized as targets of these repressors. Two main pathways have been put in place by the cells: the KRAB-ZFP (Kruppel-associated box-containing zinc finger protein) described below and piRNAs (see section 3.4.3). These two systems have the particularity of being able to evolve rapidly to adapt to the appearance of new TEs to be repressed in the genome.

3.4.1.2. *Specific targeting of TEs by KRAB-ZFP proteins*

KRAB-ZFPs belong to a family of genes that have diversified widely in tetrapods and represent the largest family of transcription factors in humans and mice (Imbeault et al. 2017). These proteins bind to DNA through a variable number of tandem copies of C2H2-type zinc fingers, each being able to interact with three nucleotides (for review, see Yang et al. 2017; Ecco et al. 2017). Through their N-terminal KRAB domain, KRAB-ZFPs recruit the universal corepressor KAP1 (also called TRIM28). In turn, KAP1 recruits histone modifying enzymes, in particular the histone deacetylase complex NuRD (nucleosome remodeling deacetylase), H3K9 HMT SETDB1, or DNMTs to favor the deposition of repressive histone marks (Schultz et al. 2001; Quenneville et al. 2011) (see Figure 3.5).

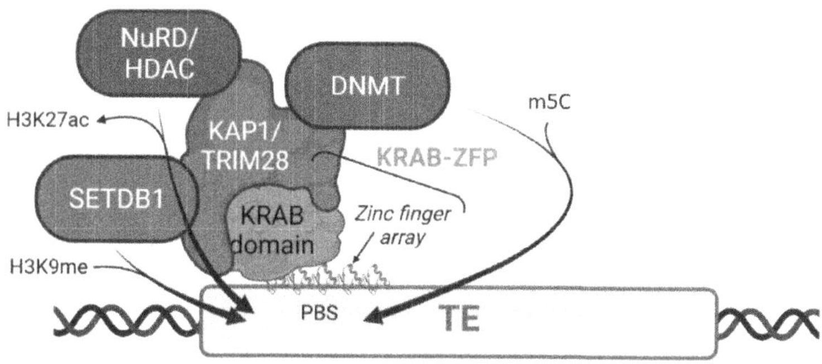

Figure 3.5. *Recognition and repression of TEs by the KRAB-ZFP complex. The KRAB proteins (in green) recognize the TE sequences, in particular their primer binding site (PBS) sequence, thanks to the zinc finger array in their C-terminal domain. Through their N-terminal KRAB domain, the KRAB proteins recruit the repressor KAP1 (also called TRIM28, in purple). KAP1 will induce heterochromatin formation by recruiting DNMT (DNA methyltransferase), which methylates DNA cytosines (m5C), NurD (nucleosome remodeling deacetylase) or HDAC (histone deacetylase), which deacetylates H3 lysine 27 (H3K27ac), and SETDB1, which methylates histone H3 lysine 9 (H3K9me). Created with BioRender.com*

Analysis of KRAB-ZFP binding profiles in genomes or loss-of-function experiments of some of these proteins or their partner KAP1 made it possible to determine that the primary function of KRAB-ZFPs is the transcriptional repression of TEs in embryonic cells (Wolf and Goff 2009; Rowe et al. 2010; Jacobs et al. 2014; Najafabadi et al. 2015; Wolf et al. 2015, 2020). The best-characterized example is probably ZFP809, which represses MLV retrovirus insertions in mouse embryonic cells (Wolf and Goff 2009). Interestingly, ZFP809 recognizes the PBS sequences not only of MLVs but also of all retroelements that share the same PBS sequence, resulting in their repression (Wolf et al. 2015). This PBS sequence is significant for retroelement biology because it is recognized by a tRNA that will be used as the primer to initiate reverse transcription. Targeting a sequence that is essential to the success of the retrotransposition cycle, and thus one of the most evolutionarily constrained of the retroviral genome, is an efficient strategy to make it more difficult for retroelements to escape repression. Indeed, KRAB-ZFPs are able to evolve following the arrival of new TEs. Nevertheless, their evolution is slow because it requires gene duplications with diversification of the zinc finger array (Wolf et al. 2020). Thus, the number of copies of KRAB-ZFP in the genomes correlates with the amount of LTR retroelements having colonized these genomes (Thomas and Schneider 2011). This arms race was illustrated in a study demonstrating that KRAB-ZFP proteins targeting LINE-1 and SVA retroelements in embryonic stem (ES) cells were able to emerge during the course of evolution as a result of the acquisition of new DNA-binding capabilities while retrotransposons themselves were subject to deletions from the regions targeted by these KRAB-ZFPs (Jacobs et al. 2014). One study thus demonstrated that KRAB domains were repeatedly fused during the course of the evolution of tetrapods to DNA-binding domains of transposases, the enzyme encoded by DNA transposons (Cosby et al. 2021). Indeed, a more detailed study of one of these potential transcriptional repressors present in the genome of bats demonstrated that the KRABINER fusion protein not only regulated the host genes, but was also able to interact with hundreds of DNA transposons of the same family as the transposase domain fused to the KRAB domain (Cosby et al. 2021).

3.4.2. *TE transcripts: choice targets for multiple restrictions*

While the majority of TEs are transcriptionally repressed, control mechanisms are not functional in all cell types. TE transcription can be activated during epigenetic reprogramming of germline cell development, or during stressful events (for review, see Maupetit-Mehouas and Vaury 2020; Horváth et al. 2017). Thus, strategies exist to restrict TEs that target post-transcriptional stages of the transposition cycle. In particular, RNAs, molecules that serve both as template for translation and for reverse transcription, are targeted by several mechanisms.

First, like DNA, TE RNAs can be chemically modified and these modifications can have an impact on their stability. In humans, uridylation by the TUT4 enzyme destabilizes LINE-1 RNA inhibiting retrotransposition (Warkocki et al. 2018). More recently, it has been shown that in mouse ES cells, the methylation of adenosine at position 6 (m^6A) of messenger RNA (mRNA) blocks ERV activity (Chelmicki et al. 2021; J. Liu et al. 2021; Xu et al. 2021). This methylation is catalyzed by the methyltransferase complex m^6A composed of METTL3 and METTL14 combined with two accessory subunits, WTAP and ZC3H13. A depletion of one of these subunits leads to an increase in the mRNA abundance of IAP and *ERV-K* elements, most likely due to an increase in ERV RNA stability (Chelmicki et al. 2021) and/or a decrease in SETDB1-mediated heterochromatin H3K9me3 marks on targeted ERVs (Liu et al. 2021; Xu et al. 2021). In contrast, in human cells (HeLa and embryonic carcinoma cells), the RNA m^6A modification of LINE-1 appears to facilitate their mobility by promoting RNA translation (Hwang et al. 2021). It is important to note that there are more than 150 different modifications cataloged on RNAs to date. It is therefore reasonable to think that the modification of RNAs adds a level of complexity to the regulation of TEs. Recently, substantial technical efforts have been made to detect RNA modifications, allowing us to believe that new discoveries will be made in the near future.

The control of retrotransposition can also be an indirect consequence of RNA modification. Indeed, one of the first demonstrations of the involvement of an enzyme capable of methylating RNAs in TE control dates back to 2009. This study suggested that DNMT2 was responsible for TE control in *Drosophila* (Phalke et al. 2009). Several years later, publications indicated that DNMT2 is indeed an RNA (not a DNA) methyltransferase (Goll et al. 2006; Raddatz et al. 2013). DNMT2 and other methyltransferases such as Nsun2 and Trm7 capable of methylating tRNAs in *Drosophila* have been identified for their roles in the activity of certain TEs (Genenncher et al. 2018; Angelova et al. 2020). However, the precise molecular mechanisms of the involvement of these tRNA methylations on the control of TEs have not been clearly elucidated to date. But one promising avenue is being explored. The loss of tRNA modifications is believed to be responsible for the fragmentation of tRNAs into fragments called tRFs. Yet, it has been shown in mouse ES cells that tRFs, corresponding to the last 18 or 22 nucleotides of tRNAs, affect transposition at different levels: through post-transcriptional repression mechanisms by inducing RNA degradation, by blocking reverse transcription or by inhibiting the translation of these RNAs (Schorn et al. 2017). Similar strategies have been found in plants capable of accumulating a tRF population corresponding to the first 19 nucleotides of tRNAs in reproductive tissues and resulting in TE RNA degradation (Martinez et al. 2017). Excessive tRNA fragmentation as a result of methylation failure could therefore explain this relationship between tRNA methylation, fragmentation and post-transcriptional control of the transposition, thus explaining the effect of tRNA methylation on the activity of some TEs. An alternative hypothesis (Genenncher et al. 2018; Angelova et al. 2020) proposes that an overabundance of tRFs could saturate the Argonaute or Dicer proteins, thus disrupting the pathways of the small, non-coding RNAs involved in TE regulation (siRNA and piRNA) (see section 3.4.3).

3.4.3. *The Swiss knives of TE restriction: piRNAs*

In 2006–2007, several groups identified a new class of small non-coding RNAs involved in TE regulation in mouse, rat and *Drosophila* gonads. These small non-coding RNAs, originally called repeat associated small interfering RNAs (rasi-RNAs), were distinguishable from the other small RNAs known at the time (small-interfering RNAs (siRNAs) and microRNAs (miRNAs)). They measure between 21 and 32 nucleotides (nt) depending on the species, their biogenesis does not involve Dicer proteins, and they are associated with Argonaute proteins of the PIWI family. Hence, these small RNAs specific to animals (Grimson et al. 2008) have been named piRNAs for PIWI interacting RNA (Aravin et al. 2006; Girard et al. 2006; Saito et al. 2006; Vagin et al. 2006; Brennecke et al. 2007; Gunawardane et al. 2007). In the remainder of this section, we will address piRNA biogenesis in three model organisms for which this pathway has been particularly studied: fruit flies (*Drosophila melanogaster*), mice (*Mus musculus*) and nematodes (*Caenorhabditis elegans*).

Regions called "piRNA clusters", enriched in fragments of TEs from different families and in heterochromatic marks, produce piRNAs (Aravin et al. 2006; Girard et al. 2006; Brennecke et al. 2007). These regions are essential for genome stability because they repress TEs, mainly in the germline. As such, piRNA clusters contain the repertoire of all the mobile sequences that need to be regulated in genomes. Around 140 piRNA clusters have been identified in the *Drosophila* genome (Brennecke et al. 2007), around a hundred in mice (Aravin et al. 2006; Girard et al. 2006) and two large piRNA clusters located on chromosome IV in nematodes (Ruby et al. 2006). In *Drosophila*, there are two types of piRNA clusters: single-strand piRNA clusters known as "uni-strand", which are mainly active in the somatic follicle cells of the gonads and for which only one of the two strands is transcribed, and double-stranded piRNA clusters known as "dual-strand", mainly active in germ cells that are transcribed on both strands (Brennecke et al. 2007). In mice and nematodes, only uni-strand piRNA clusters are found (Cecere et al. 2012; Li et al. 2013).

3.4.3.1. *Uni-strand piRNA clusters*

Uni-strand piRNA cluster transcription is similar to mRNA transcription, as transcription factors like *Cubitus interruptus* (Ci) in *Drosophila*, A-Myb in mice and Forkhead in nematodes participate in the recruitment of the RNA polymerase II to transcribe them (Cecere et al. 2012; Li et al. 2013; Goriaux et al. 2014) (see Figure 3.6).

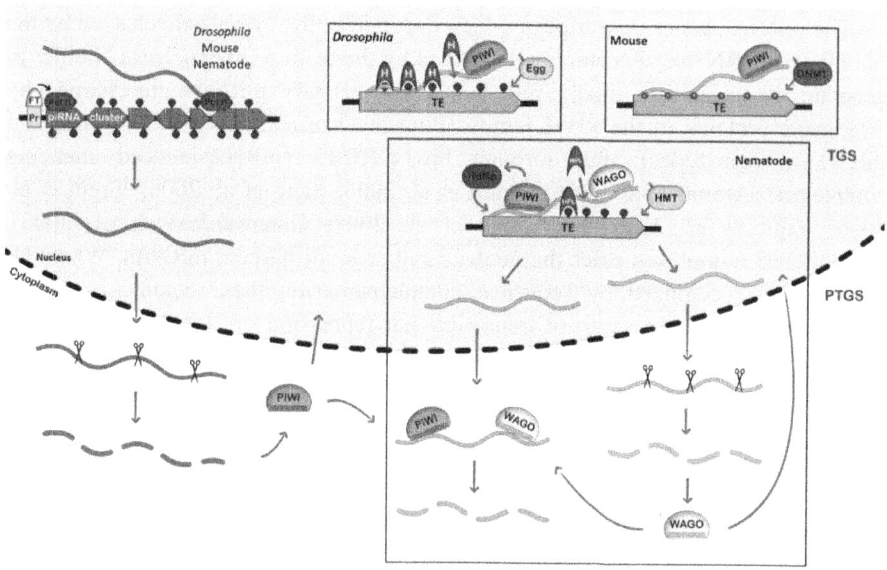

Figure 3.6. *Biogenesis of piRNAs derived from uni-strand piRNA clusters. Uni-strand piRNA clusters are transcribed by RNA polymerase II on one strand. RNA precursors are cleaved in the cytoplasm and loaded by PIWI proteins. In Drosophila, piRISC complexes scan transcripts, recruit methyltransferase Egg (Egg) that deposits H3K9me3 marks (green balls) recognized by HP1a (H). In mice, piRISC complexes scan transcripts and recruit a DNA methyltransferase (DNMT) that carries out DNA methylation of CpG islands (purple circles). In nematodes, piRISC complexes cleave TE mRNAs in the cytoplasm and can recognize transcripts in the nucleus. An RNA-dependent RNA polymerase (RdRp) is then recruited to synthesize precursor RNAs that are subsequently cleaved by WAGO into secondary siRNAs in the cytoplasm. The siRNA-WAGO complexes scan transcripts in the nucleus and recruit histone methyltransferase (HMT) to deposit H3K9me3 marks recognized by HPL2 (HPL), the homolog of HP1. TGS, transcriptional gene silencing; PTGS, post-transcriptional gene silencing; Pr, promoter; FT, transcription factor*

Surprisingly, piRNA clusters are transcribed while the sequence is enriched in heterochromatic marks, H3K9me3 in *Drosophila* and mice (Li et al. 2013; Goriaux et al. 2014; Mohn et al. 2014) and H3K27me3 in nematodes (Beltran et al. 2019). These chromatin marks are described as repressive because they are generally associated with little or no transcriptional activity. The biology of piRNA clusters is a good example that tends to indicate that heterochromatin is not systematically transcriptionally inactive.

The piRNA cluster transcripts are then exported to the cytoplasm, cleaved by the ribonuclease (RNAse) Zucchini and matured by the 3' to 5' exonuclease Nibbler to generate the so-called primary piRNAs. These primary piRNAs are charged by Argonaute proteins of the PIWI family (Piwi in *Drosophila*, MIWI2 in mice and PRG-1 in nematodes), thus forming Piwi-piRISCs (piRNA-induced silencing complexes) (Aravin et al. 2006; Girard et al. 2006; Saito et al. 2006; Vagin et al. 2006; Yigit et al. 2006; Brennecke et al. 2007; Gunawardane et al. 2007). Piwi-piRISC complexes enter the nucleus and scan all nascent mRNAs. When TE mRNAs are recognized by sequence complementarity, these complexes recruit factors involved in the setup of transcriptional repression or "transcriptional gene silencing" (TGS). In *Drosophila*, Piwi recruits Eggless, which deposits H3K9me3 marks recognized by the HP1a protein, resulting in local heterochromatinization of the locus (Rangan et al. 2011). In mice, DNA is methylated at the CpG islands by DNA methyltransferases (DNMT, see section 3.4.1) (Aravin et al. 2008). Lastly, in nematodes, PRG-1-piRNA complexes can directly cleave TE mRNAs in the cytoplasm, but also recognize transcripts during elongation and recruit an RNA-dependent RNA polymerase (RdRp) that synthesizes secondary siRNAs loaded by WAGO proteins (Yigit et al. 2006; Bagijn et al. 2012; Shirayama et al. 2012). These siRNA-WAGO complexes scan the transcripts undergoing synthesis and recruit HMTs for the deposition of H3K9me3 marks and of the HP1 homolog, HPL2, at the TE sequences (Lee et al. 2012; Shirayama et al. 2012). The siRNA-WAGO complexes are also capable of cleaving TE mRNAs in the cytoplasm (Lee et al. 2012; Shirayama et al. 2012) (see Figure 3.6).

3.4.3.2. *Dual-strand piRNA clusters*

In *Drosophila* germ cells, dual-strand piRNA clusters are enriched in H3K9me3 marks that are recognized by a germline-specific HP1-like protein called Rhino. This interaction is stabilized by the binding of Rhino to Kipferl, a DNA binding protein able to recognize guanosine-rich sequences on specific loci (Baumgartner et al., 2022). Then, Rhino interacts with cofactors involved in the recruitment of RNA polymerase II that performs transcription from the locus (Mohn et al. 2014; Andersen et al. 2017) (see Figure 3.7).

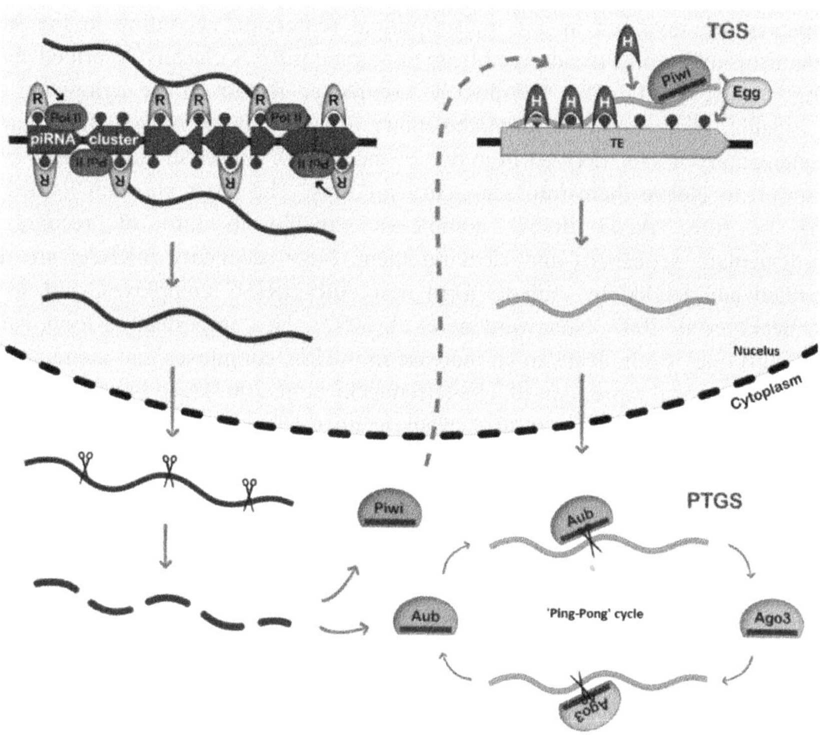

Figure 3.7. *Biogenesis of piRNAs derived from dual-strand piRNA clusters in Drosophila germ cells. Dual-strand piRNA clusters are composed of H3K9me3-enriched TE fragments, which interact with a protein complex containing Rhino (R) allowing the recruitment of RNA polymerase II (Pol II) for transcription on both strands. RNA precursors are cleaved in the cytoplasm in primary piRNAs and then loaded by Piwi or Aubergine (Aub) proteins. The Piwi-piRISC complexes go to the nucleus, scan the nascent TE transcripts and recruit the methyltransferase Eggless (Egg), which deposits H3K9me3 marks, recognized by HP1a (H), resulting in local heterochromatinization of the TE, allowing a first level of transcriptional regulation (transcriptional gene silencing [TGS]). The Aub-piRISC complexes recognize and cleave TE transcripts to generate secondary piRNAs loaded into Ago3. The Ago3-piRISC complexes in turn recognize complementary TE sequences and cleave them in secondary piRNAs loaded into Aub. This amplification loop, called "Ping-Pong", results in the increase in the number of PIWI-piRISC complexes and enables a second level of regulation which is post-transcriptional gene silencing (PTGS)*

Precursor RNAs are exported to the cytoplasm, cleaved by Zucchini and matured by Nibbler to generate primary piRNAs loaded by Piwi or Aubergine (Aub) to form Piwi-piRISC and Aub-piRISC complexes (Brennecke et al. 2007). Piwi-piRISC complexes are involved in TE transcriptional repression according to the same mechanisms as for piRNAs derived from uni-strand piRNA clusters described above (Rangan et al. 2011). The Aub-piRISC complexes remain in the cytoplasm and are capable of recognizing complementary RNAs. These Aub-piRISCs target complementary RNAs, derived from one of the piRNA cluster strands or from a TE transcript, to cleave them into secondary piRNAs. The latter are then loaded by Ago3 to form Ago3-piRISC complexes capable in turn of recognizing complementary sequences and cleaving them. New secondary piRNAs are thus generated and loaded by Aub to form new Aub-piRISC complexes, and so on (Brennecke et al. 2007; Gunawardane et al. 2007). This amplification loop, called the "Ping-Pong" cycle, leads to an increase in piRISC complexes that guarantee the post-transcriptional repression of TEs (Brennecke et al. 2007) (see Figure 3.7). Note that the Ping-Pong mechanism also exists in mice (Aravin et al. 2008) but not in nematodes, which have another amplification mechanism using an RdRp (RNA dependent RNA polymerase) (Lee et al. 2012; Shirayama et al. 2012).

Thus, thanks to their sequence complementarity with transpositionnally active TEs, the TE relics present in the piRNA clusters are essential for genome stability. It should be noted that strains carrying mutations in one of the genes involved in the piRNA pathway are for the most part sterile, partly because of the derepression of some TEs, resulting in DNA damage.

piRNA clusters are thus fundamental loci whose activation at each generation is essential for the maintenance of genome integrity. It has been shown that piRNAs synthesized from germinal piRNA clusters are transmitted to the developing oocyte allowing at each generation to redefine precisely the piRNA clusters present in the mother (de Vanssay et al. 2012). This leads to a case of paramutation, the first stable case described in animals, originally identified in maize and defined as an epigenetically conversion of one allele induced by another allele (Brink, 1956, Coe, 1959). In the absence of this maternal piRNA inheritance, loci contained in a particular genomic environment can be activated for piRNA production under specific stress conditions such as high temperature (Casier et al. 2019).

3.4.3.3. *Consequences of piRNA expression default*

In *Drosophila*, mice and *Arabidopsis Thaliana*, time windows have been observed in which the mechanisms involved in TE repression during development are released (Heard 2014). In *Drosophila*, the amount of Piwi protein is reduced during specific stages in early oogenesis. This reduction of Piwi is sufficient to open temporal windows, allowing some TEs to be transcribed and potentially to transpose (Dufourt et al. 2014; Théron et al. 2018).

The reactivation of TEs in descendants of hybrid crosses not having the same TE and piRNA repertoires has been documented in many species but the molecular mechanisms involved in these reactivations have not been determined for the most part (for review, see Serrato-Capuchina and Matute 2018). A well-known example is the case of the *P* element in *D. melanogaster* (see section 3.1.2), present in what are known as “P” strains, containing *P* copies within piRNA clusters, and absent in “M” strains. When *P* males are crossed with *M* females, the gonads of the descendants are atrophied. This phenomenon of hybrid dysgenesis is not observed in reciprocal cross (Kidwell et al. 1977). Indeed, in the gonads of *P* females, piRNAs complementary to the *P* elements are deposited in the oocyte, protecting progeny against new deleterious insertions (Brennecke et al. 2008).

3.4.4. *Reverse transcription of retroelements: a key step to inhibit*

Several of the above-mentioned mechanisms that lead to degradation of retroelement RNAs may also inhibit reverse transcription of these RNAs and thus strengthen the restriction of retrotransposition. This is the case, for example, with uridylation of LINE-1 RNA 3’ ends, which, when carried out by the TUT7 enzyme in the cytoplasm, inhibits initiation of reverse transcription after import of the ribonucleoparticles of LINE-1 into the nucleus (Warkocki et al. 2018). Similarly, while 22-nt tRFs block translation or destabilize retroelement RNAs, 18-nt tRFs, which include the last 3’-terminal nucleotide CCA of tRNA, inhibit reverse transcription initiation by interfering with the binding of initiator tRNAs on the PBS sequence of IAP and ETn LTR retrotransposons in mice (Schorn et al. 2017). Thus, this restriction mechanism targets the mobile autonomous and non-autonomous retroelements. As with KRAB-ZFPs (see section 3.4.1.2), targeting the PBS sequence is an efficient strategy to avoid retroelements escaping restriction since a mutation in the PBS would lead to a loss of retrotransposition efficiency.

In vertebrates, the AID/APOBEC deaminase family is another important line of defense against all retroviruses and retrotransposons (for review, see Goodier 2016; Orecchini et al. 2018). AID/APOBEC proteins deaminate cytosines to uracils in the first nascent DNA strand during reverse transcription. These uracils lead to a hypermutation of guanines to adenosines during second-strand DNA synthesis. While this mechanism does not prevent the integration of new copies of retroelements into the genome, it alters the integrity of their sequence and thus their level of retrotransposition in subsequent generations.

3.5. The impact of transposition events on genomes

We have seen that TE mobility is extremely controlled in germ and somatic cells ensuring that the genome is faithfully passed on from generation to generation. However, these check points can be reduced at certain stages of development or under pressure from various stress factors (environmental, aging, hybrid dysgenesis, etc. (see section 3.4.3.3)). These events could lead to bursts in the transposition activity of some TEs. While new TE insertions can be deleterious for the survival or fertility of individuals, they can also induce genetic innovations in genome evolution.

3.5.1. *The structural and functional consequences of TE activity on the genome*

TE activity can alter the structure of genomes via various active mechanisms, related to the insertion of a new TE copy, or passive mechanisms, simply related to the abundance of these repeated sequences in genomes. Changes in structure can occur locally and affect the host transcriptome and proteome or affect the organization of the genome more globally.

3.5.1.1. *Mutagenesis of the genome through insertion of a new TE copy*

The majority of TE insertions are neutral for genomes, because they are located in non-coding regions. When these insertions take place in coding or regulatory regions, they can be mutagenic. The functional consequences on the expression of the targeted gene are highly variable, depending on the location of the insertion in the gene (see Figure 3.8), as well as on the length and orientation of the inserted sequence. Insertions in coding sequences can lead to modifications in the structure of transcripts, particularly if the inserted TEs contain regulatory sequences such as promoters, transcription terminators, splicing signals or insulators (Rebollo et al. 2012a; Chuong et al. 2017). The presence in the TE of transcription terminators leads, for example, to the production of nonfunctional truncated transcripts, whereas the presence of splice-acceptor or splice-donor sites can lead to alternative splicing of transcripts or the creation of new exons (called exonization). All of these alternative transcripts will have consequences on the proteins produced from these genes. Transcripts including a TE will generate fusion proteins with additional protein domains or transcript degradation if the TE sequence introduces a premature translation stop codon. A systematic study has identified 94 fusion events during the evolution of tetrapods, the majority of which are the result of an alternative splicing between a host protein domain and a transposase domain, the enzyme encoded by DNA transposons, thus generating new fusion genes (host-transposase fusion [HTF]) most often encoding potential transcriptional regulators (Cosby et al. 2021).

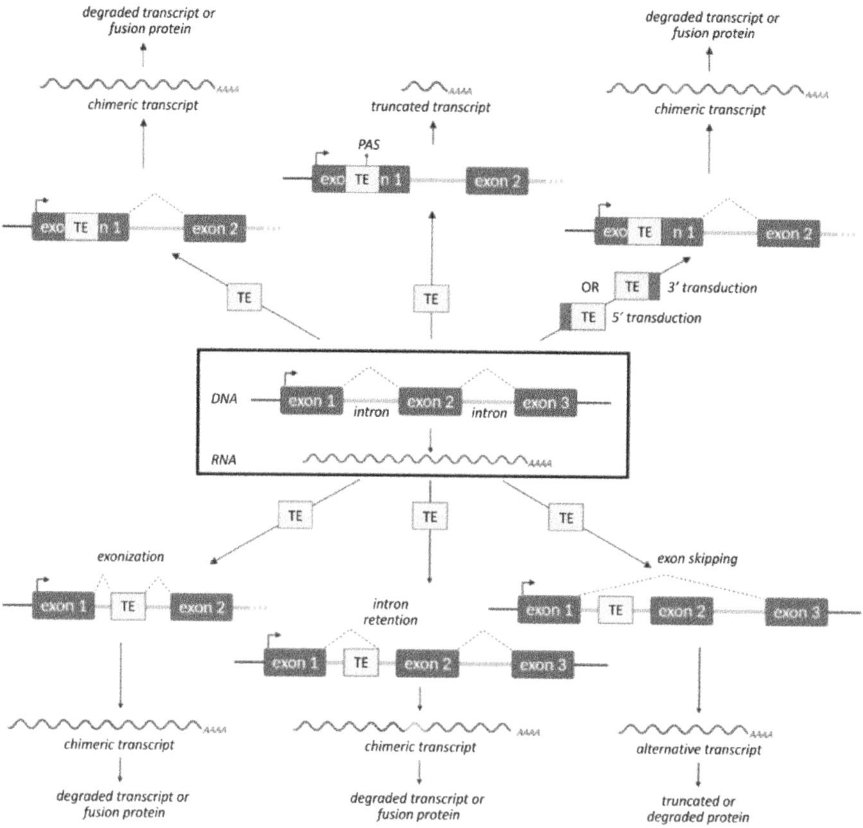

Figure 3.8. *Mutagenesis induced by the insertion of a TE in a gene. Shown at the top are several examples of the consequences of inserting a transposable element (TE in pink) in an exon (shown in green). Sequences from the TE locus of origin are shown in blue. Shown at the bottom are the main consequences of alternative splices resulting from the insertion of a TE in an intron (shown in yellow). PAS: polyadenylation signal. Created with BioRender.com*

When TEs are inserted into regulatory sequences, this results in modifications of the gene expression level: repression, overexpression and even new regulations (see Figure 3.9). A very well-characterized example of the regulation of a gene by a TE is that of the *Agouti* gene in mice, at the 5’ end of which an IAP element is found.

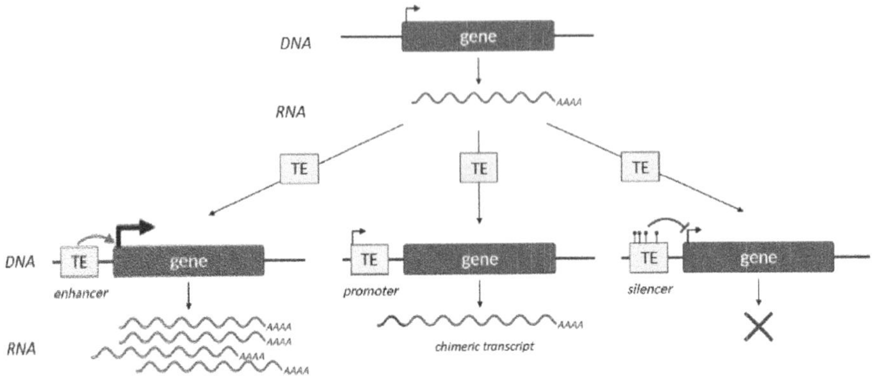

Figure 3.9. *Functional consequences of the insertion of a TE in the regulatory sequences of a gene. The insertion of a TE (in pink) upstream of a coding region (gene in green) can activate its transcription by serving as an enhancer or activator (on the left), producing new chimeric transcripts from an alternative promoter located in the TE (in the middle) or repressing its transcription because of the TE heterochromatinization (on the right). Created with BioRender.com*

The phenotypic consequences of IAP regulation on the *Agouti* expression can be directly observed in the mouse coat color: if the IAP is completely repressed the coat is yellow, whereas it is dark brown when the IAP element is active (Michaud et al. 1994). Similarly, inter-individual differences in human gene expression have been associated with polymorphic insertions of LINE-1, Alu and SVA (Wang et al. 2017). The promoters present in the TEs can also induce transcription of nearby non-coding sequences normally repressed or lead to the appearance of chimeric TE-gene transcripts (see Figure 3.9) (Cordaux and Batzer 2009; Rebollo et al. 2012a; Dubin et al. 2018). A noteworthy example highlighted in human cell lineages concerns the miRNAs of the cluster present on chromosome 19 (*C19CM*) that are transcribed by RNA polymerase III from the *Alu* retrotransposon promoters located upstream (Borchert et al. 2006). Many other examples of genes functionally affected by TE regulatory elements have been identified in mammals, a certain number of which have been cataloged on the *C-GATE* (Catalog of Genes Affected by TE) site (Rebollo et al. 2012b).

During transposition, TEs can sometimes carry genomic sequences adjacent to their initial copy to the new insertion site (see Figure 3.8). This phenomenon, called transduction, is based on the use of an alternative promoter present upstream of the retrotransposon (transduction 5') (Damert et al. 2009) or on the weakness of the polyadenylation signal of the retrotransposon and the use of an alternative polyadenylation signal located at the 3' end of the retroelement (3' transduction). These additional genomic sequences will be reverse transcribed and integrated into a new genome locus at the same time as the retroelement (Maxwell and Curcio 2007; Cordaux and Batzer 2009). It is estimated that about 10% of LINE-1 and SVA insertions in the human genome are associated with a 3' transduction of genomic sequences (Goodier et al. 2000; Pickeral et al. 2000; Wang et al. 2005). Reverse transcriptases can also accidentally use cellular mRNAs as a template, thus producing, after integration, pseudogenes present in many genomes. Through the acquisition of new regulatory sequences, certain pseudogenes, initially non-functional, are at the origin of new genes (Kaessmann et al. 2009).

Lastly, TE insertion can result in another significant structural modification of the genome, consisting of the deletion of sequences adjacent to the new insertion site, reaching up to several kilobases. This phenomenon has been observed in particular following the insertion of retrotransposons without LTRs, such as LINE-1 or *Alu* (Cordaux and Batzer 2009). As an example of the significance of the phenomenon, it is now estimated that during the evolution of primates, over 45,000 deletions induced by insertions resulted in the elimination of more than 30 Mb of genomic sequences (Cordaux 2008).

3.5.1.2. *Rearrangements of the genome by non-allelic homologous recombination between TEs*

Independently of the ability of TEs to move into genomes, the presence of multiple copies of TEs with high sequence homology can lead to large chromosomal rearrangements following the repair of double-stranded DNA breaks by non-allelic homologous recombination (NAHR) between two TE copies (see Figure 3.10). The first effect of these ectopic recombinations is to eliminate TE copies from the genome (Bourgeois and Boissinot 2019) even if, paradoxically, the transcriptional repression of TEs by the heterochromatinization of their sequence (see section 3.4.1) protects TEs from recombination (Kent et al. 2017).

However, the consequences of NAHR events on the genome can also be very deleterious. If both TE copies are located on different chromosomes, NAHR can lead to the deletion, duplication or translocation of part of the chromosomes. If both copies are located on the same chromosome, recombination can result in the deletion, duplication or inversion of a part of the chromosome (see Figure 3.10) (Klein and O'Neill 2018). In humans, for example, NAHR events have been observed, essentially between Alu retrotransposons, and are often associated with deletions or inversions (Collins and Rubin 1984; Gilbert et al. 2002; Sen et al. 2006; Han et al. 2008; Lee et al. 2008; Franke et al. 2009). Thus, it is estimated that since the divergence of humans and chimpanzees, these deletions associated with recombinations involving Alu or LINE-1 elements resulted in the elimination of almost 1 Mb of genomic sequences, illustrating the evolutionary significance of these phenomena (Sen et al. 2006; Han et al. 2008).

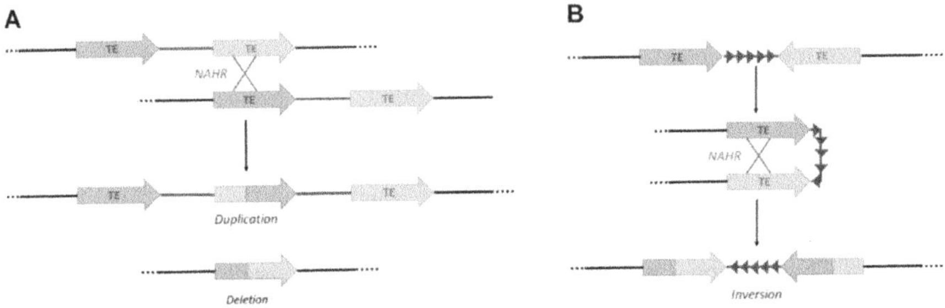

Figure 3.10. *Consequences of recombination between TEs of the same family (NAHR). (A) The recombination between two related TEs present in tandem on the same strand or two strands of DNA can lead to duplications or deletions. The duplicated or deleted sequence is shown in red. (B) The recombination between two inverted TEs on the same strand of DNA results in the inversion of the red sequence*

Retrotransposon sequences are thus known to be fragile sites of eukaryotic genomes (for review, see Curcio 2019; Scheifele and Boeke 2008). The many studies using the Ty1 retrotransposon of the yeast *S. cerevisiae* as a model have revealed that these chromosomal rearrangements appear in particular in response to selection pressures and are essential for cell survival in the event of exposure of cells to environmental or replicative stresses, but can be reversible when the selection pressure disappears. The multiple copies of Ty1 and the NAHR events that they promote thus appear as an essential step to rapidly adapt to changes in environmental conditions (Curcio 2019).

3.5.2. *Pathologies associated with TE activity*

3.5.2.1. *Pathologies associated with new TE insertions*

When the structural and functional modifications described above appear in germ cells, they are passed on to the next generations. In humans, it is estimated that a new insertion occurs during one meiosis out of 20 for Alu, one meiosis out of 20 or 200 for LINE-1 and one meiosis out of 900 for SVA (Kazazian 1999; Cordaux et al. 2006; Xing et al. 2009). These insertions therefore contribute to genome evolution and to inter-individual or family variations while posing a risk to genomic integrity. Indeed, some have been associated with the emergence of genetic diseases.

The first case of a pathology induced by a somatic neo-insertion of TE in humans was demonstrated in 1988 through the study of patients with type A hemophilia (Kazazian Jr. et al. 1988). The authors were able to identify in two unrelated patients insertions of the LINE-1 retrotransposon in exon 14 of the *Factor VIII* coagulation gene. Indeed, this is how LINE-1 was identified. On the other hand, one example of a family disease induced by a NAHR event between TEs is a family hypercholesterolaemia, associated with homologous recombination between *Alu* retroelements inserted into the gene encoding the low-density lipoprotein receptor resulting in a deletion of approximately 5 kb in the transmembrane domain of the receptor (Lehrman et al. 1985).

A link between the activity of TEs in somatic cells and cancers has been proposed in different models (*Drosophila*, nematodes, zebrafish and human cell lineages). In humans, the generalized DNA hypomethylation that characterizes many cancers results in increased expression of retrotransposons in tumor cells. Thus, the overexpression of the ORF1p protein of LINE-1 retrotransposons is a frequent manifestation of human cancers (more than 90% of breast or ovarian cancers and nearly 90% of pancreatic cancers) (for review, see Burns 2017). Somatic LINE-1 retrotransposition events have also been detected in many types of cancers, particularly in epithelial cancers such as those of the gastrointestinal tract. However, the number of somatic insertions varies widely between epithelial tumors (from one to more than 50) (Kazazian and Moran 2017). In these tumors, the majority of mutations induced by these insertions are called "passenger", that is, secondary consequences of genome hypomethylation and activation of the LINE-1 expression following cell transformation. However, some examples of insertions in tumor-suppressor genes such as APC or PTEN demonstrate that mutations induced by TEs can also be drivers of tumorigenesis, notably promoting the development of colon cancer in two reported cases of insertion in the APC gene (Miki et al. 1992; Helman et al. 2014; Scott et al. 2016). Lastly, it has been shown that TE activity is higher when the *p53* gene is mutated (Wylie et al. 2016). *p53* is a central gene that is widely studied in cancerology. Its loss of function is known to be involved in the development of a large number of cancers, earning the name "tumor-suppressor gene" or "guardian of the genome". For example, cancer cases associated with a *p53* gene mutation are accompanied by the derepression of certain TEs, leading to genomic instability and poorly repaired DNA breaks.

Today in humans there are more than 100 genetic diseases or cancers associated with retrotransposition events of LINE-1 or SINE elements, some examples of which are presented in Table 3.1 (for review, see Kazazian and Moran 2017).

Disease/syndrome	TE	Type(s) of modification(s)	References
Hemophilia A	LINE-1	Insertion in exon 14 of the *factor VIII* gene.	Kazazian et al. (1988)
Colon cancer	LINE-1	Insertion in the last exon of the *APC* gene, which results in its inactivation.	Miki et al. (1992)
Duchenne muscular dystrophy	LINE-1	Insertion in exon 44 of the *dystrophin* gene, leading to exon excision.	Narita et al. (1993)
Alport syndrome	Alu	Insertion in intron 5 of the *COL4A3* gene, resulting in splicing defects: excision of exon 5 and appearance of a splicing site.	Knebelmann et al. (1995)
Fukuyama muscular dystrophy	LINE-1	For the majority of patients: insertion in the 3'UTR region of the *FCMD* gene that destabilizes the mRNA; other insertions identified in several exons and in intron 7 of the *FCMD* gene resulting in nonsense and missense mutations, reading shifts or an exon excision.	Kondo-Iida et al. (1999)
Autoimmune lymphoproliferative syndrome	Alu	Insertion between intron 7 and 5' of exon 8 of the *Fas* gene, resulting in loss of exon 8.	Tighe et al. (2002)
Hemophilia A	Alu	Insertion in intron 18 of the *Factor VIII* gene results in excision of exon 19 and the formation of a nonfunctional truncated protein.	Ganguly et al. (2003)
Hemophilia B	LINE-1	Insertion in the *Factor IX* gene.	Mukherjee et al. (2004)
Cystic fibrosis	Alu	Insertions in exons 16 and 17 of the *CFTR* gene, resulting in splicing defects.	Chen et al. (2008)

Lung and colon cancers	LINE-1	Multiple insertions resulting in transcription of TE-gene chimeric transcripts from the TE promoter.	Cruickshanks and Tufarelli (2009)
Parkinson's disease	Alu	Insertions in the *PARK2* gene, resulting in different exon deletions.	Morais et al. (2016)
Schizophrenia	LINE-1	Multiple insertions in different genes.	Doyle et al. (2017)
Lung cancer	Alu	Insertion 20 kb upstream of the *LIN28B* oncogene, resulting in its overexpression.	Jang et al. (2019)

Table 3.1. *Examples of "transposopathies" for which TE insertion is associated with the emergence of various syndromes or diseases arranged in chronological order with respect to their discovery (Wylie et al. 2016). These examples represent only a small fraction of published studies on the involvement of TEs in disease development. There are numerous reviews covering a large number of examples of diseases associated with TE activity and discussing the molecular mechanisms involved (Konkel and Batzer 2010; Hancks and Kazazian 2012, 2016; Guffanti et al. 2014; Anwar et al. 2017; Saleh et al. 2019; Grundy et al. 2021)*

3.5.2.2. *Pathologies associated with abnormal TE expression*

Around a hundred copies of LINE-1 are mobile among the thousands of copies present in the human genome. Nevertheless, numerous studies describe the activation of expression of retrotransposons (whether mobile or not) in many inflammatory, autoimmune or neurodegenerative diseases. This observation suggests that the products of their expression (RNA, proteins or cDNA) participate in the pathogenesis of various diseases, including Aicardi-Goutières syndrome and systemic lupus erythematosus (Saleh et al. 2019; Burns 2020; Mustelin and Ukadike 2020). Indeed, most of these diseases are characterized by a high level of production of type-I interferon, which is involved in the anti-viral immune response. Thus, several studies have proposed that nucleic acids derived from retroelements (double-stranded RNA, DNA: RNA hybrids, extrachromosomal DNA) would be detected by cytoplasmic nucleic acid sensors such as RIG-I (retinoic acid inducible gene I), MDA5 (melanoma differentiation-associated gene 5), or c-GAS-STING (cyclic guanosine monophosphate–adenosine monophosphate-stimulator of interferon genes), which would cause activation of interferon-type inflammatory signaling pathways.

Paradoxically, these inflammatory properties of cells induced by the expression of retrotransposons are being explored as pathways for new anti-cancer treatments. Inhibitors of DNA methyltransferases, such as azacitidine or decitabine, have thus long been used to treat myeloplastic syndromes or acute myeloid leukemia. Yet, it has been demonstrated that the action mechanism of these treatments requires reactivation of the expression of retrotransposons that produce RNAs with secondary structures of double-stranded RNA and thus activate the cytoplasmic sensors and an anti-tumor immune response (Chiappinelli et al. 2015; Roulois et al. 2015).

3.5.3. *The impact of TEs on the evolution of the host*

3.5.3.1. *Molecular domestication*

Among the *de novo* transposition events of TEs, particular insertions can confer an advantage to survive in a given environment. If these insertions have taken place in the germline, they can then be fixed in the population over generations and become predominant. Insertions conferring a selective advantage on their host have been called "molecular domestications" (Miller et al. 1997). These domestications correspond to cases where a TE sequence is maintained in a genome across generations to fulfill a function that benefits its host. This sequence can include all or part of a TE. Table 3.2 cites some fascinating examples of molecular domestication. Although this aspect has been discussed on numerous occasions in reviews (e.g. Volff 2006; Jangam et al. 2017; Cosby et al. 2019), we wished to focus on two key examples that enable us, on the one hand, to effectively fight pathogens and, on the other hand, that enable us to be what we are: placental mammals.

Transposable elements and immune system: the extraordinary ability of our immune system to respond to the attack of a wide variety of pathogens is due in particular to the ability of immunoglobulin (antibody) genes to be somatically rearranged. The first step in these rearrangements is provided by a heterodimer of proteins called RAG1 and RAG2 (recombinant activating gene), which will associate different domains of the V, D and J regions of these genes, in a "patchwork" manner. The molecular mechanisms of this recombination, called VDJ recombination, involve a recognition step of specific sequences at the ends of the fragment to be excised, followed by a ligation at the immunoglobulin gene locus. The parallel with the first steps of transposase-mediated "cut-and-paste" transposition of class-II elements was proposed in 1996 (van Gent et al. 1996a). Phylogenetic analysis confirmed this hypothesis and identified that the RAG proteins are derived from ancient transposases of Transib elements (Kapitonov and Jurka 2005; for review, see Carmona and Schatz 2017).

Transposable elements and placenta: the envelope protein (ENV) of viruses confers infectivity to viral particles through a domain of interaction with cell receptors. This interaction induces fusion of the cell membrane with the envelope of the viral particle, allowing the viral nucleocapsid to be released into the cytoplasm of the infected cell. While most ERVs have accumulated mutations over the course of evolution, some ERVs have maintained the gene encoding the ENV protein. For example, the expression of the ERV ENV protein HERV-W (human ERV-W) in cell culture induces the formation of syncytia (cells containing multiple nuclei) (Mi et al. 2000). Because of this fusogenic property, this protein was named "syncytin". One of the remarkable characteristics of syncytin is its strong expression in the placenta, resulting from the formation of syncytiotrophoblast, which enables exchanges between the blood of the mother and the fetus. Systematic studies then revealed a second human syncytin, unrelated to the first (Blond et al. 2000; Blaise et al. 2003; Frendo et al. 2003), three in rodents (Dupressoir et al. 2005; Redelsperger et al. 2014), one in rabbits (Heidmann et al. 2009), ruminants (Koshi et al. 2011; Cornelis et al. 2013), carnivores (Funk et al. 2019), Afrotheria (Cornelis et al. 2014) and marsupials (Cornelis et al. 2015). All of these syncytins are the result of endogenization events of unrelated retroviruses expressed in the placenta. This observation would explain the existence of different types of placenta highlighting the fact that the placenta was "reinvented" on many occasions by the endogenization of different retroviruses. To confirm the role of syncytin in early development, a complete deletion of the mouse gene encoding syncytin A was tested. In these experiments, the embryos were not able to develop due to a disorganization of the syncytiotrophoblast altering the exchanges between mother and fetus (Dupressoir et al. 2009). In addition, another ENV protein expressed in the placenta, called Suppressyn, has been shown to bind to the Syncytin-1 membrane receptor, preventing the latter from attaching to it and apparently ensuring fine regulation of the role of Syncytin-1 in humans (Sugimoto et al. 2013, 2019).

	Description of the TE	Cellular functions	Organism	References
		Full-length TEs		
Het-A, TART, TAHRE	LINE	The *Drosophila* telomeres are made by the three tandemly repeated TEs.	*Drosophila*	Levis et al. (1993); Abad et al. (2004)
EnJSRV (endogenous Jaagsiekte virus)	Endogenous retrovirus	Protection against infection of exogenous JSRV viruses involved in forms of lung cancer.	Sheep	Mura et al. (2004); Armezzani et al. (2014)

TE fragments				
ENV protein				
Syncytin	Endogenous retrovirus	Function of the placenta (see above).	Placental mammals	Mi et al. (2000); Blond et al. (2000); Blaise et al. (2003); Frendo et al. (2003); Dupressoir et al. (2005); Funk et al. (2019)
Suppressyn	Endogenous retrovirus	Function of the placenta, Syncytin-1 functional repressor (see above).	Human	Sugimoto et al. (2013, 2019)
Fv-4 (Friend virus susceptibility-4)	Endogenous retrovirus	The *Fv-4* locus blocks the murine leukemia virus (MLV) receptor through the synthesis of an endogenous gp70: a phenomenon of receptor interference.	Mouse	Ikeda and Sugimura (1989)
GAG protein				
Fv-1	Endogenous retrovirus of the HERV-L family	Resistance to MLV infection by interacting with viral capsid.	Mouse	Best et al. (1996)
Transposase				
RAG1, RAG2	DNA transposon	VDJ recombination (see above).	Jawed vertebrates	(van Gent et al. 1996b; Kapitonov and Jurka 2005; Carmona and Schatz 2017)
PiggyMac	DNA transposon	Programmed genome rearrangement.	Paramecia	Baudry et al. (2009)
KRAB zinc finger protein	DNA transposon	Formation of new transcription factors and/or new chromatin-structuring factors involved in TE repression (see section 3.4.1.2).	Vertebrates	Cosby et al. (2021)

Any region				
piRNA	All TE classes	Regulation of TEs (see section 3.4.3).	Metazoans	Brennecke et al. (2007); Gunawardane et al. (2007)

Table 3.2. *Some notable or iconic examples of domestication of different TE encoding sequences (Henssen et al. 2017)*

Beyond these examples revealing how genomes have domesticated TE functions during species evolution (Table 3.2 and Jangam et al. 2017), human activity has also led, deliberately or not, to the selection of insertion events, among others of agronomic or "recreational" interest (Table 3.3).

Organism	TE	TE class	Position and impact of TE insertion	References
Insertions of agronomic interest				
Tomato	Rider	LTR retrotransposon	The insertion of Rider into the Sun locus is at the origin of a duplication that induces an increase in its expression, which is responsible for the elongated shape of tomatoes.	Xiao et al. (2008)
Orange	Tsc2	LTR retrotransposon	Different insertions of Tsc2 in the Ruby gene (involved in anthocyanin biosynthesis) produce blood oranges under the influence of cold.	Butelli et al. (2012)
Grape	Gret1	LTR retrotransposon	The insertion of Gret1 in the Vvmby1A gene, involved in anthocyanin biosynthesis, in Cabernet grapes, induces the loss of pigmentation characteristic of Chardonnay.	This et al. (2007)
Silk worm	Taguchi	LINE	The insertion of Taguchi in the *EO* gene, which controls molting during development, facilitates silkworm breeding.	Sun et al. (2014)

“Recreational” insertions				
Merle dog	Merle	SINE	Insertion of the SINE in the *SILV/PMEL* gene induces locus instability that could lead to alternative splicings resulting in a mottled coat.	Clark et al. (2006); Murphy et al. (2018)
Four o’clock flower	dTmj1	DNA transposon	The red variegation on yellow background observed on the petals is due to excision of the Tmj1 element present in the *CYP76AD3* gene involved in betalain biosynthesis.	Suzuki et al. (2014)
Morning glory flower	Tip100, Ac/Ds family	DNA transposon	The variegation in petal pigmentation is due to the insertion of Tip100 in the *CHS-D* gene involved in anthocyanin biosynthesis.	Habu et al. (1998)
Maize	Ds and Spm family	DNA transposon	The gene expression involved in anthocyanin biosynthesis is affected by the insertion of these TEs. Their excisions can restore its expression. The earlier this event occurs during development, the larger the size of the pigmented sector will be on the maize grain.	Wessler (1988)
Insertion selected by human activity				
Moth	carb-TE	DNA transposon	The insertion of carb-TE into the *cortex* gene induces a dark pigmentation that protects the butterfly from its predators in environments polluted by industrialization.	Van’t Hof et al. (2016)

Table 3.3. *Examples of natural transposition events, selected by humans*

In addition, understanding the molecular bases of transposition mechanisms has recently enabled the development of biotechnological tools for the analysis and modification of TE-derived genomes (Table 3.4).

TE	TE class	Biotechnological applications	References
P element	DNA transposon	The *P* element is used in *Drosophila* as a vector for *transgenesis*. Note: Other DNA transposons have since been used as a basis for transgenes in different organisms.	Rubin and Spradling (1983)
Sleeping Beauty, PiggyBac, Tol2	Tc1/mariner, PiggyBac and hAT	Transgenesis: Modification of the genome and creation of stable cell lineages for agronomic (breeding) or medical (gene therapy) purposes.	Sandoval-Villegas et al. (2021)
Tn5	DNA transposon	Tagmentation: Optimization of the preparation of genomic DNA libraries for high-throughput sequencing.	Adey et al. (2010)
Tn5	DNA transposon	ATAC-seq (Assay for Transposase-Accessible Chromatin with high-throughput sequencing): Identification of low-condensed chromatin regions, nucleosome position and transcription factors.	Buenrostro et al. (2013)

Table 3.4. *Examples of molecular biology tools developed from TE*

3.5.3.2. *TEs and adaptation to stress*

TEs, considered primarily as invaders threatening genomes, are also essential drivers of evolution. Therefore, the repression of TEs must be released for specific and short periods of time (see section 3.4.3.3). The ability of TEs to induce mutations, allowing new products (transcripts, proteins, etc.) and thus new functions (see section 3.5.1.), can indeed represent an asset under conditions of stress. TEs have stress response sequences (environment response element (Rebollo et al. 2012a)) and many are biotic (viral or bacterial infections), abiotic (temperature, radiation and pollutants), nutrient or replicative stresses capable of increasing the expression and even mobilization of TEs (Curcio et al. 2007; García Guerreiro 2012; Servant et al. 2012; Miousse et al. 2015; Morales et al. 2015; Kofler et al. 2018; Esnault et al. 2019; de Oliveira et al. 2021; Liang et al. 2021). Note, however, that from one TE family to another the response to a particular stress can differ: it may increase, decrease or not affect their expression (Miousse et al. 2015; de Oliveira et al. 2021; Liang et al. 2021). Sustained environmental change is forcing populations to adapt rapidly. Yet, the increase in the rate of transposition leads to an increase in genetic variability, resulting in a greater probability of survival of the species and an acceleration of the adaptive process (Piacentini et al. 2014). When facing stress, TEs can also modify the expression of genes in the vicinity of which they are inserted. This is notably the case with the *bari* transposon inserted near the *Jheh1*, *Jheh2* and *Jheh3* genes in *D. melanogaster*. In the absence of stress, these genes are repressed and their loci enriched with the histone mark H3K9me3. Under conditions of oxidative stress, we observe an enrichment of the H3K4me3 activating marks and the H3K27me3 repressive marks associated with the modification of the expression of the three *Jheh* genes. In the absence of *bari*, those modifications in the region are not observed, illustrating the role of the *bari* TE in the regulation of the *Jheh* genes (Guio et al. 2018). In the yeast *S. pombe*, exposure to cobalt chloride ($CoCl_2$) results in increased transcription and transposition of the *Tf1* retrotransposon and the emergence of stress-resistant cells, some of which have acquired a new copy of *Tf1* in the cobalt-resistance gene promoter (Esnault et al. 2019).

These results indicate that TEs constitute a reserve of genetic diversity useful for overcoming the stresses encountered by an organism. As discussed above, mutations that provide a selective advantage are likely to be fixed in populations (McDonald 1995). Conversely, deleterious mutations do not persist over generations, because they tend to be counterselected. Thus, the deleterious effects associated with TE activity are short-term effects that do not persist in populations, while the beneficial effects (domestication, adaptation to stress) can be selected and detected in the longer term.

3.5.3.3. *The role of TEs in brain evolution of primates*

The presence of a neocortex with a particular neuronal organization, composition and connectivity is perhaps the characteristic that has most contributed to the evolutionary success of mammals. Hominids, and in particular humans, also present an extraordinarily extensive cortical volume, an increase in the neuron size associated with an enrichment of the repertoire of neural cell types and more sophisticated models of neuronal connectivity (Ferrari et al. 2021).

The scientific community has long sought to understand the genetic basis of cognitive differences between primates, focusing particularly on the human brain. Although all types of mutations have worked in concert with the forces of evolution to generate today's human brain, TEs, in particular non-LTR retrotransposons, have played a fundamental role in the current organization of the primate brain (Linker et al. 2017). During the evolution of primates, multiple waves of retrotransposition of LINE-1, SINE, Alu and SVA occurred, eventually affecting the organization of the human brain (Linker et al. 2017).

These TEs are also essential for neurogenesis in humans and participate in neuronal plasticity during human brain development (Singer et al. 2010), as well as during neurogenesis, which also occurs in adults (Bodea et al. 2018). These mobilizations and their consequences are so powerful that they can go as far as modifying and customizing the genome of certain types of neurons of the same individual (Gage and Muotri 2012). As a result, it is now recognized that temporal deregulation of these very powerful evolutionary mechanisms can induce a variety of human neurodevelopmental and neurodegenerative pathologies (Ferrari et al. 2021 and Table 3.1).

3.6. Conclusion

Coding sequences have long been considered to be the key elements of the differences between organisms. Our ever-increasing knowledge of the functions provided by TEs in the structure, organization, expression and evolution of genomes has forced us to reconsider this reductive/erroneous view. In this chapter, we have thus illustrated how TEs are major players in the variability between species and between individuals of the same species. We have also described how complex interactions between TEs and their host, ranging from parasitism to cooperation, have led to the architecture of contemporary genomes and the emergence of sophisticated mechanisms for regulating the spread of TEs (see Figure 3.11).

However, modern tools for genome manipulation, microscopy, next-generation sequencing and bioinformatics tools that are increasingly adapted to repeated sequences are rapidly developing. Thus, we can anticipate that the next decade will bring new discoveries as to the unexpected ways in which TEs have shaped our genomes and contributed to the evolution of organisms. Almost 70 years after their revolutionary discovery by Barbara McClintock and 40 years after she received the Nobel Prize in Physiology or Medicine for her discoveries of mobile genetic elements, the study of TEs is probably entering its golden age and is likely to continue to challenge well-established concepts of our current knowledge.

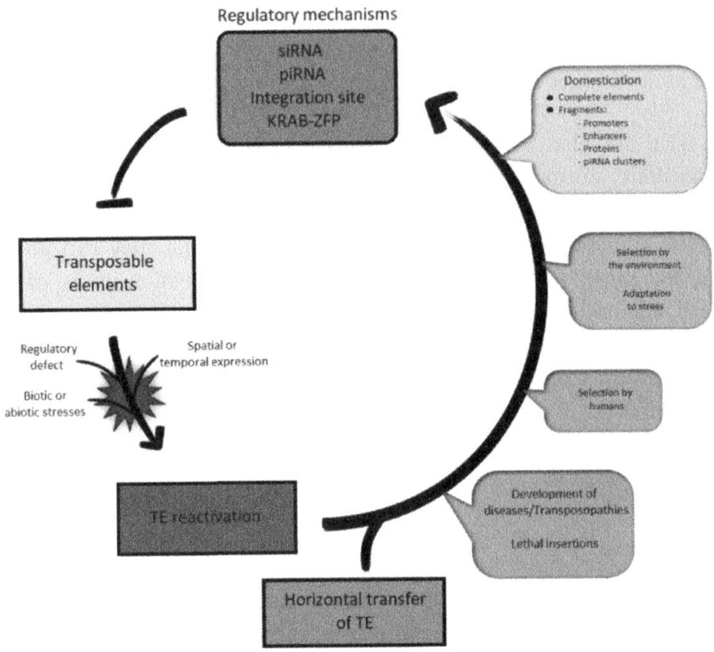

Figure 3.11. *Complex interactions between TEs and their host. TE activity is regulated by different mechanisms that vary according to the organism and/or cell type but can, however, be reactivated under specific conditions (stress, mutations in factors involved in regulation, differential expression from one tissue to another or depending on the stage of development). The consequences of the reactivation of TEs are numerous in the short or long term: disease development, selection, adaptation or domestication of elements. Over generations, certain insertions are selected, especially when they provide a selective advantage to the host*

3.7. References

Abad, J.P., De Pablos, B., Osoegawa, K., De Jong, P.J., Martín-Gallardo, A., Villasante, A. (2004). TAHRE, a novel telomeric retrotransposon from *Drosophila melanogaster*, reveals the origin of *Drosophila telomeres*. *Molecular Biology and Evolution*, 21(9), 1620–1624.

Adey, A., Morrison, H.G., Xu Xun, A., Kitzman, J.O., Turner, E.H., Stackhouse, B., MacKenzie, A.P., Caruccio, N.C., Zhang, X., Shendure, J. (2010). Rapid, low-input, low-bias construction of shotgun fragment libraries by high-density in vitro transposition. *Genome Biology*, 11(12), R119.

Amselem, J., Cornut, G., Choisne, N., Alaux, M., Alfama-Depauw, F., Jamilloux, V., Maumus, F., Letellier, T., Luyten, I., Pommier, C. et al. (2019). RepetDB: A unified resource for transposable element references. *Mobile DNA*, 10(1), 6.

Andersen, P.R., Tirian, L., Vunjak, M., Brennecke, J. (2017). A heterochromatin-dependent transcription machinery drives piRNA expression. *Nature*, 549(7670), 54–59.

Angelova, M.T., Dimitrova, D.G., Da Silva, B., Marchand, V., Jacquier, C., Achour, C., Brazane, M., Goyenvalle, C., Bourguignon-Igel, V., Shehzada, S. et al. (2020). tRNA 2′-O-methylation by a duo of TRM7/FTSJ1 proteins modulates small RNA silencing in Drosophila. *Nucleic Acids Research*, 48(4), 2050–2072.

Anwar, S.L., Wulaningsih, W., Lehmann, U. (2017). Transposable elements in human cancer: Causes and consequences of deregulation. *International Journal of Molecular Sciences*, 18(5). https://doi.org/10.3390/ijms18050974.

Anxolabéhère, D., Kidwell, M.G., Periquet, G. (1988). Molecular characteristics of diverse populations are consistent with the hypothesis of a recent invasion of *Drosophila melanogaster* by mobile P elements. *Molecular Biology and Evolution*, 5(3), 252–269.

Aravin, A., Gaidatzis, D., Pfeffer, S., Lagos-Quintana, M., Landgraf, P., Iovino, N., Morris, P., Brownstein, M.J., Kuramochi-Miyagawa, S., Nakano, T. et al. (2006). A novel class of small RNAs bind to MILI protein in mouse testes. *Nature*, 442(7099), 203–207.

Aravin, A., Sachidanandam, R., Bourc'his, D., Schaefer, C., Pezic, D., Toth, K.F., Bestor, T., Hannon, G.J. (2008). A piRNA pathway primed by individual transposons is linked to *de novo* DNA methylation in mice. *Molecular Cell*, 31(6), 785–799.

Arkhipova, I.R. (2017). Using bioinformatic and phylogenetic approaches to classify transposable elements and understand their complex evolutionary histories. *Mobile DNA*, 8(December), 19.

Armezzani, A., Varela, M., Spencer, T.E., Palmarini, M., Arnaud, F. (2014). "Ménage à Trois": The evolutionary interplay between JSRV, enJSRVs and domestic sheep. *Viruses*, 6(12), 4926–4945.

Asif-Laidin, A., Conesa, C., Bonnet, A., Grison, C., Adhya, I., Menouni, R., Fayol, H., Palmic, N., Acker, J., Lesage, P. (2020). A small targeting domain in Ty1 integrase is sufficient to direct retrotransposon integration upstream of tRNA genes. *The EMBO Journal*, 39(17), e104337.

Augé-Gouillou, C., Hamelin, M.H., Demattei, M.V., Periquet, M., Bigot, Y. (2001). The wild-type conformation of the Mos-1 inverted terminal repeats is suboptimal for transposition in bacteria. *Molecular Genetics and Genomics: MGG*, 265(1), 51–57.

Bagijn, M.P., Goldstein, L.D., Sapetschnig, A., Weick, E.-M., Bouasker, S., Lehrbach, N.J., Simard, M.J., Miska, E.A. (2012). Function, targets, and evolution of *Caenorhabditis elegans* piRNAs. *Science*, 337(6094), 574–578.

Baller, J.A., Gao, J., Voytas, D.F. (2011). Access to DNA establishes a secondary target site bias for the yeast retrotransposon Ty5. *Proceedings of the National Academy of Sciences of the United States of America*, 108(51), 20351–20356.

Baller, J.A., Gao, J., Stamenova, R., Curcio, M.J., Voytas, D.F. (2012). A nucleosomal surface defines an integration hotspot for the *Saccharomyces cerevisiae* Ty1 retrotransposon. *Genome Research*, 22(4), 704–713.

Barau, J., Teissandier, A., Zamudio, N., Roy, S., Nalesso, V., Hérault, Y., Guillou, F., Bourc'his, D. (2016). The DNA methyltransferase DNMT3C protects male germ cells from transposon activity. *Science*, 354(6314), 909–912.

Baudry, C., Malinsky, S., Restituito, M., Kapusta, A., Rosa, S., Meyer, E., Bétermier, M. (2009). PiggyMac, a domesticated PiggyBac transposase involved in programmed genome rearrangements in the ciliate *Paramecium tetraurelia*. *Genes & Development*, 23(21), 2478–2483.

Baumgartner L., Handler D., Platzer S.W., Yu C., Duchek P., Brennecke J. (2022) The Drosophila ZAD zinc finger protein Kipferl guides Rhino to piRNA clusters. *eLife*, 11:e80067.

Beauregard, A., Curcio, M.J., Belfort, M. (2008). The take and give between retrotransposable elements and their hosts. *Annual Review of Genetics*, 42, 587–617.

Belfort, M. (2017). Mobile self-splicing introns and inteins as environmental sensors. *Current Opinion in Microbiology*, 38(August), 51–58.

Beltran, T., Barroso, C., Birkle, T.Y., Stevens, L., Schwartz, H.T., Sternberg, P.W., Fradin, H., Gunsalus, K., Piano, F., Sharma, G. et al. (2019). Comparative epigenomics reveals that RNA polymerase II pausing and chromatin domain organization control nematode piRNA biogenesis. *Developmental Cell*, 48(6), 793–810.e6.

Best, S., Tissier, P.L., Towers, G., Stoye, J.P. (1996). Positional cloning of the mouse retrovirus restriction gene Fv1. *Nature*, 382(6594), 826–829.

Biémont, C. (2010). A brief history of the status of transposable elements: From junk DNA to major players in evolution. *Genetics*, 186(4), 1085–1093.

Blaise, S., de Parseval, N., Bénit, L., Heidmann, T. (2003). Genomewide screening for fusogenic human endogenous retrovirus envelopes identifies Syncytin 2, a gene conserved on primate evolution. *Proceedings of the National Academy of Sciences of the United States of America*, 100(22), 13013–13018.

Blond, J.L., Lavillette, D., Cheynet, V., Bouton, O., Oriol, G., Chapel-Fernandes, S., Mandrand, B., Mallet, F., Cosset. F.L. (2000). An envelope glycoprotein of the human endogenous retrovirus HERV-W is expressed in the human placenta and fuses cells expressing the type D mammalian retrovirus receptor. *Journal of Virology*, 74(7), 3321–3329.

Bodea, G.O., McKelvey, E.G.Z., Faulkner, G.J. (2018). Retrotransposon-induced mosaicism in the neural genome. *Open Biology*, 8(7). https://doi.org/10.1098/rsob.180074.

Bonnet, A. and Lesage, P. (2021). Light and shadow on the mechanisms of integration site selection in yeast Ty retrotransposon families. *Current Genetics*, 67(3), 347–357.

Borchert, G.M., Lanier, W., Davidson, B.L. (2006). RNA polymerase III transcribes human microRNAs. *Nature Structural & Molecular Biology*, 13(12), 1097–1101.

Bourgeois, Y. and Boissinot, S. (2019). On the population dynamics of junk: A review on the population genomics of transposable elements. *Genes*, 10(6). https://doi.org/10.3390/genes10060419.

Bourque, G., Burns, K.H., Gehring, M., Gorbunova, V., Seluanov, A., Hammell, M., Imbeault, M., Izsvák, Z., Levin, H.L., Macfarlan, T.S. et al. (2018). Ten things you should know about transposable elements. *Genome Biology*, 19(1), 199.

Brennecke, J., Aravin, A.A., Stark, A., Dus, M., Kellis, M., Sachidanandam, R., Hannon, G.J. (2007). Discrete small RNA-generating loci as master regulators of transposon activity in Drosophila. *Cell*, 128(6), 1089–1103.

Brennecke, J., Malone, C.D., Aravin, A.A., Sachidanandam, R., Stark, A., Hannon, G.J. (2008). An epigenetic role for maternally inherited piRNAs in transposon silencing. *Science*, 322(5906), 1387–1392.

Bridier-Nahmias, A., Tchalikian-Cosson, A., Baller, J.A., Menouni, R., Fayol, H., Flores, A., Saïb, A., Werner, M., Voytas, D.F., Lesage, P. (2015). Retrotransposons. An RNA polymerase III subunit determines sites of retrotransposon integration. *Science*, 348(6234), 585–588.

Brink, R.A. (1956). A genetic change associated with the R locus in maize which is directed and potentially reversible. *Genetics*, 41, 872–889

Buenrostro, J.D., Giresi, P.G., Zaba, L.C., Chang, H.Y., Greenleaf, W.J. (2013). Transposition of native chromatin for fast and sensitive epigenomic profiling of open chromatin, DNA-binding proteins and nucleosome position. *Nature Methods*, 10(12), 1213–1218.

Bulut-Karslioglu, A., De La Rosa-Velázquez, I.A., Ramirez, F., Barenboim, M., Onishi-Seebacher, M., Arand, J., Galán, C., Winter, G.E., Engist, B., Gerle, B. et al. (2014). Suv39h-dependent H3K9me3 marks intact retrotransposons and silences LINE elements in mouse embryonic stem cells. *Molecular Cell*, 55(2), 277–290.

Burns, K.H. (2017). Transposable elements in cancer. *Nature Reviews Cancer*, 17(7), 415–424.

Burns, K.H. (2020). Our conflict with transposable elements and its implications for human disease. *Annual Review of Pathology*, 15(January), 51–70.

Butelli, E., Licciardello, C., Zhang, Y., Liu, J., Mackay, S., Bailey, P., Reforgiato-Recupero, G., Martin, C. (2012). Retrotransposons control fruit-specific, cold-dependent accumulation of anthocyanins in blood oranges. *The Plant Cell*, 24(3), 1242–1255.

Carmona, L.M., and Schatz, D.G. (2017). New insights into the evolutionary origins of the recombination-activating gene proteins and V(D)J recombination. *The FEBS Journal*, 284(11), 1590–1605.

Casacuberta, E. (2017). Drosophila: Retrotransposons making up telomeres. *Viruses*, 9(7). https://doi.org/10.3390/v9070192.

Casier, K., Delmarre, V., Gueguen, N., Hermant, C., Viodé, E., Vaury, C., Ronsseray, S., Brasset, E., Teysset, L., Boivin, A. (2019). Environmentally-induced epigenetic conversion of a piRNA Cluster. *eLife*, 8(March). https://doi.org/10.7554/eLife.39842.

Cecere, G., Zheng, G.X.Y., Mansisidor, A.R., Klymko, K.E., Grishok, A. (2012). Promoters recognized by forkhead proteins exist for individual 21U-RNAs. *Molecular Cell*, 47(5), 734–745.

Chelmicki, T., Roger, E., Teissandier, A., Dura, M., Bonneville, L., Rucli, S., Dossin, F., Fouassier, C., Lameiras, S., Bourc'his, D. (2021). m6A RNA methylation regulates the fate of endogenous retroviruses. *Nature*, 591(7849), 312–316.

Chen, J.-M., Masson, E., Macek Jr., M., Raguénès, O., Piskackova, T., Fercot, B., Fila, L., Cooper, D.N., Audrézet, M.-P., Férec, C. (2008). Detection of two alu insertions in the CFTR gene. *Journal of Cystic Fibrosis: Official Journal of the European Cystic Fibrosis Society*, 7(1), 37–43.

Cherepanov, P., Maertens, G., Proost, P., Devreese, B., Beeumen, J.V., Engelborghs, Y., De Clercq, E., Debyser, Z. (2003). HIV-1 integrase forms stable tetramers and associates with LEDGF/p75 protein in human cells. *The Journal of Biological Chemistry*, 278(1), 372–381.

Chiappinelli, K.B., Strissel, P.L., Desrichard, A., Li, H., Henke, C., Akman, B., Hein, A., Rote, N.S., Cope, L.M., Snyder, A. et al. (2015). Inhibiting DNA methylation causes an interferon response in cancer via dsRNA including endogenous retroviruses. *Cell*, 162(5), 974–986.

Christensen, S., Pont-Kingdon, G., Carroll, D. (2000). Target specificity of the endonuclease from the *Xenopus laevis* non-long terminal repeat retrotransposon, Tx1L. *Molecular and Cellular Biology*, 20(4), 1219–1226.

Chung, T., Siol, O., Dingermann, T., Winckler, T. (2007). Protein interactions involved in tRNA gene-specific integration of dictyostelium discoideum non-long terminal repeat retrotransposon TRE5-A. *Molecular and Cellular Biology*, 27(24), 8492–8501.

Chuong, E.B., Elde, N.C., Feschotte, C. (2017). Regulatory activities of transposable elements: From conflicts to benefits. *Nature Reviews Genetics*, 18(2), 71–86.

Clark, L.A., Wahl, J.M., Rees, C.A., Murphy, K.E. (2006). Retrotransposon insertion in SILV is responsible for Merle patterning of the domestic dog. *Proceedings of the National Academy of Sciences of the United States of America*, 103(5), 1376–1381.

Coe Jr., E.H. (1959). A regular and continuing conversion-type phenomenon at the B locus in maize. *Proc. Natl. Acad. Sci. USA*, 45, 828–832.

Collins, M. and Rubin, G.M. (1984). Structure of chromosomal rearrangements induced by the FB transposable element in Drosophila. *Nature*, 308(5957), 323–327.

Cordaux, R. (2008). The human genome in the LINE of fire. *Proceedings of the National Academy of Sciences of the United States of America*, 105(49), 19033–19034.

Cordaux, R. and Batzer, M.A. (2009). The impact of retrotransposons on human genome evolution. *Nature Reviews Genetics*, 10(10), 691–703.

Cordaux, R., Hedges, D.J., Herke, S.W., Batzer, M.A. (2006). Estimating the retrotransposition rate of human alu elements. *Gene*, 373(May), 134–137.

Cornelis, G., Heidmann, O., Degrelle, S.A., Vernochet, C., Lavialle, C., Letzelter, C., Bernard-Stoecklin, S., Hassanin, A., Mulot, B., Guillomot, M. et al. (2013). Captured retroviral envelope syncytin gene associated with the unique placental structure of higher ruminants. *Proceedings of the National Academy of Sciences of the United States of America*, 110(9), E828–E837.

Cornelis, G., Vernochet, C., Malicorne, S., Souquere, S., Tzika, A.C., Goodman, S.M., Catzeflis, F., Robinson, T.J., Milinkovitch, M.C., Pierron, G. et al. (2014). Retroviral envelope syncytin capture in an ancestrally diverged mammalian clade for placentation in the primitive Afrotherian tenrecs. *Proceedings of the National Academy of Sciences of the United States of America*, 111(41), E4332–E4341.

Cornelis, G., Vernochet, C., Carradec, Q., Souquere, S., Mulot, B., Catzeflis, F., Nilsson, M.A., Menzies, B.R., Renfree, M.B., Pierron, G. et al. (2015). Retroviral envelope gene captures and syncytin exaptation for placentation in marsupials. *Proceedings of the National Academy of Sciences of the United States of America*, 112(5), E487–E496.

Cosby, R.L., Chang, N.-C., Feschotte, C. (2019). Host-transposon interactions: Conflict, cooperation, and cooption. *Genes & Development*, 33(17–18), 1098–1116.

Cosby, R.L., Judd, J., Zhang, R., Zhong, A., Garry, N., Pritham, E.J., Feschotte, C. (2021). Recurrent evolution of vertebrate transcription factors by transposase capture. *Science*, 371(6531). https://doi.org/10.1126/science.abc6405.

Cottee, M.A., Beckwith, S.L., Letham, S.C., Kim, S.J., Young, G.R., Stoye, J.P., Garfinkel, D.J., Taylor, I.A. (2021). Structure of a Ty1 restriction factor reveals the molecular basis of transposition copy number control. *Nature Communications*, 12(1), 5590.

Cruickshanks, H.A. and Tufarelli, C. (2009). Isolation of cancer-specific chimeric transcripts induced by hypomethylation of the LINE-1 antisense promoter. *Genomics*, 94(6), 397–406.

Curcio, M.J. (2019). Border collies of the genome: Domestication of an autonomous retrovirus-like transposon. *Current Genetics*, 65(1), 71–78.

Curcio, M.J. and Derbyshire, K.M. (2003). The outs and ins of transposition: From mu to kangaroo. *Nature Reviews. Molecular Cell Biology*, 4(11), 865–877.

Curcio, M.J., Kenny, A.E., Moore, S., Garfinkel, D.J., Weintraub, M., Gamache, E.R., Scholes. D.T. (2007). S-phase checkpoint pathways stimulate the mobility of the retrovirus-like transposon Ty1. *Molecular and Cellular Biology*, 27(24), 8874–8885.

Dai, J., Xie, W., Brady, T.L., Gao, J., Voytas, D.F. (2007). Phosphorylation regulates integration of the yeast Ty5 retrotransposon into heterochromatin. *Molecular Cell*, 27(2), 289–299.

Damert, A., Raiz, J., Horn, A.V., Löwer, J., Wang, H., Xing, J., Batzer, M.A., Löwer, R., Schumann, G.G. (2009). 5'-Transducing SVA retrotransposon groups spread efficiently throughout the human genome. *Genome Research*, 19(11), 1992–2008.

Daniels, S.B., Peterson, K.R., Strausbaugh, L.D., Kidwell, M.G., Chovnick, A. (1990). Evidence for horizontal transmission of the P transposable element between Drosophila species. *Genetics*, 124(2), 339–355.

De Rijck, J., De Kogel, C., Demeulemeester, J., Vets, S., El Ashkar, S., Malani, N., Bushman, F.D., Landuyt, B., Husson, S.J., Busschots, K. et al. (2013). The BET family of proteins targets Moloney Murine Leukemia Virus integration near transcription start sites. *Cell Reports*, 5(4), 886–894.

Dietrich, C.R., Cui, F., Packila, M.L., Li, J., Ashlock, D.A., Nikolau, B.J., Schnable, P.S. (2002). Maize Mu transposons are targeted to the 5' untranslated region of the gl8 gene and sequences flanking Mu target-site duplications exhibit nonrandom nucleotide composition throughout the genome. *Genetics*, 160(2), 697–716.

Doyle, G.A., Crist, R.C., Karatas, E.T., Hammond, M.J., Ewing, A.D., Ferraro, T.N., Hahn, C.-G., Berrettini, W.H. (2017). Analysis of LINE-1 elements in DNA from postmortem brains of individuals with schizophrenia. *Neuropsychopharmacology: Official Publication of the American College of Neuropsychopharmacology*, 42(13), 2602–2611.

Dubin, M.J., Scheid, O.M., Becker, C. (2018). Transposons: A blessing curse. *Current Opinion in Plant Biology*, 42(April), 23–29.

Dufourt, J., Dennis, C., Boivin, A., Gueguen, N., Théron, E., Goriaux, C., Pouchin, P., Ronsseray, S., Brasset, E., Vaury, C. (2014). Spatio-temporal requirements for transposable element piRNA-mediated silencing during Drosophila oogenesis. *Nucleic Acids Research*, 42(4), 2512–2524.

Dupressoir, A., Marceau, G., Vernochet, C., Bénit, L., Kanellopoulos, C., Sapin, V., Heidmann, T. (2005). Syncytin-A and Syncytin-B, two fusogenic placenta-specific murine envelope genes of retroviral origin conserved in Muridae. *Proceedings of the National Academy of Sciences of the United States of America*, 102(3), 725–730.

Dupressoir, A., Vernochet, C., Bawa, O., Harper, F., Pierron, G., Opolon, P., Heidmann, T. (2009). Syncytin-A knockout mice demonstrate the critical role in placentation of a fusogenic, endogenous retrovirus-derived, envelope gene. *Proceedings of the National Academy of Sciences of the United States of America*, 106(29), 12127–12132.

Ecco, G., Imbeault, M., Trono, T. (2017). KRAB zinc finger proteins. *Development*, 144(15), 2719–2729.

Edith H. and Martienssen, R.A. (2014). Transgenerational epigenetic inheritance: Myths and mechanisms. *Cell*, 157(1), 95–109.

El Baidouri, M., Carpentier, M.-C., Cooke, R., Gao, D., Lasserre, E., Llauro, C., Mirouze, M., Picault, N., Jackson, S.A., Panaud, O. (2014). Widespread and frequent horizontal transfers of transposable elements in plants. *Genome Research*, 24(5), 831–838.

Esnault, C., Lee, M., Ham, C., Levin, H.L. (2019). Transposable element insertions in fission yeast drive adaptation to environmental stress. *Genome Research*, 29(1), 85–95.

Ferrari, R., Grandi, N., Tramontano, E., Dieci, G. (2021). Retrotransposons as drivers of mammalian brain evolution. *Life*, 11(5). https://doi.org/10.3390/life11050376.

Feschotte, C. and Pritham, E.J. (2007). DNA transposons and the evolution of eukaryotic genomes. *Annual Review of Genetics*, 41, 331–368.

Finnegan, D.J. (1989). Eukaryotic transposable elements and genome evolution. *Trends in Genetics: TIG*, 5(4), 103–107.

Finnegan, D.J. (1992). Transposable elements. *Current Opinion in Genetics & Development*, 2(6), 861–867.

Flasch, D.A., Macia, A., Sánchez, L., Ljungman, M., Heras, S.R., García-Pérez, J.L., Wilson, T.E., Moran, J.V. (2019). Genome-wide *de novo* L1 retrotransposition connects endonuclease activity with replication. *Cell*, 177(4), 837–851.e28.

Franke, G., Bausch, B., Hoffmann, M.M., Cybulla, M., Wilhelm, C., Kohlhase, J., Scherer, G., Neumann, H.P.H. (2009). Alu-Alu recombination underlies the vast majority of large VHL germline deletions: Molecular characterization and genotype-phenotype correlations in VHL patients. *Human Mutation*, 30(5), 776–786.

Frendo, J.-L., Olivier, D., Cheynet, V., Blond, J.-L., Bouton, O., Vidaud, M., Rabreau, M., Evain-brion, D., Mallet, F. (2003). Direct involvement of HERV-W Env glycoprotein in human trophoblast cell fusion and differentiation. *Molecular and Cellular Biology*, 23(10), 3566–3574.

Fujiwara, H. (2015). Site-specific non-LTR retrotransposons. *Microbiology Spectrum*, 3(2), MDNA3–0001–2014.

Fujiwara, H., Osanai, M., Matsumoto, T., Kojima, K.K. (2005). Telomere-specific non-LTR retrotransposons and telomere maintenance in the silkworm. *Bombyx mori. Chromosome Research: An International Journal on the Molecular, Supramolecular and Evolutionary Aspects of Chromosome Biology*, 13(5), 455–467.

Fuller, A.M., Cook, E.G., Kelley, K.J., Pardue, M.-L. (2010). Gag proteins of Drosophila telomeric retrotransposons: Collaborative targeting to chromosome ends. *Genetics*, 184(3), 629–636.

Funk, M., Cornelis, G., Vernochet, C., Heidmann, O., Dupressoir, A., Conley, A., Glickman, S., Heidmann, T. (2019). Capture of a hyena-specific retroviral envelope gene with placental expression associated in evolution with the unique emergence among carnivorans of hemochorial placentation in Hyaenidae. *Journal of Virology*, 93(4). https://doi.org/10.1128/JVI.01811-18.

Gage, F. and Muotri, A. (2012). Des gènes sauteurs dans le cerveau. *Pour La Science*, 417, 70–75.

Gai, X. and Voytas, D.F. (1998). A single amino acid change in the yeast retrotransposon Ty5 abolishes targeting to silent chromatin. *Molecular Cell*, 1(7), 1051–1055.

Ganguly, A., Dunbar, T., Chen, P., Godmilow, L., Ganguly, T. (2003). Exon skipping caused by an intronic insertion of a young Alu Yb9 element leads to severe hemophilia A. *Human Genetics*, 113(4), 348–352.

Gao, D., Chu, Y., Xia, H., Xu, C., Heyduk, K., Abernathy, B., Ozias-Akins, P., Leebens-Mack, J.-H., Jackson, S.A. (2018). Horizontal transfer of non-LTR retrotransposons from arthropods to flowering plants. *Molecular Biology and Evolution*, 35(2), 354–364.

García Guerreiro, M.P. (2012). What makes transposable elements move in the Drosophila genome? *Heredity*, 108(5), 461–468.

Garcia-Vallvé, S., Romeu, A., Palau, J. (2000). Horizontal gene transfer in bacterial and archaeal complete genomes. *Genome Research*, 10(11), 1719–1725.

Geis, F.K. and Goff, S.P. (2020). Silencing and transcriptional regulation of endogenous retroviruses: An overview. *Viruses*, 12(8). https://doi.org/10.3390/v12080884.

Genenncher, B., Durdevic, Z., Hanna, K., Zinkl, D., Mobin, M.B., Senturk, N., Da Silva, B., Legrand, C., Carré, C., Lyko, F. et al. (2018). Mutations in cytosine-5 tRNA methyltransferases impact mobile element expression and genome stability at specific DNA repeats. *Cell Reports*, 22(7), 1861–1874.

van Gent, D.C., Mizuuchi, K., Gellert, M. (1996a). Similarities between initiation of V(D)J recombination and retroviral integration. *Science*, 271(5255), 1592–1594.

van Gent, D.C., Ramsden, D.A., Gellert, M. (1996b). The RAG1 and RAG2 proteins establish the 12/23 rule in V(D)J recombination. *Cell*, 85(1), 107–113.

Gilbert, C. and Cordaux, R. (2017). Viruses as vectors of horizontal transfer of genetic material in eukaryotes. *Current Opinion in Virology*, 25(August), 16–22.

Gilbert, N., Lutz-Prigge, S., Moran, J.V. (2002). Genomic deletions created upon LINE-1 retrotransposition. *Cell*, 110(3), 315–325.

Girard, A., Sachidanandam, R., Hannon, G.J., Carmell, M.A. (2006). A germline-specific class of small RNAs binds mammalian Piwi proteins. *Nature*, 442(7099), 199–202.

Goerner-Potvin, P. and Bourque, G. (2018). Computational tools to unmask transposable elements. *Nature Reviews. Genetics*, 19(11), 688–704.

Göke, J., Lu, X., Chan, Y.-S., Ng, H.H., Ly, L.-M., Sachs, F., Szczerbinska, L. (2015). Dynamic transcription of distinct classes of endogenous retroviral elements marks specific populations of early human embryonic cells. *Cell Stem Cell*, 16(2), 135–141.

Goll, M.G., Kirpekar, F., Maggert, K.A., Yoder, J.A., Hsieh, C.-H., Zhang, X., Golic, K.G., Jacobsen, S.E., Bestor, T.H. (2006). Methylation of tRNAAsp by the DNA methyltransferase homolog Dnmt2. *Science*, 311(5759), 395–398.

Goodier, J.L. (2016). Restricting retrotransposons: A review. *Mobile DNA*, 7(August), 16.

Goodier, J.L., Ostertag, E.M., Kazazian Jr., H.H. (2000). Transduction of 3'-flanking sequences is common in L1 retrotransposition. *Human Molecular Genetics*, 9(4), 653–657.

Goriaux, C., Desset, S., Renaud, Y., Vaury, C., Brasset, E. (2014). Transcriptional properties and splicing of the flamenco piRNA cluster. *EMBO Reports*, 15(4), 411–418.

Grimson, A., Srivastava, M., Fahey, B., Woodcroft, B.J., Chiang, H.R., King, N., Degnan, B.M., Rokhsar, D.S., Bartel, D.P. (2008). Early origins and evolution of microRNAs and Piwi-interacting RNAs in animals. *Nature*, 455(7217), 1193–1197.

Grundy, E.E., Diab, N., Chiappinelli, K.B. (2021). Transposable element regulation and expression in cancer. *The FEBS Journal*. https://doi.org/10.1111/febs.15722.

Guffanti, G., Gaudi, S., Fallon, J.H., Sobell, J., Potkin, S.G., Pato, C., Macciardi, F. (2014). Transposable elements and psychiatric disorders. *American Journal of Medical Genetics. Part B, Neuropsychiatric Genetics: The Official Publication of the International Society of Psychiatric Genetics*, 165B(3), 201–216.

Guio, L., Vieira, C., González, J. (2018). Stress affects the epigenetic marks added by natural transposable element insertions in *Drosophila melanogaster*. *Scientific Reports*, 8(1), 12197.

Gunawardane, L.S., Saito, K., Nishida, K.M., Miyoshi, K., Kawamura, Y., Nagami, T., Siomi, H., Siomi, M.C. (2007). A slicer-mediated mechanism for repeat-associated siRNA 5' end formation in Drosophila. *Science*, 315(5818), 1587–1590.

Guo, Y. and Levin, H.L. (2010). High-throughput sequencing of retrotransposon integration provides a saturated profile of target activity in *Schizosaccharomyces pombe*. *Genome Research*, 20(2), 239–248.

Gupta, S.S., Maetzig, T., Maertens, G.N., Sharif, A., Rothe, M., Weidner-Glunde, M., Galla, M., Schambach, A., Cherepanov, P., Schulz, T.F. (2013). Bromo- and extraterminal domain chromatin regulators serve as cofactors for murine leukemia virus integration. *Journal of Virology*, 87(23), 12721–12736.

Habu, Y., Hisatomi, Y., Iida, S. (1998). Molecular characterization of the mutable flaked allele for flower variegation in the common morning glory. *The Plant Journal: For Cell and Molecular Biology*, 16(3), 371–376.

Han, K., Lee, J., Meyer, T.J., Remedios, P., Goodwin, L., Batzer, M.A. (2008). L1 recombination-associated deletions generate human genomic variation. *Proceedings of the National Academy of Sciences of the United States of America*, 105(49), 19366–19371.

Hancks, D.C. and Kazazian Jr., H.H. (2012). Active human retrotransposons: Variation and disease. *Current Opinion in Genetics & Development*, 22(3), 191–203.

Hancks, D.C. and Kazazian Jr., H.H. (2016). Roles for retrotransposon insertions in human disease. *Mobile DNA*, 7(May), 9.

Havecker, E.R., Gao, X., Voytas, D.F. (2004). The diversity of LTR retrotransposons. *Genome Biology*, 5(6), 225.

Heidmann, O., Vernochet, C., Dupressoir, A., Heidmann, T. (2009). Identification of an endogenous retroviral envelope gene with fusogenic activity and placenta-specific expression in the rabbit: A new "syncytin" in a third order of mammals. *Retrovirology*, 6(November), 107.

Helman, E., Lawrence, M.S., Stewart, C., Sougnez, C., Getz, G., Meyerson, M. (2014). Somatic retrotransposition in human cancer revealed by whole-genome and exome sequencing. *Genome Research*, 24(7), 1053–1063.

Henssen, A.G., Koche, R., Zhuang, J., Jiang, E., Reed, C., Eisenberg, A., Still, E., MacArthur, I.C., Rodríguez-Fos, E., Gonzalez, S. et al. (2017). PGBD5 promotes site-specific oncogenic mutations in human tumors. *Nature Genetics*, 49(7), 1005–1014.

Hickey, A., Esnault, C., Majumdar, A., Chatterjee, A.G., Iben, J.R., McQueen, P.G., Yang, A.X., Mizuguchi, T., Grewal, S.I.S., Levin, H.L. (2015). Single-nucleotide-specific targeting of the Tf1 retrotransposon promoted by the DNA-binding protein Sap1 of *Schizosaccharomyces pombe*. *Genetics*, 201(3), 905–924.

Horváth, V., Merenciano, M., González, J. (2017). Revisiting the relationship between transposable elements and the eukaryotic stress response. *Trends in Genetics: TIG*, 33(11), 832–841.

Hughes, S.H. (2015). Reverse transcription of retroviruses and LTR retrotransposons. *Microbiology Spectrum*, 3(2), MDNA3–0027–2014.

Hwang, S.-Y., Jung, H., Mun, S., Lee, S., Park, K., Baek, S.C., Moon, H.C., Kim, H., Kim, B., Choi, Y. et al. (2021). L1 retrotransposons exploit RNA m6A modification as an evolutionary driving force. *Nature Communications*, 12(1), 880.

Ikeda, H. and Sugimura, H. (1989). Fv-4 resistance gene: A truncated endogenous murine leukemia virus with ecotropic interference properties. *Journal of Virology*, 63(12), 5405–5412.

Imbeault, M., Helleboid, P.-Y., Trono, D. (2017). KRAB zinc-finger proteins contribute to the evolution of gene regulatory networks. *Nature*, 543(7646), 550–554.

Jacobs, F.M.J., Greenberg, D., Nguyen, N., Haeussler, M., Ewing, A.D., Katzman, S., Paten, B., Salama, S.R., Haussler, D. (2014). An evolutionary arms race between KRAB zinc-finger genes ZNF91/93 and SVA/L1 retrotransposons. *Nature*, 516(7530), 242–245.

Jacobs, J.Z., Rosado-Lugo, J.D., Cranz-Mileva, S., Ciccaglione, K.M., Tournier, V., Zaratiegui, M. (2015). Arrested replication forks guide retrotransposon integration. *Science*, 349(6255), 1549–1553.

Jang, H.S., Shah, N.M., Du, A.Y., Dailey, Z.Z., Pehrsson, E.C., Godoy, P.M., Zhang, D., Li, D., Xing, X., Kim, S. et al. (2019). Transposable elements drive widespread expression of oncogenes in human cancers. *Nature Genetics*, 51(4), 611–617.

Jangam, D., Feschotte, C., Betrán, E. (2017). Transposable element domestication as an adaptation to evolutionary conflicts. *Trends in Genetics: TIG*, 33(11), 817–831.

Jiang, N. and Wessler, S.R. (2001). Insertion preference of maize and rice miniature inverted repeat transposable elements as revealed by the analysis of nested elements. *The Plant Cell*, 13(11), 2553–2564.

Kaessmann, H., Vinckenbosch, N., Long, M. (2009). RNA-based gene duplication: Mechanistic and evolutionary insights. *Nature Reviews. Genetics*, 10(1), 19–31.

Kapitonov, V.V. and Jurka, J. (2005). RAG1 core and V(D)J recombination signal sequences were derived from transib transposons. *PLoS Biology*, 3(6), e181.

Kato, Y., Kaneda, M., Hata, K., Kumaki, K., Hisano, M., Kohara, Y., Okano, M., Li, E., Nozaki, M., Sasaki, H. (2007). Role of the Dnmt3 family in *de novo* methylation of imprinted and repetitive sequences during male germ cell development in the mouse. *Human Molecular Genetics*, 16(19), 2272–2280.

Kato, M., Takemoto, K., Shinkai, Y. (2018). A somatic role for the histone methyltransferase Setdb1 in endogenous retrovirus silencing. *Nature Communications*, 9(1), 1683.

Kazazian Jr., H.H. (1999). An estimated frequency of endogenous insertional mutations in humans. *Nature Genetics*, 22(2), 130.

Kazazian Jr., H.H. and Moran, J.V. (2017). Mobile DNA in health and disease. *The New England Journal of Medicine*, 377(4), 361–370.

Kazazian Jr., H.H., Wong, C., Youssoufian, H., Scott, A.F., Phillips, D.G., Antonarakis, S.E. (1988). Haemophilia A resulting from *de novo* insertion of L1 sequences represents a novel mechanism for mutation in man. *Nature*, 332 (6160), 164–166.

Kent, T.V., Uzunović, J., Wright, S.I. (2017). Coevolution between transposable elements and recombination. *Philosophical Transactions of the Royal Society of London. Series B, Biological Sciences*, 372(1736). https://doi.org/10.1098/rstb.2016.0458.

Kidwell, M.G., Kidwell, J.F., Sved, J.A. (1977). Hybrid dysgenesis in *Drosophila melanogaster*: A syndrome of aberrant traits including mutation, sterility and male recombination. *Genetics*, 86(4), 813–833.

Kim, A., Terzian, C., Santamaria, P., Pélisson, A., Purd'homme, N., Bucheton, A. (1994). Retroviruses in invertebrates: The gypsy retrotransposon is apparently an infectious retrovirus of *Drosophila melanogaster*. *Proceedings of the National Academy of Sciences of the United States of America*, 91(4), 1285–1289.

Kirchner, J., Connolly, C.M., Sandmeyer, S.B. (1995). Requirement of RNA polymerase III transcription factors for in vitro position-specific integration of a retroviruslike element. *Science*, 267(5203), 1488–1491.

Klein, S.J. and O'Neill, RJ. (2018). Transposable elements: Genome innovation, chromosome diversity, and centromere conflict. *Chromosome Research: An International Journal on the Molecular, Supramolecular and Evolutionary Aspects of Chromosome Biology*, 26(1–2), 5–23.

Knebelmann, B., Forestier, L., Drouot, L., Quinones, S., Chuet, C., Benessy, F., Saus, J., Antignac, C. (1995). Splice-mediated insertion of an alu sequence in the COL4A3 mRNA causing autosomal recessive Alport syndrome. *Human Molecular Genetics*, 4(4), 675–679.

Kofler, R., Senti, K.-A., Nolte, V., Tobler, R., Schlötterer, C. (2018). Molecular dissection of a natural transposable element invasion. *Genome Research*, 28(6), 824–835.

Kojima, K.K. (2020). Structural and sequence diversity of eukaryotic transposable elements. *Genes & Genetic Systems*, 94(6), 233–252.

Kondo-Iida, E., Kobayashi, K., Watanabe, M., Sasaki, J., Kumagai, T., Koide, H., Saito, K., Osawa, M., Nakamura, Y., Toda, T. (1999). Novel mutations and genotype-phenotype relationships in 107 families with Fukuyama-type congenital muscular dystrophy (FCMD). *Human Molecular Genetics*, 8(12), 2303–2309.

Konkel, M.K. and Batzer, M.A. (2010). A mobile threat to genome stability: The impact of non-LTR retrotransposons upon the human genome. *Seminars in Cancer Biology*, 20(4), 211–221.

Koshi, K., Ushizawa, K., Kizaki, K., Takahashi, T., Hashizume, K. (2011). Expression of endogenous retrovirus-like transcripts in bovine trophoblastic cells. *Placenta*, 32(7), 493–499.

Lampe, D.J., Akerley, B.J., Rubin, E.J., Mekalanos, J.J., Robertson, H.M. (1999). Hyperactive transposase mutants of the Himar1 mariner transposon. *Proceedings of the National Academy of Sciences of the United States of America*, 96(20), 11428–11433.

Laski, F.A., Rio, D.C., Rubin, G.M. (1986). Tissue specificity of Drosophila P element transposition is regulated at the level of mRNA splicing. *Cell*, 44(1), 7–19.

Laureau, R., Dyatel, A., Dursuk, G., Brown, S., Adeoye, H., Yue, J.-X., Chiara, M.D., Harris, A., Ünal, E., Liti, G. et al. (2021). Meiotic cells counteract programmed retrotransposon activation via RNA-binding translational repressor assemblies. *Developmental Cell*, 56(1), 22–35.e7.

Leblanc, P., Dastugue, B., Vaury, C. (1999). The integration machinery of ZAM, a retroelement from *Drosophila melanogaster*, acts as a sequence-specific endonuclease. *Journal of Virology*, 73(8), 7061–7064.

Lee, J., Han, K., Meyer, T.J., Kim, H.-S., Batzer, M.A. (2008). Chromosomal inversions between human and chimpanzee lineages caused by retrotransposons. *PloS One*, 3(12), e4047.

Lee, H.-C., Gu, W., Shirayama, M., Youngman Jr., E., D.C., Mello, C.C. (2012). *C. elegans* piRNAs mediate the genome-wide surveillance of germline transcripts. *Cell*, 150(1), 78–87.

Lehrman, M.A., Schneider, W.J., Südhof, T.C., Brown, M.S., Goldstein, J.L., Russell, D.W. (1985). Mutation in LDL receptor: Alu-Alu recombination deletes exons encoding transmembrane and cytoplasmic domains. *Science*, 227(4683), 140–146.

Lelek, M., Casartelli, N., Pellin, D., Rizzi, E., Souque, P., Severgnini, M., Serio, C.D., Fricke, T., Diaz-Griffero, F., Zimmer, C. et al. (2015). Chromatin organization at the nuclear pore favours HIV replication. *Nature Communications*, 6(March), 6483.

Levis, R.W., Ganesan, R., Houtchens, K., Tolar, L.A., Sheen, F.M. (1993). Transposons in place of telomeric repeats at a Drosophila telomere. *Cell*, 75(6), 1083–1093.

Li, X.Z., Roy, C.K., Dong, X., Bolcun-Filas, E., Wang, J., Han, B.W., Xu, J., Moore, M.J., Schimenti, J.C., Weng, Z. et al. (2013). An ancient transcription factor initiates the burst of piRNA production during early meiosis in mouse testes. *Molecular Cell*, 50(1), 67–81.

Liang, Z., Anderson, S.N., Noshay, J.M., Crisp, P.A., Enders, T.A., Springer, N.M. (2021). Genetic and epigenetic variation in transposable element expression responses to abiotic stress in maize. *Plant Physiology*. https://doi.org/10.1093/plphys/kiab073.

Lin, X., Faridi, N., Casola, C. (2016). An ancient transkingdom horizontal transfer of Penelope-like retroelements from arthropods to conifers. *Genome Biology and Evolution*, 8(4), 1252–1266.

Linker, S.B., Marchetto, M.C., Narvaiza, I., Denli, A.M., Gage, F.H. (2017). Examining non-LTR retrotransposons in the context of the evolving primate brain. *BMC Biology*, 15(1), 68.

Liu, D. and Chalmers, R. (2014). Hyperactive mariner transposons are created by mutations that disrupt allosterism and increase the rate of transposon end synapsis. *Nucleic Acids Research*, 42(4), 2637–2645.

Liu, G., NISC Comparative Sequencing Program, Zhao, S., Bailey, J.A., Sahinalp, S.C., Alkan, C., Tuzun, E., Green, E.D., Eichler, E.E. (2003). Analysis of primate genomic variation reveals a repeat-driven expansion of the human genome. *Genome Research*, 13(3), 358–368.

Liu, S., Yeh, C.-T., Ji, T., Ying, K., Wu, H., Tang, H.M., Fu, F., Nettleton, D., Schnable, P.S. (2009). Mu transposon insertion sites and meiotic recombination events co-localize with epigenetic marks for open chromatin across the maize genome. *PLoS Genetics*, 5(11), e1000733.

Liu, S., Brind'Amour, J., Karimi, M.M., Shirane, K., Bogutz, A., Lefebvre, L., Sasaki, H., Shinkai, Y., Lorincz, M.C. (2014). Setdb1 is required for germline development and silencing of H3K9me3-marked endogenous retroviruses in primordial germ cells. *Genes & Development*, 28(18), 2041–2055.

Liu, J., Gao, M., He, J., Wu, K., Lin, S., Jin, L., Chen, Y., Liu, H., Shi, J., Wang, X. et al. (2021). The RNA m6A reader YTHDC1 silences retrotransposons and guards ES cell identity. *Nature*, 591(7849), 322–326.

Llano, M., Saenz, D.T., Meehan, A., Wongthida, P., Peretz, M., Walker, W.H., Teo, W., Poeschla, E.M. (2006). An essential role for LEDGF/p75 in HIV integration. *Science*, 314(5798), 461–464.

Löber, U., Hobbs, M., Dayaram, A., Tsangaras, K., Jones, K., Alquezar-Planas, D.E., Ishida, Y., Meers, J., Mayer, J., Quedenau, C. et al. (2018). Degradation and remobilization of endogenous retroviruses by recombination during the earliest stages of a germ-line invasion. *Proceedings of the National Academy of Sciences of the United States of America*, 115(34), 8609–8614.

Lohe, A.R. and Hartl, D.L. (1996). Autoregulation of mariner transposase activity by overproduction and dominant-negative complementation. *Molecular Biology and Evolution*, 13(4), 549–555.

Loiseau, V., Herniou, E.A., Moreau, Y., Lévêque, N., Meignin, C., Daeffler, L., Federici, B., Cordaux, R., Gilbert, C. (2020). Wide spectrum and high frequency of genomic structural variation, including transposable elements, in large double-stranded DNA viruses. *Virus Evolution*, 6(1), vez060.

Loreto, E.L.S., Carareto, C.M.A., Capy, P. (2008). Revisiting horizontal transfer of transposable elements in Drosophila. *Heredity*, 100(6), 545–554.

Lusic, M. and Siliciano, R.F. (2017). Nuclear landscape of HIV-1 infection and integration. *Nature Reviews. Microbiology*, 15(2), 69–82.

Majumdar, S. and Rio, D.C. (2015). P transposable elements in Drosophila and other eukaryotic organisms. *Microbiology Spectrum*, 3(2), MDNA3–0004–2014.

Maksakova, I.A., Thompson, P.J., Goyal, P., Jones, S.J., Singh, P.B., Karimi, M.M., Lorincz, M.C. (2013). Distinct roles of KAP1, HP1 and G9a/GLP in silencing of the two-cell-specific retrotransposon MERVL in mouse ES cells. *Epigenetics & Chromatin*, 6(1), 15.

Malik, H.S., Henikoff, S., Eickbush, T.H. (2000). Poised for contagion: Evolutionary origins of the infectious abilities of invertebrate retroviruses. *Genome Research*, 10(9), 1307–1318.

Manhas, S., Ma, L., Measday, V. (2018). The yeast Ty1 retrotransposon requires components of the nuclear pore complex for transcription and genomic integration. *Nucleic Acids Research*, 46(7), 3552–3578.

Marini, B., Kertesz-Farkas, A., Ali, H., Lucic, B., Lisek, K., Manganaro, L., Pongor, S., Luzzati, R., Recchia, A., Mavilio, F. et al. (2015). Nuclear architecture dictates HIV-1 integration site selection. *Nature*, 521(7551), 227–231.

Martinez, G., Choudury, S.G., Slotkin, R.K. (2017). tRNA-derived small RNAs target transposable element transcripts. *Nucleic Acids Research*, 45(9), 5142–5152.

Mátés, L., Chuah, M.K.L., Belay, E., Jerchow, B., Manoj, N., Acosta-Sanchez, A., Grzela, D.P., Schmitt, A., Becker, K., Matrai, J. et al. (2009). Molecular evolution of a novel hyperactive sleeping beauty transposase enables robust stable gene transfer in vertebrates. *Nature Genetics*, 41(6), 753–761.

Maupetit-Mehouas, S. and Vaury, C. (2020). Transposon reactivation in the germline may be useful for both transposons and their host genomes. *Cells*, 9(5), 1172.

Maxwell, P.H. and Curcio, M.J. (2007). Retrosequence formation restructures the yeast genome. *Genes & Development*, 21(24), 3308–3318.

McClintock, B. (1950). The origin and behavior of mutable loci in maize. *Proceedings of the National Academy of Sciences of the United States of America*, 36(6), 344–355.

McDonald, J.F. (1995). Transposable elements: Possible catalysts of organismic evolution. *Trends in Ecology & Evolution*, 10(3), 123–126.

Mendoza, A., Bonnet, A., Vargas-Landin, D.B., Ji, N., Li, H., Yang, F., Li, L., Hori, K., Pflueger, J., Buckberry, S. et al. (2018). Recurrent acquisition of cytosine methyltransferases into eukaryotic retrotransposons. *Nature Communications*, 9(1), 1341.

Merlin, C. and Toussaint, A. (1999). Les éléments transposables bactériens. *Medecine Sciences: M/S*, 15(8–9), I.

Mi, S., Lee, X., Li, X., Veldman, G.M., Finnerty, H., Racie, L., LaVallie, E., Tang, X.Y., Edouard, P., Howes, S. et al. (2000). Syncytin is a captive retroviral envelope protein involved in human placental morphogenesis. *Nature*, 403(6771), 785–789.

Michaud, E.J., van Vugt, M.J., Bultman, S.J., Sweet, H.O., Davisson, M.T., Woychik, R.P. (1994). Differential expression of a new dominant agouti allele (Aiapy) is correlated with methylation state and is influenced by parental lineage. *Genes & Development*, 8(12), 1463–1472.

Miki, Y., Nishisho, I., Horii, A., Miyoshi, Y., Utsunomiya, J., Kinzler, K.W., Vogelstein, B., Nakamura, Y. (1992). Disruption of the APC gene by a retrotransposal insertion of L1 sequence in a colon cancer. *Cancer Research*, 52(3), 643–645.

Miller, W.J., McDonald, J.F., Pinsker, W. (1997). Molecular domestication of mobile elements. *Genetica*, 100(1–3), 261–270.

Miousse, I.R., Chalbot, M.-C.G., Lumen, A., Ferguson, A., Kavouras, I.G., Koturbash, I. (2015). Response of transposable elements to environmental stressors. *Mutation Research/Reviews in Mutation Research*, 765(July), 19–39.

Mita, P., Wudzinska, A., Sun, X., Andrade, J., Nayak, S., Kahler, D.J., Badri, S., LaCava, J., Ueberheide, B., Yun, C.Y. et al. (2018). Line-1 protein localization and functional dynamics during the cell cycle. *eLife*, 7(1).

Mohn, F., Sienski, G., Handler, D., Brennecke, J. (2014). The rhino-deadlock-cutoff complex licenses noncanonical transcription of dual-strand piRNA clusters in Drosophila. *Cell*, 157(6), 1364–1379.

Molaro, A. and Malik, H.S. (2016). Hide and seek: How chromatin-based pathways silence retroelements in the mammalian germline. *Current Opinion in Genetics & Development*, 37(April), 51–58.

Morais, S., Bastos-Ferreira, R., Sequeiros, J., Alonso, I. (2016). Genomic mechanisms underlying PARK2 large deletions identified in a cohort of patients with PD. *Neurology Genetics*, 2(3), e73.

Morales, M.E., Servant, G., Ade, C., Roy-Engel, A.M. (2015). Altering genomic integrity: Heavy metal exposure promotes transposable element-mediated damage. *Biological Trace Element Research*, 166(1), 24–33.

Moran, J.V., Holmes, S.E., Naas, T.P., DeBerardinis, R.J., Boeke, J.D., Kazazian Jr., H.H. (1996). High frequency retrotransposition in cultured mammalian cells. *Cell*, 87(5), 917–927.

Mukherjee, S., Mukhopadhyay, A., Banerjee, D., Chandak, G.R., Ray, K. (2004). Molecular pathology of haemophilia B: Identification of five novel mutations including a LINE 1 insertion in Indian patients. *Haemophilia: The Official Journal of the World Federation of Hemophilia*, 10(3), 259–263.

Mularoni, L., Zhou, Y., Bowen, T., Gangadharan, S., Wheelan, S.J., Boeke, J.D. (2012). Retrotransposon Ty1 integration targets specifically positioned asymmetric nucleosomal DNA segments in tRNA hotspots. *Genome Research*, 22(4), 693–703.

Mura, M., Murcia, P., Caporale, M., Spencer, T.E., Nagashima, K., Rein, A., Palmarini, M. (2004). Late viral interference induced by transdominant gag of an endogenous retrovirus. *Proceedings of the National Academy of Sciences of the United States of America*, 101(30), 11117–11122.

Murphy, S.C., Evans, J.M., Tsai, K.L., Clark, L.A. (2018). Length variations within the Merle retrotransposon of canine PMEL: Correlating genotype with phenotype. *Mobile DNA*, 9(August), 26.

Mustelin, T. and Ukadike, K.C. (2020). How retroviruses and retrotransposons in our genome may contribute to autoimmunity in rheumatological conditions. *Frontiers in Immunology*, 11(November), 593891.

Naito, K., Zhang, F., Tsukiyama, T., Saito, H., Hancock, C.N., Richardson, A.O., Okumoto, Y., Tanisaka, T., Wessler, S.R. (2009). Unexpected consequences of a sudden and massive transposon amplification on rice gene expression. *Nature*, 461(7267), 1130–1134.

Najafabadi, H.S., Mnaimneh, S., Schmitges, F.W., Garton, M., Lam, K.N., Yang, A., Albu, M., Weirauch, M.T., Radovani, E., Kim, P.M. et al. (2015). C2H2 zinc finger proteins greatly expand the human regulatory lexicon. *Nature Biotechnology*, 33(5), 555–562.

Narita, N., Nishio, H., Kitoh, Y., Ishikawa, Y., Ishikawa, Y., Minami, R., Nakamura, H., Matsuo, M. (1993). Insertion of a 5' truncated L1 element into the 3' end of exon 44 of the dystrophin gene resulted in skipping of the exon during splicing in a case of Duchenne muscular dystrophy. *The Journal of Clinical Investigation*, 91(5), 1862–1867.

de Oliveira, D.S., Rosa, M.T., Vieira, C., Loreto, E.L.S. (2021). Oxidative and radiation stress induces transposable element transcription in *Drosophila melanogaster*. *Journal of Evolutionary Biology*, 34(4), 628–638.

Orecchini, E., Frassinelli, L., Galardi, S., Ciafrè, S.A., Michienzi, A. (2018). Post-transcriptional regulation of LINE-1 retrotransposition by AID/APOBEC and ADAR deaminases. *Chromosome Research: An International Journal on the Molecular, Supramolecular and Evolutionary Aspects of Chromosome Biology*, 26(1–2), 45–59.

Peccoud, J., Loiseau, V., Cordaux, R., Gilbert, C. (2017). Massive horizontal transfer of transposable elements in insects. *Proceedings of the National Academy of Sciences of the United States of America*, 114(18), 4721–4726.

Penton, E.H. and Crease, T.J. (2004). Evolution of the transposable element pokey in the ribosomal DNA of species in the subgenus Daphnia (Crustacea: Cladocera). *Molecular Biology and Evolution*, 21(9), 1727–1739.

Phalke, S., Nickel, O., Walluscheck, D., Hortig, F., Onorati, M.C., Reuter, G. (2009). Retrotransposon silencing and telomere integrity in somatic cells of Drosophila depends on the cytosine-5 methyltransferase DNMT2. *Nature Genetics*, 41(6), 696–702.

Piacentini, L., Fanti, L., Specchia, V., Bozzetti, M.P., Berloco, M., Palumbo, G., Pimpinelli, S. (2014). Transposons, environmental changes, and heritable induced phenotypic variability. *Chromosoma*, 123(4), 345–354.

Pickeral, O.K., Makałowski, W., Boguski, M.S., Boeke, J.D. (2000). Frequent human genomic DNA transduction driven by LINE-1 retrotransposition. *Genome Research*, 10(4), 411–415.

Poulter, R.T.M. and Butler, M.I. (2015). Tyrosine recombinase retrotransposons and transposons. *Microbiology Spectrum*, 3(2), MDNA3–0036–2014.

Qi, X. and Sandmeyer, S. (2012). In vitro targeting of strand transfer by the Ty3 retroelement integrase. *The Journal of Biological Chemistry*, 287(22), 18589–18595.

Qi, X., Daily, K., Nguyen, K., Wang, H., Mayhew, D., Rigor, P., Forouzan, S., Johnston, M., Mitra, R.D., Baldi, P. et al. (2012). Retrotransposon profiling of RNA polymerase III initiation sites. *Genome Research*, 22(4), 681–692.

Quadrana, L., Etcheverry, M., Gilly, A., Caillieux, E., Madoui, M.-A., Guy, J., Silveira, A.B., Engelen, S., Baillet, V., Wincker, P. et al. (2019). Transposition favors the generation of large effect mutations that may facilitate rapid adaption. *Nature Communications*, 10(1), 3421.

Quenneville, S., Verde, G., Corsinotti, A., Kapopoulou, A., Jakobsson, J., Offner, S., Baglivo, I., Pedone, P.V., Grimaldi, G., Riccio, A. et al. (2011). In embryonic stem cells, ZFP57/KAP1 recognize a methylated hexanucleotide to affect chromatin and DNA methylation of imprinting control regions. *Molecular Cell*, 44(3), 361–372.

Raddatz, G., Guzzardo, P.M., Olova, N., Fantappié, M.R., Rampp, M., Schaefer, M., Reik, W., Hannon, G.J., Lyko, F. (2013). Dnmt2-dependent methylomes lack defined DNA methylation patterns. *Proceedings of the National Academy of Sciences of the United States of America*, 110(21), 8627–8631.

Rangan, P., Malone, C.D., Navarro, C., Newbold, S.P., Hayes, P.S., Sachidanandam, R., Hannon, G.J., Lehmann, R. (2011). piRNA production requires heterochromatin formation in Drosophila. *Current Biology: CB*, 21(16), 1373–1379.

Rashkova, S., Karam, S.E., Kellum, R., Pardue, M.-L. (2002). Gag proteins of the two Drosophila telomeric retrotransposons are targeted to chromosome ends. *The Journal of Cell Biology*, 159(3), 397–402.

Rebollo, R., Farivar, S., Mager, D.L. (2012a). C-GATE – Catalogue of genes affected by transposable elements. *Mobile DNA*, 3(1), 9.

Rebollo, R., Romanish, M.T., Mager, D.L. (2012b). Transposable elements: An abundant and natural source of regulatory sequences for host genes. *Annual Review of Genetics*, 46(August), 21–42.

Redelsperger, F., Cornelis, G., Vernochet, C., Tennant, B.C., Catzeflis, F., Mulot, B., Heidmann, O., Heidmann, T., Dupressoir, A. (2014). Capture of Syncytin-Mar1, a fusogenic endogenous retroviral envelope gene involved in placentation in the rodentia squirrel-related clade. *Journal of Virology*, 88(14), 7915–7928.

Roberts, D., Hoopes, B.C., McClure, W.R., Kleckner, N. (1985). IS10 transposition is regulated by DNA adenine methylation. *Cell*, 43(1), 117–130.

Rohrmann, G.F. and Karplus, P.A. (2001). Relatedness of baculovirus and gypsy retrotransposon envelope proteins. *BMC Evolutionary Biology*, 1(February), 1.

Ronsseray, S., Lehmann, M., Anxolabéhère, D. (1989). Copy number and distribution of P and I mobile elements in *Drosophila melanogaster* populations. *Chromosoma*, 98(3), 207–214.

Ros, F. and Kunze, R. (2001). Regulation of activator/dissociation transposition by replication and DNA methylation. *Genetics*, 157(4), 1723–1733.

Roulois, D., Yau, H.L., Singhania, R., Wang, Y., Danesh, A., Shen, S.Y., Han, H., Liang, G., Jones, P.A., Pugh, T.J. et al. (2015). DNA-demethylating agents target colorectal cancer cells by inducing viral mimicry by endogenous transcripts. *Cell*, 162(5), 961–973.

Rowe, H.M., Jakobsson, J., Mesnard, D., Rougemont, J., Reynard, S., Aktas, T., Maillard, P.V., Layard-Liesching, H., Verp, S., Marquis, J. et al. (2010). KAP1 controls endogenous retroviruses in embryonic stem cells. *Nature*, 463(7278), 237–240.

Rubin, G.M. and Spradling, A.C. (1983). Vectors for P element-mediated gene transfer in Drosophila. *Nucleic Acids Research*, 11(18), 6341–6351.

Ruby, J.G., Jan, C., Player, C., Axtell, M.J., Lee, W., Nusbaum, C., Ge, H., Bartel, D.P. (2006). Large-scale sequencing reveals 21U-RNAs and additional microRNAs and endogenous siRNAs in *C. elegans*. *Cell*, 127(6), 1193–1207.

Saha, A., Mitchell, J.A., Nishida, Y., Hildreth, J.E., Ariberre, J.A., Gilbert, W.V., Garfinkel, D.J. (2015). A trans-dominant form of gag restricts Ty1 retrotransposition and mediates copy number control. *Journal of Virology*, 89(7), 3922–3938.

Saito, K., Nishida, K.M., Mori, T., Kawamura, Y., Miyoshi, K., Nagami, T., Siomi, H., Siomi, M.C. (2006). Specific association of Piwi with rasiRNAs derived from retrotransposon and heterochromatic regions in the Drosophila genome. *Genes & Development*, 20(16), 2214–2222.

Saleh, A., Macia, A., Muotri, A.R. (2019). Transposable elements, inflammation, and neurological disease. *Frontiers in Neurology*, 10(August), 894.

Sandoval-Villegas, N., Nurieva, W., Amberger, M., Ivics, Z. (2021). Contemporary transposon tools: A review and guide through mechanisms and applications of *Sleeping Beauty, piggyBac* and *Tol2* for genome engineering. *International Journal of Molecular Sciences*, 22(10). https://doi.org/10.3390/ijms22105084.

Scheifele, L.Z. and Boeke, J.D. (2008). From the shards of a shattered genome, diversity. *Proceedings of the National Academy of Sciences of the United States of America*. 105(33), 11593–11594.

Schorn, A.J., Gutbrod, M.J., LeBlanc, C., Martienssen, R. (2017). LTR-retrotransposon control by tRNA-derived small RNAs. *Cell*, 170(1), 61–71.e11.

Schultz, D.C., Friedman, J.R., Rauscher 3rd, F.J. (2001). Targeting histone deacetylase complexes via KRAB-zinc finger proteins: The PHD and bromodomains of KAP-1 form a cooperative unit that recruits a novel isoform of the Mi-2alpha subunit of NuRD. *Genes & Development*, 15(4), 428–443.

Scott, E.C., Gardner, E.J., Masood, A., Chuang, N.T., Vertino, P.M., Devine, S.E. (2016). A hot L1 retrotransposon evades somatic repression and initiates human colorectal cancer. *Genome Research*, 26(6), 745–755.

Seifarth, W., Frank, O., Zeilfelder, U., Spiess, B., Greenwood, A.D., Hehlmann, R., Leib-Mösch, C. (2005). Comprehensive analysis of human endogenous retrovirus transcriptional activity in human tissues with a retrovirus-specific microarray. *Journal of Virology*, 79(1), 341–352.

Sen, S.K., Han, K., Wang, J., Lee, J., Wang, H., Callinan, P.A., Dyer, M., Cordaux, R., Liang, P., Batzer, M.A. (2006). Human genomic deletions mediated by recombination between alu elements. *American Journal of Human Genetics*, 79(1), 41–53.

Serrao, E., Ballandras-Colas, A., Cherepanov, P., Maertens, G.N., Engelman, A.N. (2015). Key determinants of target DNA recognition by retroviral intasomes. *Retrovirology*, 12(1), 39.

Serrato-Capuchina, A. and Matute, D.R. (2018). The role of transposable elements in speciation. *Genes*, 9(5). https://doi.org/10.3390/genes9050254.

Servant, G., Pinson, B., Tchalikian-Cosson, A., Coulpier, F., Lemoine, S., Pennetier, C., Bridier-Nahmias, A., Todeschini, A.L., Fayol, F., Daignan-Fornier, B. et al. (2012). Tye7 regulates yeast Ty1 retrotransposon sense and antisense transcription in response to adenylic nucleotides stress. *Nucleic Acids Research*, 40(12), 5271–5282.

Sharma, A., Larue, R.C., Plumb, M.R., Malani, N., Male, F., Slaughter, A., Kessl, J.J., Shkriabai, N., Coward, E., Aiyer, S.S. et al. (2013). BET proteins promote efficient murine leukemia virus integration at transcription start sites. *Proceedings of the National Academy of Sciences of the United States of America*, 110(29), 12036–12041.

Shirayama, M., Seth, M., Lee, H.-C., Gu, W., Ishidate, T., Conte Jr., D., Mello, C.C. (2012). piRNAs initiate an epigenetic memory of nonself RNA in the *C. elegans* germline. *Cell*, 150(1), 65–77.

Singer, T., McConnell, M.J., Marchetto, M.C.N., Coufal, N.G., Gage, F.H. (2010). LINE-1 retrotransposons: Mediators of somatic variation in neuronal genomes? *Trends in Neurosciences*, 33(8), 345–354.

Song, S.U., Gerasimova, T., Kurkulos, M., Boeke, J.D., Corces, V.G. (1994). An Env-like protein encoded by a Drosophila retroelement: Evidence that gypsy is an infectious retrovirus. *Genes & Development*, 8(17), 2046–2057.

Spaller, T., Kling, E., Glöckner, G., Hillmann, F., Winckler, T. (2016). Convergent evolution of tRNA gene targeting preferences in compact genomes. *Mobile DNA*, 7(1), 17.

Spradling, A.C., Stern, D.M., Kiss, I., Roote, J., Laverty, T., Rubin, G.M. (1995). Gene disruptions using P transposable elements: An integral component of the Drosophila genome project. *Proceedings of the National Academy of Sciences of the United States of America*, 92(24), 10824–10830.

Sugimoto, J., Sugimoto, M., Bernstein, H., Jinno, Y., Schust, D. (2013). A novel human endogenous retroviral protein inhibits cell-cell fusion. *Scientific Reports*, 3, 1462.

Sugimoto, J., Schust, D.J., Kinjo, T., Aoki, Y., Jinno, Y., Kudo, Y. (2019). Suppressyn localization and dynamic expression patterns in primary human tissues support a physiologic role in human placentation. *Scientific Reports*, 9(1), 19502.

Sultana, T., Zamborlini, A., Cristofari, G., Lesage, P. (2017). Integration site selection by retroviruses and transposable elements in eukaryotes. *Nature Reviews Genetics*, 18(5), 292–308.

Sultana, T., van Essen, D., Siol, O., Bailly-Bechet, M., Philippe, C., Aabidine, A.Z.E., Pioger, L., Nigumann, P., Saccani, S., Andrau, K.C. et al. (2019). The landscape of L1 retrotransposons in the human genome is shaped by pre-insertion sequence biases and post-insertion selection. *Molecular Cell*, 74(3), 555– 570.e7.

Sun, W., Shen, Y.-H., Han, M.-J., Cao, Y.-F., Zhang, Z. (2014). An adaptive transposable element insertion in the regulatory region of the EO gene in the domesticated silkworm, *Bombyx mori. Molecular Biology and Evolution*, 31(12), 3302–3313.

Suzuki, M., Miyahara, T., Tokumoto, H., Hakamatsuka, T., Goda, Y., Ozeki, Y., Sasaki, N. (2014). Transposon-mediated mutation of CYP76AD3 affects betalain synthesis and produces variegated flowers in four o'clock (*Mirabilis jalapa*). *Journal of Plant Physiology*, 171(17), 1586–1590.

Teysset, L., Burns, J.C., Shike, H., Sullivan, B.L., Bucheton, A., Terzian, C. (1998). A moloney murine leukemia virus-based retroviral vector pseudotyped by the insect retroviral gypsy envelope can infect Drosophila cells. *Journal of Virology*, 72(1), 853–856.

Théron, E., Maupetit-Mehouas, S., Pouchin, P., Baudet, L., Brasset, E., Vaury, C. (2018). The interplay between the argonaute proteins piwi and aub within Drosophila germarium is critical for oogenesis, piRNA biogenesis and TE silencing. *Nucleic Acids Research*, 46(19), 10052–10065.

This, P., Lacombe, T., Cadle-Davidson, M., Owens, C.L. (2007). Wine grape (*Vitis vinifera* L.) color associates with allelic variation in the domestication gene VvmybA1. *TAG: Theoretical and Applied Genetics*, 114(4), 723–730.

Thomas, J.H. and Schneider, S. (2011). Coevolution of retroelements and tandem zinc finger genes. *Genome Research*, 21(11), 1800–1812.

Tighe, P.J., Stevens, S.E., Dempsey, S., Deist, F.L., Rieux-Laucat, F., Edgar, J.D.M. (2002). Inactivation of the fas gene by Alu insertion: Retrotransposition in an intron causing splicing variation and autoimmune lymphoproliferative syndrome. *Genes and Immunity*, 3(Suppl. 1), S66–S70.

Tokuyama, M., Kong, Y., Song, E., Jayewickreme, T., Kang, I., Iwasaki, A. (2018). ERVmap analysis reveals genome-wide transcription of human endogenous retroviruses. *Proceedings of the National Academy of Sciences of the United States of America*, 115(50), 12565–12572.

Vagin, V.V., Sigova, A., Li, C., Seitz, H., Gvozdev, V., Zamore, P.D. (2006). A distinct small RNA pathway silences selfish genetic elements in the germline. *Science*, 313(5785), 320–324.

de Vanssay, A., Bougé, A.-L., Boivin, A., Hermant, C., Teysset, L., Delmarre, V., Antoniewski, C., Ronsseray, S. (2012). Paramutation in Drosophila linked to emergence of a piRNA-producing locus. *Nature*, 490(7418), 112–115.

Van't Hof, A.E., Campagne, P., Rigden, D.J., Yung, C.J., Lingley, J., Quail, M.A., Hall, N., Darby, A.C., Saccheri, I.J. (2016). The industrial melanism mutation in British peppered moths is a transposable element. *Nature*, 534(7605), 102–105.

Vigdal, T., Christopher, J., Kaufman, D., Izsvák, Z., Voytas, D.F., Ivics, Z. (2002). Common physical properties of DNA affecting target site selection of sleeping beauty and other Tc1/mariner transposable elements. *Journal of Molecular Biology*, 323(3), 441–452.

Volff, J.-N. (2006). Turning junk into gold: Domestication of transposable elements and the creation of new genes in eukaryotes. *BioEssays: News and Reviews in Molecular, Cellular and Developmental Biology*, 28(9), 913–922.

Walsh, C.P., Chaillet, J.R., Bestor, T.H. (1998). Transcription of IAP endogenous retroviruses is constrained by cytosine methylation. *Nature Genetics*, 20(2), 116–117.

Wang, H., Xing, J., Grover, D., Hedges, D.J., Han, K., Walker, J.K., Batzer, M.A. (2005). SVA elements: A hominid-specific retroposon family. *Journal of Molecular Biology*, 354(4), 994–1007.

Wang, L., Rishishwar, L., Mariño-Ramírez, L., Jordan, I.K. (2017). Human population-specific gene expression and transcriptional network modification with polymorphic transposable elements. *Nucleic Acids Research*, 45(5), 2318–2328.

Warkocki, Z., Krawczyk, P.S., Adamska, D., Bijata, K., Garcia-Perez, J.L., Dziembowski, A. (2018). Uridylation by TUT4/7 restricts retrotransposition of human LINE-1s. *Cell*, 174(6), 1537–1548.e29.

Wells, J.N. and Feschotte, C. (2020). A field guide to eukaryotic transposable elements. *Annual Review of Genetics*, 54(November), 539–561.

Wessler, S.R. (1988). Phenotypic diversity mediated by the maize transposable elements Ac and Spm. *Science*, 242(4877), 399–405.

Wicker, T., Sabot, F., Hua-Van, A., Bennetzen, J.L., Capy, P., Chalhoub, B., Flavell, A., Leroy, P., Morgante, M., Panaud, O. et al. (2007). A unified classification system for eukaryotic transposable elements. *Nature Reviews Genetics*, 8(12), 973–982.

Wolf, D. and Goff, S.P. (2009). Embryonic stem cells use ZFP809 to silence retroviral DNAs. *Nature*, 458(7242), 1201–1204.

Wolf, G., Yang, P., Füchtbauer, A.C., Füchtbauer, E-M., Silva, A.D., Park, C., Wu, W., Nielsen, A.L., Pedersen, F.S., Macfarlan, T.S. (2015). The KRAB zinc finger protein ZFP809 is required to initiate epigenetic silencing of endogenous retroviruses. *Genes & Development*, 29(5), 538–554.

Wolf, G., de Iaco, A., Sun, M.-A., Bruno, M., Tinkham, M., Hoang, D., Mitra, A., Ralls, S., Trono, D., Macfarlan, T.S. (2020). KRAB-zinc finger protein gene expansion in response to active retrotransposons in the murine lineage. *eLife*, 9. https://doi.org/10.7554/eLife.56337.

Wylie, A., Jones, A.E., Abrams, J.M. (2016). p53 in the game of transposons. *BioEssays: News and Reviews in Molecular, Cellular and Developmental Biology*, 38(11), 1111–1116.

Xiao, H., Jiang, N., Schaffner, E., Stockinger, E.J., van der Knaap, E. (2008). A retrotransposon-mediated gene duplication underlies morphological variation of tomato fruit. *Science*, 319(5869), 1527–1530.

Xie, W., Gai, X., Zhu, Y., Zappulla, D.C., Sternglanz, R., Voytas, D.F. (2001). Targeting of the yeast Ty5 retrotransposon to silent chromatin is mediated by interactions between integrase and Sir4p. *Molecular and Cellular Biology*, 21(19), 6606–6614.

Xing, J., Zhang, Y., Han, K., Salem, A.H., Sen, S.K., Huff, C.D., Zhou, Q., Kirkness, E.F., Levy, S., Batzer, M.A. et al. (2009). Mobile elements create structural variation: Analysis of a complete human genome. *Genome Research*, 19(9), 1516–1526.

Xu, W., Li, J., He, C., Wen, J., Ma, H., Rong, B., Diao, J., Wang, L., Wang, K., Wu, F. et al. (2021). METTL3 regulates heterochromatin in mouse embryonic stem cells. *Nature*, 591(7849), 317–321.

Yang, P., Wang, Y., Macfarlan, T.S. (2017). The role of KRAB-ZFPs in transposable element repression and mammalian evolution. *Trends in Genetics: TIG*, 33(11), 871–881.

Ye, J., Pérez-González, C.E., Eickbush, D.G., Eickbush, T.H. (2005). Competition between R1 and R2 transposable elements in the 28S rRNA genes of insects. *Cytogenetic and Genome Research*, 110(1–4), 299–306.

Yieh, L., Hatzis, H., Kassavetis, G., Sandmeyer, S.B. (2002). Mutational analysis of the transcription factor IIIB-DNA target of Ty3 retroelement integration. *The Journal of Biological Chemistry*, 277(29), 25920–25928.

Yigit, E., Batista, P.J., Bei, Y., Pang, K.M., Chen, C.-C.G., Tolia, N.H., Joshua-Tor, L., Mitani, S., Simard, M.J., Mello, C.C. (2006). Analysis of the *C. elegans* argonaute family reveals that distinct argonautes act sequentially during RNAi. *Cell*, 127(4), 747–757.

Yoder, J.A., Walsh, C.P., Bestor, T.H. (1997). Cytosine methylation and the ecology of intragenomic parasites. *Trends in Genetics: TIG*, 13(8), 335–340.

Zaratiegui, M. (2017). Cross-regulation between transposable elements and host DNA replication. *Viruses*, 9(3). https://doi.org/10.3390/v9030057.

Zhang, L., Beaucher, M., Cheng, Y., Rong, Y.S. (2014). Coordination of transposon expression with DNA replication in the targeting of telomeric retrotransposons in Drosophila. *The EMBO Journal*, 33(10), 1148–1158.

Zhang, H.-H., Peccoud, J., Xu, M.-R.-S., Zhang, X.-G., Gilbert, C. (2020). Horizontal transfer and evolution of transposable elements in vertebrates. *Nature Communications*, 11(1), 1362.

4

Insights Into the Evolutionary Diversity of Centromeres

Nuria CORTES-SILVA, Aruni P. SENARATNE and Ines A. DRINNENBERG
Evolution des centromères et ségrégation des chromosomes, Institut Curie, CNRS UMR3664, Paris, France

4.1. The centromere

4.1.1. *Definition and historical background*

The centromere is a specialized chromosomal region, essential for faithful transmission of genetic material into daughter cells during cell division. This region was first observed by light microscopy in dividing salamander cells and was hand-drawn by Walther Flemming in 1882 (see Figure 4.1(A)). In his drawings, Flemming represented nuclear threads (or *Mitosen* as he named) arranged in the center of the cell during *Metakinese* (or metaphase) with spindle fibers attached onto a single region (i.e. primary constriction). Furthermore, he observed that these threads were split longitudinally and he predicted that one half of this longitudinally split pair was destined to go to one daughter cell while the other half should go to the other daughter cell (Flemming 1882; Paweletz 2001). These nuclear threads were later referred to as *Chromosomen* (for "stainable bodies", now known as chromosomes) by Waldeyer in 1888 (Waldeyer 1888).

For a color version of all figures in this chapter, see http://www.iste.co.uk/richard/function.zip.

Function and Evolution of Repeated DNA Sequences,
coordinated by Guy-Franck RICHARD. © ISTE Ltd 2023.

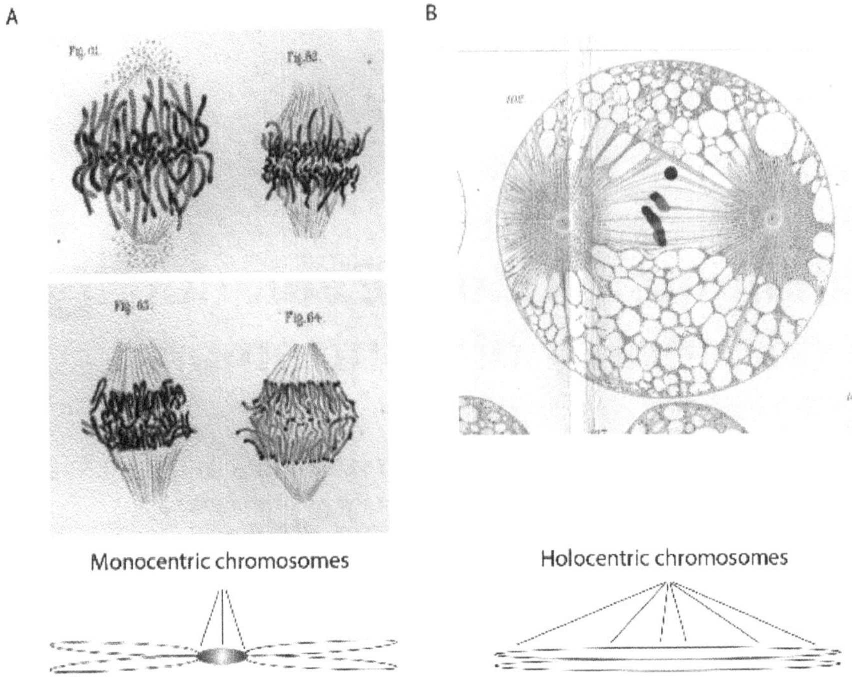

Figure 4.1. *(A) Illustrations of salamander cells with chromosomes (monocentric) in mitosis published in Walter Flemming's book Zellsubstanz, Kern und Zelltheilung ('Cell substance, nucleus and cell division') from 1882 (Flemming 1882). (B) Illustrations of cell division in the nematode Ascaris megalocephala bivalens (holocentric) drawn by Theodor Boveri in the beginning of the 20th century as part of his six comprehensive monographs Zellen-Studien ("Cell studies") (Boveri 2019) (kindly provided by Markus Engstler, Theodor-Boveri-Institut, Universität Wuerzburg). The drawing shows chromosomes, spindles, two centrosomes and two centrioles within each centrosome*

It was in 1936 that Cyril Dean Darlington used the term "centromere" to designate the primary constriction of a chromosome or "spindle fiber attachments chromosome" (Darlington 1936b). For Darlington, "the centromeres are specialized chromomeres and determine chromosome movement", as they "arrange themselves in an equatorial plate across the spindle. They are repelled by the centrosomes and by one another" (Darlington 1936b). Furthermore, by studying meiosis in triploid plants (organisms with three homologous chromosomes), he concluded that centromeres "modify the structural orientation of the spindle, and at anaphase they control the spindle [...]", in these ways "the centromeres are analogous to the centrosomes, and for this reason I have adopted the term centromere to describe them" (Darlington 1936a).

Decades later, studies aimed to describe the genetic nature of the centromere. In 1970, Mary Lou Pardue and Joseph G. Gall applied hybridization techniques using radioactive nucleic acid probes to cytologically localize specific sequences in the genome and found that sequences of satellite DNA were located at the centromeres of mouse chromosomes. Further studies in other organisms found that satellite DNA is also present at the centromeres of *Rhynchosciara hollaenderi* (sciarid fly), *Triturus viridescens* and *Drosophila melanogaster*. In addition to these hybridization techniques, modifications of Giemsa staining, notably G-banding (Caspersson et al. 1970; Arrighi and Hsu 1971; Drets and Shaw 1971; Seabright 1972) and C-banding (McKay 1973) allowed the distinction between heterochromatic regions of chromosomes (G-banding) and constitutive heterochromatin present at centromeres (C-banding). These techniques were also widely used to count and characterize the visual appearance of chromosomes, for example, by distinguishing chromosomes based on their distribution of the C band position.

Even though the word "centromere" from the Greek *kentron* for "center" and *meros* for "part" (McKay 1973) implies a position at the center of the chromosomes, this region can be located at diverse positions along a chromosome. Namely, in metacentric chromosomes the centromere is located in the middle, in acrocentric chromosomes the centromere is positioned closer to one end thus creating a short arm and a long arm and in telocentric chromosomes the centromere is terminally positioned, giving rise to a single chromosomal arm (Musacchio and Desai 2017). These distinctions apply to monocentromeres, which is one of two types of centromeric architectures that will be described in the following section (see Figure 4.1).

4.1.2. *Two main types of centromeric architectures*

Monocentromeres are restricted to one region on each chromosome. This architecture is found in most eukaryotes and has first been visualized and described as primary constrictions in Flemming's early drawings of salamander cell mitosis (Flemming 1882) (see Figure 4.1(A)). Only 6 years later, Theodor Boveri published an illustration of meiotic cells of the parasitic nematode Ascaris, a model that he studied together with his wife Marcella O'Grady. As opposed to Flemming's drawings, his illustrations depicted spindle fiber attachments that extended along the entire length of Ascaris chromosomes (Boveri 2018) (see Figure 4.1(B)). These two independent observations from the late 19th century can collectively be considered as the first evidence proving that centromeres can indeed exhibit different morphological forms. It was however only decades later that this difference in spindle fiber attachment came to be acknowledged and that the latter centromere type (i.e. with chromosome-wide spindle fiber attachments) was termed as the holocentromere (or holokinetochore) by Franz Schrader (Schrader 1935). Monocentromeres and holocentromeres are the two common centromeric architectures found among eukaryotes today.

In this chapter, we will discuss several types of mono- and holocentric architectures that have been described. In addition to a brief overview on monocentric architectures in common animal and plant model organisms, we will mainly focus on the diversity of monocentromeres in fungi. Furthermore, we will review known holocentric architectures, in particular those which have been analyzed using modern microscope-based and genomic approaches.

4.2. Monocentromeres

4.2.1. *The diversity of monocentric architectures across fungi*

With centromeres of more than 60 fungal species analyzed, centromeric organizations and regulations display a large degree of diversity across fungi (see Figure 4.2). The simplest monocentromere is the point centromere, which evolved in the Saccharomycetaceae clade of the budding yeasts (Saccharomycotina), containing the model yeast *Saccharomyces cerevisiae* (Dujon 2010). The point centromere consists of unique short DNA sequences and captures a single microtubule (Peterson and Ris 1976). The centromeric DNA (CEN) of *S. cerevisiae* was isolated by Clarke and Carbon using a technique called overlap hybridization screening or chromosome "walking" (Chinault and Carbon 1979). In brief, the authors first cloned fragments containing two centromere III-linked genes (*LEU2* and *CDC10*) located on opposite sides of *CEN3* (chromosome III centromeric DNA) (Clarke and Carbon 1980a). These genes were then used to "walk" through the DNA sequence in between. Then, the isolated DNA sequences were reintroduced into yeast using a shuttle plasmid carrying a yeast replication origin to test which of these sequences conferred stability to the shuttle vector over mitotic and meiotic divisions. Using this methodology, Clarke and Carbon (1980a) narrowed down the region to a segment of 1.6 kb containing the *CEN3*.

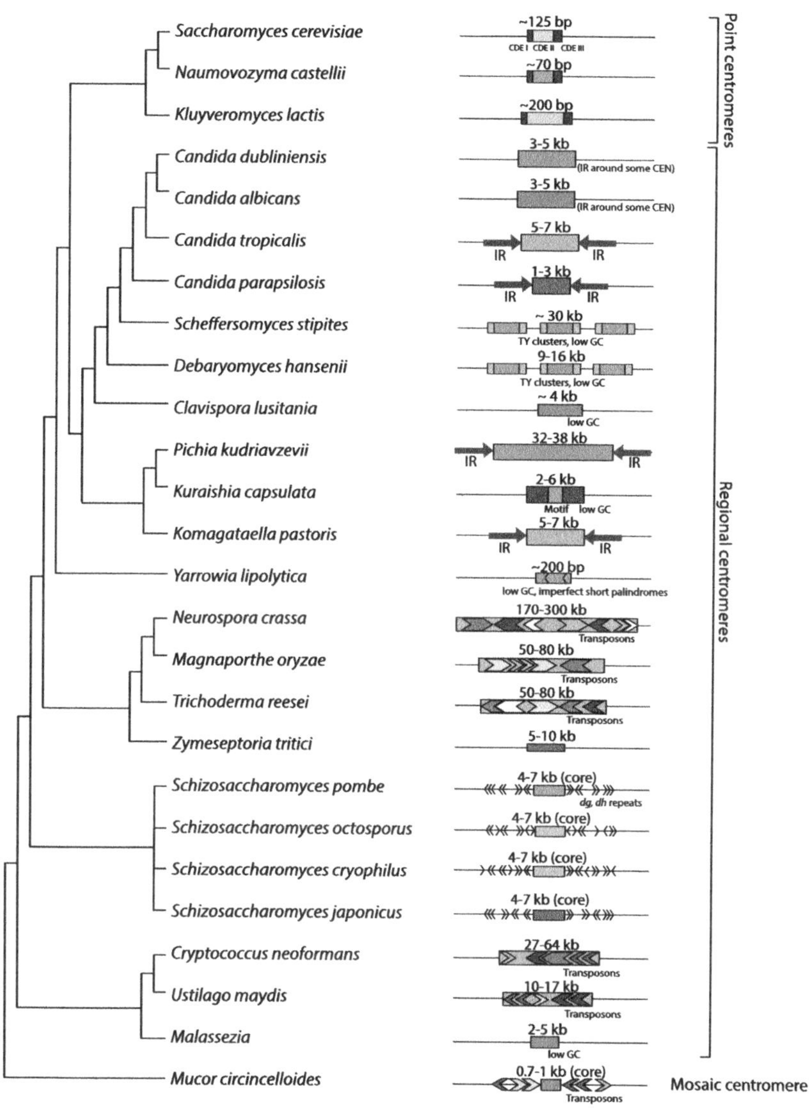

Figure 4.2. *Phylogeny of several fungal organisms along with their respective centromeric architectures. Centromere lengths are not drawn to scale*

Later studies showed that the minimal functional budding yeast centromere encompasses three conserved DNA elements (CDE): CDEI (8 bp palindromic sequence), CDEII (78–86 bp) and CDEIII (26 bp), giving a total of ~120 bp sequence (Fitzgerald-Hayes et al. 1982; Pluta et al. 1995; Clarke 1998). Among these, the 8 bp CDEI motif is also found in the promoters of a number of genes and bound by the transcriptional activator Cpf1 (Bram and Kornberg 1987; Baker et al. 1989; Cai and Davis 1990; Mellor et al. 1990). CDEIII is bound by the point centromere-specific CBF3 protein complex which is absolutely essential for centromere function. In fact, a single base mutation in the CCG motif of CDEIII impairs CBF3 binding, thereby abolishing centromere function (Lechner and Carbon 1991). CDEII is an AT-rich sequence that wraps the nucleosome containing the centromere-specific histone H3 variant CenH3 (Allshire and Karpen 2008) (first described as centromeric protein A [CENP-A] in humans; Earnshaw and Rothfield 1985). DNA sequence recognition by the point centromere-specific protein complexes contributes to the genetic identity of this type of centromere, enabling this sequence to be both necessary and sufficient for chromosome segregation.

Additional species of the Saccharomycetaceae clade contain CDE-like elements often at syntenic locations of their chromosomes (Gordon et al. 2011). The major difference among their centromere structures is the length of CDEII that can be variable among species. The point centromere organization is probably an evolutionarily derived state from a regional centromeric organization (see below) and is likely to have originated once in a common ancestor of all Saccharomycetaceae. While the CDE-like organization can be found in other Saccharomycetes, an unconventional point centromere has been recently described in *Naumovozyma castellii* and *Naumovozyma diarenensis*. These centromeres were found in a different syntenic location suggesting a unique evolutionary origin. Nevertheless, these unconventional point centromeres are still bound by the CBF3-complex, which does however recognize diverged centromere DNA sequences (Kobayashi et al. 2015).

In contrast to the genetically defined point centromere, epigenetic factors are crucial for the identity and inheritance of regional centromeres (Allshire, Karpen 2008). Regional centromeres capture multiple microtubules and are found in most other eukaryotes including the related budding yeast clades Pichiaceae and CUG-Ser1. The CUG-Ser1 clade within the Saccharomycotina contains many *Candida* and other species, distinguished by a coding change of the CUG codon, from Leucine to Serine (Ohama et al. 1993). The centromeres of *Candida albicans* and *Candida dubliniensis* are described as "small regional". They are located in large open reading frame (ORF)-free regions of 4–18 kb, which includes a 3–5 kb core region of CenH3-containing nucleosomes. While synteny is conserved among species, centromere sequences are diverged sharing no common sequence motif or repeats either within one organism or between them. On a few chromosomes, centromeres are flanked by small inverted repeats (IRs) and long terminal repeats

(LTRs) (Sanyal et al. 2004; Mishra et al. 2007; Padmanabhan et al. 2008). In contrast, centromeres of chromosomes in species belonging to the *Candida parapsilosis* sensu lato clade and *Candida tropicalis* are all surrounded by IRs (Chatterjee et al. 2016; Ola et al. 2020). Based on the phylogenetic relationship among *Candida* species, it has been suggested that the ancestral centromere of this clade was flanked by pericentric IRs, while these were lost in *C. albicans* and *C. dubliniensis* during evolution (Chatterjee et al. 2016). A few centromeres in the more distantly related species of the CUG-Ser1 clade have also been analyzed. Centromeres in *Clavispora lusitania* were found to be small (around 4 kb), low in GC-content with no sequence conservation (Kapoor et al. 2015). Regions with low GC-content also mark putative centromeres in *Debaryomyces hansenii* and *Scheffersomyces stipites* (Lynch et al. 2010), which however consist of clusters of retrotransposons in both organisms (Lynch et al. 2010; Coughlan and Wolfe 2019).

Centromeres of organisms belonging to the Pichiaceae clade including *Pichia kudriavzevii* and *Komagataella pastoris* are also flanked by IRs (Douglass et al. 2018; Coughlan and Wolfe 2019). In contrast, similar to the situation of *C. albicans* and *C. dubliniensis*, IRs surrounding centromeres have likely been lost in the related *Kuraishia capsulata*, where centromeres are located in 2–6 kb regions with low GC-content and a 200 bp motif that is conserved across some chromosomes (Morales et al. 2013).

Insights into centromere structures of organisms ancestral to the Pichiceae, CUG-Ser1 and Saccharomycetaceae clades are sparse. An exception is *Yarrowia lipolytica* in which centromeres also coincide with regions of poor GC-content and encompass 17 to 21 bp long imperfect palindromes (Yamane et al. 2008). Nevertheless, to get insights into the ancestral centromere structure of Saccharomycotina centromeres, additional outgroups will have to be determined.

Centromere structures in most other fungal organisms are usually longer in comparison to those in the budding yeasts. Exceptions are those in the basidiomycete Malassezia that possesses highly AT-rich short regional centromeres of about 2–5 kb (Sankaranarayanan et al. 2020) and the ascomycete *Zymeseptoria tritici*, which contains 5–10 kb CenH3-rich regions without distinct sequence patterns (Schotanus et al. 2015). As in most other eukaryotes with regional centromeres, the centromeric core of fungal regional centromeres is flanked by regions of heterochromatin, named pericentric heterochromatin (with the exception of those regional centromeres of the Saccharomycotina that have lost the lysine methylation machinery (Malik and Henikoff 2009)).

The centromere of the fission yeast, *Schizosaccharomyces pombe*, is among the first regional centromeres to be characterized. This regional centromere encompasses a ~40–110 kb region which includes a non-repetitive central core region (4–7 kb) flanked by repetitive DNA (Clarke et al. 1993). The heterochromatic pericentric regions consist of dg and dh classes of repeats (Clarke and Carbon 1980b). The central core region functions as the binding site for centromere-specific proteins including CenH3 (Polizzi and Clarke 1991; Thakur et al. 2015), while the flanking repeat regions are involved in the proper recruitment of cohesin for binding sister chromatids (Bernard et al. 2001; Allshire and Ekwall 2015). Additional species of the Schizosaccharomyces including *S. octosporus*, *S. japonicus* and *S. cryophilus* have also been studied, and their genomes have been sequenced. While the synteny around centromeric regions flanked by repeat elements is conserved, the central core is not conserved and repeat structures and arrangements have diverged between species (Rhind et al. 2011).

Long regional centromeres, rich in transposable elements, have also been described in other organisms belonging to the Ascomycota and the Basidiomycota. Among the Ascomycota, *Neurospora crassa* contains 170–300 kb long regional centromeres with high AT-content and enriched in inactive retrotransposons. CenH3 was found to be enriched in a centromeric core that is surrounded by 2–45 kb regions of pericentric heterochromatin (Smith et al. 2011). The assembly of centromeric and pericentric regions in *N. crassa* was aided by the presence of repeat-induced point (RIP) mutation, which affects duplicated sequences as short as 150 bp leading to centromere sequence divergence (Selker 1990; Gladyshev and Kleckner 2016). Organisms closely related to *Neurospora* have shorter regional centromeres. For instance, *Magnaporthe oryzae* contains AT-rich centromeres with active and inactive retrotransposons ranging from 57 to 109 kb (Thon et al. 2006; Yadav et al. 2019).

Among the Basidiomycota, centromeres in the *Cryptococcus* species complex have been analyzed by profiling centromere-associated proteins CenH3 and CENP-C (Loftus et al. 2005; Janbon et al. 2014; Yadav et al. 2018). All species were found to contain regional centromeres featuring retrotransposons of the same kind or their remnants. Interestingly, centromeres of RNAi-deficient species of this complex were shorter than those of RNAi-proficient species and contained no full-length transposable elements. The authors proposed a model in which the loss of RNAi (and DNA methylation) ancestral to some species enabled the transient activation and amplification of retroelements that integrated into the genome including the centromeres. This triggered increased recombination (facilitated by the absence of DNA methylation) particularly at the centromeres where those elements are in close proximity. Increased recombination in turn shortens the centromeres and renders retroelements inactive. They further expanded their analyses to the *Ustilago* species complex, which is the only other group of basidiomycetes known to contain species that have lost RNAi and DNA methylation. They found that putative centromeric regions are also shorter in RNAi-deficient *Ustilago* species in contrast to RNAi-proficient ones (Yadav et al. 2018).

Centromere structures in organisms belonging to clades ancestral to Ascomycota and Basidiomycota have only been analyzed in the early diverging subphylum Mucoromycotina. Interestingly, organisms belonging to this group of species have also lost the centromere-specifying factor CenH3 which is otherwise conserved in most other eukaryotes (Hooff et al. 2017; Navarro-Mendoza et al. 2019). The loss of CenH3 in Mucoromycotina thus raises the question of how centromeres are defined and organized in these organisms. In other organisms, CenH3 specifies the site for the assembly of the kinetochore protein complex, which in turn mediates the attachment of chromosomes to spindle microtubules (Cheeseman 2014). A recent study visualized and profiled kinetochore components along chromosomes in *Mucor circinelloides,* an opportunistic human pathogen that causes an infectious disease called mucormycosis. By expressing fluorescence-fusion proteins, the authors found that both outer kinetochore components, Mis12 and Dsn1 and the inner kinetochore component CENP-T localized to a single chromosomal region leading to the conclusion that *M. circinelloides* chromosomes are monocentric. Pulling down Mis12- and Dsn1-associated centromeric DNA further revealed that *M. circinelloides* centromeres display features of both point and regional centromeres. Indeed, *M. circinelloides* centromeres have short kinetochore-binding regions, a DNA sequence motif specific to centromeres and an AT-rich core region, features similar to point centromeres. On the other hand, the pericentromeres are also associated with retrotransposons reminiscent to regional centromeres. Owing to these mixed features, the authors named the *M. circinelloides* centromere as a "mosaic centromere". Moreover, other histone variants did not show centromere-specific binding. Thus, it appears that the function of CenH3 at centromeric chromatin has not been replaced by other histone variants in this organism (Navarro-Mendoza et al. 2019). Future studies on *M. circinelloides* will aim to determine the role of its centromere-specific DNA motif in order to reveal whether or not *M. circinelloides* centromeres follow a genetic definition.

In conclusion, fungi present a range of different centromeric architectures. Furthermore, even the regulation of centromere function differs among organisms of this kingdom. To get further insights into this diversity, more recently developed approaches including Hi-C methods in addition to standard genome sequencing efforts can increasingly be used in future studies to identify centromeres as regions with the strongest inter-chromosomal interactions (Marie-Nelly et al. 2014).

Hi-C-based methods will be faster and more accessible for organisms in which molecular-genetic tools including endogenous tagging of CenH3 or transformations are limited. Insights into the organization of fungal centromeres, in particular those represented by only a few organisms as well as those belonging to basal fungal clades, will enable us to trace the evolutionary trajectories of centromere divergence and provide clues into the ancestral organization of centromeres in fungi.

4.2.2. *Animal and plant models contain long repetitive regional centromeres*

The regional centromere is the most prevalent centromere organization across eukaryotes including well-established model organisms. Known regional centromeres of animals and plants consist of large stretches of repetitive DNA in the range of kilobases to megabases and capturing several microtubules (Manuelidis and Wu 1978; Copenhaver et al. 1999; Sun 2003; Aldrup-Macdonald and Sullivan 2014). The long, repetitive nature of animal and plant regional centromeres poses significant challenges to genome assembly projects leading to assembly gaps in centromeric regions. Similar to the shorter regional centromeres found in fungi, regional centromeres in animals and plants are thought to be epigenetically defined by the presence of the centromere-specifying epigenetic marker protein CenH3 (Allshire and Karpen 2008). Alternatively, a genetic basis of centromere identity in most organisms has also been proposed through the recognition of tertiary DNA structures by the CenH3 incorporation machinery, with the centromere-specific incorporation of CenH3 itself being a by-product of this process (Kasinathan and Henikoff 2018).

Human centromeric core DNA encompasses AT-rich alpha-satellite repeats (Jones and Corneo 1971; Jones et al. 1974; Manuelidis 1978b, 1978a) based on ~171 bp monomer repeat unit arranged in a head-to-tail fashion (Manuelidis and Wu 1978; Willard 1985; Willard and Waye 1987). Subsequent studies showed that these monomers assemble into a higher-order-repeat (HOR) pattern, which are up to >90% identical to each other at the center of the satellite repeat array (Waye and Willard 1989). Flanking the HOR array, which is bound by kinetochore-specific proteins, monomers are more randomly arranged. The length of centromeric DNA varies between chromosomes, creating chromosome-specific arrays of 0.25–5 mb and representing 3–5% of the chromosome (Willard 1985; Choo et al. 1991; Aldrup-Macdonald and Sullivan 2014; Fukagawa and Earnshaw 2014; Kursel and Malik 2016). Due to high sequence similarities within the satellite repeat arrays, it has been challenging to assemble and characterize the HOR structures of human centromeres. Nevertheless, more recent efforts using long-read sequencing completed centromeric assemblies for chromosomes 8, X and Y (Schueler et al. 2001; Nusbaum et al. 2006; Miga et al. 2014; Jain et al. 2018).

Centromere structures in several additional animal model organisms have also been investigated. Fly centromeres for instance are associated with large regions of pericentric heterochromatin ranging from 200 to 500 kb and map to regions within large blocks of tandem repeats (Le et al. 1995; Blower et al. 2002; Sun 2003; Garavís et al. 2015). An early study on an X-derived minichromosome mapped a minimal centromeric sequence to a 420 kb region containing 5–12 bp tandem repeats interspersed by regions of complex sequences (Sun et al. 1997; Sun 2003). More recently, combining long-read sequencing, chromatin immunoprecipitation (ChIP) for the centromeric histone CenH3 and high-resolution chromatin fiber imaging allowed for the determination of fly centromere architecture. This study found that CenH3 primarily occupies islands of complex DNA enriched in retrotransposable elements including non-LTR Jockey elements that are flanked by large blocks of satellites. The authors thus suggest that retroelements specify centromere function in flies (Chang et al. 2019). In contrast, another study investigating sequences identified using CenH3-associated immunoprecipitations suggests that simple satellites are the functional centromeric sequences in *D. melanogaster* (Talbert et al. 2018).

Investigations of centromeric structures in plants have lagged behind those performed in animals and yeasts. Nevertheless, insights into centromere sequence and organization have been obtained in particular for the model organisms *Arabidopsis thaliana* and Graminaceae species. As in humans, known plant centromeres are composed of megabase-sized arrays of tandem repetitive DNA sequences. In addition, transposable elements are also abundant in their centromeric regions. In *A. thaliana*, the five chromosomes contain centromeres composed of a 180 bp centromere-specific satellite that was found to be enriched in CenH3 immunoprecipitations (Nagaki et al. 2003). In addition, non-centromere specific repeats were also found including fragments of Athila retrotransposon elements (Arabidopsis Genome Initiative 2000). Related *Arabidopsis* species also contain satellites that can be different in length and sequence to the 180 bp repeat (Nagaki et al. 2009). Sequences of centromeres in Graminaceae are variable in both size and organization. Nevertheless, the repeats can be organized into two groups: satellites and centromere-specific retrotransposons. Centromeres in *Oryza sativa* chromosomes, for instance, contain 155 bp CentO satellites arranged into a total array size of 65 kb to 2 Mb (Cheng et al. 2002). Satellites have also been reported at the centromeres of other Graminacaea including maize (CentC, 156 bp) (Ananiev et al. 1998; Zhong et al. 2002) and sugarcane (SCEN, 140 bp) (Nagaki et al. 1998; Nagaki and Murata 2005). The centromere-specific retrotransposons (CR) in

Graminacaea belong to the Ty-gypsy family of retrotransposons that are flanked by LTRs on each side. These retrotransposons include CRR elements in rice (Dong et al. 1998; Nonomura and Kurata 1999; Cheng et al. 2002; Nagaki et al. 2004), CRM elements in maize (Ananiev et al. 1998; Zhong et al. 2002) and CRS elements in sugarcane (Nagaki and Murata 2005). Given the abundance of solo LTRs and the lack of orthology among anchored LTRs in related species, CR elements appear to be active components of the centromere of Graminacaea (Ma and Jackson 2006).

Note that some regional centromeres are composed of complex non-repetitive DNA. In fact, chickens (Shang et al. 2010), horses (Piras et al. 2010) or orangutans (Locke et al. 2011) were found to contain chromosomes with non-repetitive regional centromeres that have been hypothesized to represent more recent evolutionary states (Nergadze et al. 2018).

4.3. Holocentromeres

In contrast to the monocentric architecture, several organisms have holocentromeres in which the centromeric microtubule-binding regions span in large portions or even the full length of the chromosomes (see Figure 4.1). This type of centromeric architecture was first described by Franz Schrader based on cytological analyses of chromosome segregation in Hemiptera (true bugs) during cell division. From his observations, he concluded that "in Hemiptera [...] the entire pole-ward surface of the chromosome serves as a base for the half spindle component" (Schrader 1935).

Since then, the criteria used to characterize holocentric chromosomes have been based on cytological remarks (Hughes-Schrader and Ris 1941).

These include the following:

– The lack of a primary constriction on mitotic chromosomes.

– The migration pattern and shape of sister chromatids during cell division. Indeed, during anaphase, holocentric sister chromatids migrate to the spindle poles as parallel lines (since microtubule-pulling forces are distributed along the chromosomes), while monocentric sister chromatids adopt a "V"-shape since microtubule-pulling forces are located at one chromosomal region.

– The ability to recover from chromosomal breaks; for example, those induced by X-ray irradiation. Holocentric organisms could have a greater ability to resist the harmful effects of chromosomal breakage than monocentric organisms. This is due to the fact that holocentric chromosomal fragments resulting from breakage events have a high probability of possessing at least one centromeric site, therefore enabling its continued mitotic potency over cell cycles. This is in contrast to the deleterious loss of chromosomal fragments in the monocentric condition owing to the presence of only one centromere site.

– The presence of structural adaptations during meiosis. Franz and Sally-Hughes Schrader noted that in spermatocyte chromosomes of stink bugs (sub-order: Heteroptera), there is a temporary restriction of kinetic activity, where the diffuse spindle seen at mitosis is instead confined to a terminal region of the chromosome at meiosis, thus temporarily behaving like a monocentric chromosome. Additionally, when studying the chromosome behaviors in some coccid species (scale insects) collected in Mexico, Sally Hughes-Schrader noted the apparent loss of all cohesion on sister chromatids in meiosis I (Hughes-Schrader and Ris 1941; Hughes-Schrader 1944) – a feature that is usually not observed until meiosis II. These observations are examples of meiotic adaptations in holocentric organisms, the molecular mechanisms of which are currently not well understood.

Holocentricity is not a rare event during evolution. Instead, it has been described at least 13 diverse groups of animals and plants (Melters et al. 2012). The nested occurrence of these holocentric organisms within larger monocentric groups indicates that the holocentric architecture is the derived state from monocentric ancestors. However, as cytologically based methods have been the main driver to identify holocentric species, it is possible that the prevalence of holocentromeres is underestimated as there are many species that are difficult to study cytologically. On the contrary, limited often by the small size of chromosomes in certain species and poor cytological resolution of early light microscopes, to this day, much remains to be verified about the inferred holocentricity in some clades. Most recently, genomic approaches and advanced microscopy have been applied for the study of holocentric architectures at the molecular level in a few organisms.

4.3.1. *Nematodes*

The holocentric organism best characterized is the nematode *Caenorhabditis elegans*. Several of the earlier described characteristics associated with the holocentric architecture in Hemiptera (i.e. lack of a primary constriction and presence of diffuse microtubule attachments on mitotic chromosomes as well as the mitotic stability of radiation-induced chromosomal fragments) were first demonstrated in *C. elegans* using light and electron microscopy (Albertson and Thomson 1982). Later on, along with the discovery of homologs of the centromere-specific histone CenH3 in various eukaryotes, immunostaining using an antibody against the *C. elegans* CenH3 protein revealed a localization pattern as longitudinal bands that span the polar length of condensed *C. elegans* chromosomes (Buchwitz et al. 1999). Two more recent studies have profiled CenH3 localization patterns along *C. elegans* chromosomes using DNA microarray technology and high-throughput sequencing, respectively (Gassmann et al. 2012; Steiner and Henikoff 2014) (see Figure 4.3, top).

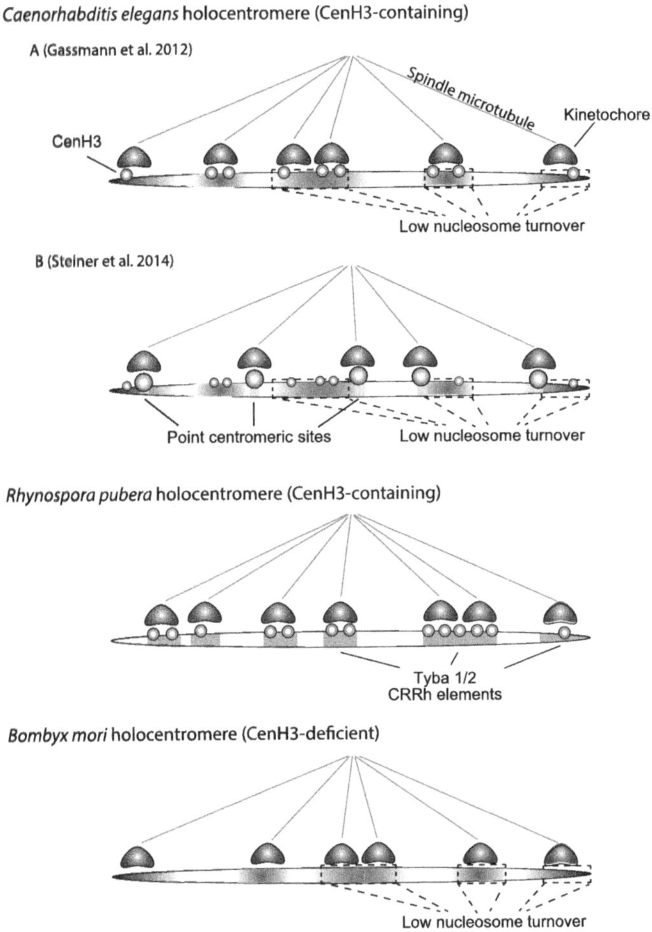

Figure 4.3. *Schematics of models for different holocentric architectures of C. elegans (top two) (Gassmann et al. 2012; Steiner and Henikoff 2014), R. pubera (middle) (Marques et al. 2015) and B. mori (bottom) (Senaratne et al. 2020). Note that the number of CenH3 molecules, kinetochores and microtubules was chosen arbitrarily. Larger circles of CenH3 in C. elegans Model B indicate high abundance CenH3 peaks in ChIP-seq datasets*

Among those, the first study by the Desai lab that was conducted in the later embryonic stages revealed that CenH3 incorporates at low abundance into large domains of complex DNA which encompass half of the genome. Moreover, the authors found that those centromeric domains are enriched over transcriptionally silent regions of low RNA polymerase II occupancy levels leading to the hypothesis that centromere formation requires low chromatin activity (Gassmann et al. 2012). The second study by the Henikoff lab added that CenH3 incorporation into these transcriptionally silent chromatin regions is likely a result of low nucleosome turnover, which delimited CenH3 to these sites by way of exclusion from regions with high nucleosome turnover (Steiner and Henikoff 2014). In very early, transcriptionally inactive embryonic stages, this simple inverse relationship to transcription might however be more complex. Here, cues transmitted from the female germline related to its transcriptional status may contribute to determining centromeric profiles in these early embryonic stages (Gassmann et al. 2012).

In addition, the study by the Henikoff lab also identified around 700 discrete and dispersed loci along the chromosomes with high CenH3 abundance. The chromatin organization of and around these CenH3 peaks resembles the organization of the budding yeast point centromere, where one CenH3-containing nucleosome is flanked by two well-positioned H3-containing nucleosomes. The authors therefore conclude that the CenH3 peaks rather than the CenH3 domains represent the actual centromeric sites in *C. elegans* (Steiner and Henikoff 2014). The architecture of the *C. elegans* holocentromere and the relevance of centromeric domains or peaks to contribute to functional centromeric sites in this organism is still an open question in the field.

4.3.2. *Plants*

In plants, holocentric chromosomes were found to have evolved at least four times in lineages of flowering plants (Angiosperms) (Melters et al. 2012). Fluorescence analyses visualizing CenH3 in two monocots belonging to the Juncaceae family (*Luzula elegans* and *Luzula nivea*) along with electron microscopy analyses of their chromosomes revealed band-like distributions of CenH3 signal along the longitudinal groove of sister chromatids (Nagaki et al. 2005; Heckmann et al. 2011).

The holocentromere architecture is also characterized in one member of the monocot Cypereaceae family, the beak-sedge (*Rhyncospora pubera*). In contrast to *C. elegans*, immunofluorescence-fluorescence *in situ* hybridization (IF-FISH) and CenH3-specific chromatin immunoprecipitation followed by high-throughput sequencing (ChIP-seq) in *R. pubera* revealed that centromeric sites are located in repeat-rich regions. The authors identified 172 bp long satellite DNAs (Tyba-1 and Tyba-2 elements) that are arranged into 3–16 kb long arrays. In addition, LTR retrotransposons belonging to the Ty3/gypsy clade (termed as CRRh) were also found to be associated with CenH3 containing chromatin (Marques et al. 2015) (see Figure 4.3, middle). To what extent those repeats play an active role in recruiting centromeric proteins is still unclear.

4.3.3. *Insects*

The presence of a holocentric architecture is common in insects. Analyses of insect chromosomes over many decades revealed that holocentric chromosomes have evolved in several insect orders including the Hemiptera (true bugs), Phthiraptera (lice), Lepidoptera (butterflies and moths), Dermaptera (earwigs) and Odonata (dragonflies). The phylogenetic distribution of these orders among other monocentric insect orders supports at least four independent transitions from mono- to holocentric chromosomes in insects. The recurrent rise to holocentric chromosomes together with the availability of genomic or transcriptomic data from insects belonging to multiple mono- and holocentric orders allows for the application of comparative genomics to test for genetic changes associated with the two types of centromeric architectures. Taking advantage of this opportunity, a recent computational survey across several insect orders revealed the recurrent loss of CenH3 in insects belonging to several holocentric orders listed above. In contrast, CenH3 homologs could be identified in monocentric insects including Diptera (flies and mosquitoes), Coleoptera (beetles), Hymenoptera (wasps, bees, and ants), Blattodea (cockroaches), Phasmatodea (stick insects), Orthoptera (crickets) and Ephemeroptera (mayflies). Besides the analyses of insect genomes, transcriptome assemblies of additional mono- and holocentric insects further supported the presence of CenH3 in several monocentric insects, but its loss in several holocentric insects. The loss of CenH3 in some holocentric insects was unlikely due to an insufficient coverage of the transcriptome assemblies nor to a lower CenH3 expression compared to monocentric ones. In fact, a similar or even higher number of predicted proteins in the holocentric assemblies supported the fact that transcriptome coverages were comparable in both types of insects. Moreover, CenH3 transcripts were not in low abundance in monocentric insects. Taken together, these findings showed that CenH3 was lost multiple times in insects and that this loss is associated with the presence of holocentric chromosomes (Drinnenberg et al. 2014).

The loss of CenH3 in holocentric insects raises the question of how centromeres are specified in these organisms. Interestingly, despite the loss of CenH3, kinetochore components that are conserved in other eukaryotes continue to be present in these CenH3-lacking insects and can therefore be used to map centromeres (i.e. sites for kinetochore assembly) along chromosomes (Cortes-Silva et al. 2020). Using cell lines derived from the silk moth (*Bombyx mori)* as a holocentric insect model system and leveraging microscopy and genomics approaches, as well as antibodies against the conserved kinetochore component CENP-T, a recent study has profiled and characterized holocentromere formation along *B. mori* chromosomes. It was observed that the CENP-T-specific IF signal formed broad localization patterns that extend along the entire polar length of mitotic sister chromatids, a pattern that has recurrently been observed for kinetochore staining in other holocentric organisms (Buchwitz et al. 1999). Furthermore, ChIP-seq analyses targeting CENP-T in unsynchronized *B. mori* cell populations were performed to obtain genomic resolution maps of CENP-T distribution. Here, the mapping results revealed broad domains of enrichment for CENP-T protein genome-wide. CENP-T domains vary considerably in size, with a median size of ~36 kb and at times extending up to approximately 1 Mb. Domains are scattered evenly across the genome and are relatively depleted in genes. Based on cut-offs, it was estimated that ~54% of the *B. mori* genome is covered by CENP-T domains. Considering that the ChIP-seq datasets represent entire cell populations, this implies that approximately half of the *B. mori* genome in a single cell is permissive for CENP-T localization. Further analyses of the CENP-T domains revealed that those are not specifically enriched for repeats, instead these domains are as repeat-rich as the rest of the genome.

Furthermore, genomics approaches to profile the underlying chromatin environment revealed several correlations in the CENP-T distribution with respect to chromatin marks governing chromatin activity status. CENP-T profiles were anticorrelated with active histone modifications, RNA polymerase II and nucleosome turnover, but positively correlated with silent histone marks. These results supported the hypothesis that *B. mori* centromere profiles are being shaped by chromatin activity (see Figure 4.3, bottom). To test this hypothesis, transcriptional perturbation experiments were subsequently done to demonstrate that centromeres become excluded from active chromatin regions but can form de novo in regions where chromatin activity is low. Taken together, these results revealed a new mode of centromere regulation that is dependent on the chromosome-wide chromatin landscape rather than the presence of CenH3 (Senaratne et al. 2020). Furthermore, this CenH3-independent mode of centromere formation in Lepidoptera is conceptually similar to the CenH3-dependent mode of centromere formation along large chromosomal domains anticorrelated to active nucleosome turnover in *C. elegans* (Gassmann et al. 2012). This resemblance thus highlights the relevance of chromatin composition for holocentromere organization in the presence or absence of CenH3.

4.4. Open questions

What drives centromere divergence among organisms? To what extent is centromere drive implicated in changes in centromere length and sequence?

How did the budding yeast point centromere emerge?

What enabled or drove the transition from mono- to holocentric chromosomes?

How was the centromere in the last eukaryotic common ancestor organized and defined?

4.5. Acknowledgments

NCS receives salary support from Sorbonne Université (2781/2017) and the Fondation ARC pour la recherche sur le cancer (ARCDOC42020020001567). APS receives salary support from Institut Curie and the Fondation Recherche Médicale. IAD receives salary support from the CNRS. This work is supported by the Labex DEEP ANR-11-LABX-0044 part of the IDEX Idex PSL ANR-10-IDEX-0001-02 PSL, the Institut Curie and the ERC (CENEVO-758757).

4.6. References

Albertson, D.G. and Thomson, J.N. (1982). The kinetochores of *Caenorhabditis elegans*. *Chromosoma*, 86(3), 409–428.

Aldrup-Macdonald, M.E. and Sullivan, B.A. (2014). The past, present, and future of human centromere genomics. *Genes*, 5(1), 33–50.

Allshire, R.C. and Ekwall, K. (2015). Epigenetic regulation of chromatin states in *Schizosaccharomyces pombe*. *Cold Spring Harbor Perspectives in Biology*, 7(7), a018770.

Allshire, R.C. and Karpen, G.H. (2008). Epigenetic regulation of centromeric chromatin: Old dogs, new tricks? *Nature Reviews. Genetics*, 9(12), 923–937.

Ananiev, E.V., Phillips, R.L., Rines, H.W. (1998). Chromosome-specific molecular organization of maize (*Zea mays* L.) centromeric regions. *Proceedings of the National Academy of Sciences*, 95(22), 13073–13078.

Arabidopsis Genome Initiative (2000). Analysis of the genome sequence of the flowering plant *Arabidopsis thaliana*. *Nature*, 408(6814), 796–815.

Arrighi, F.E. and Hsu, T.C. (1971). Localization of heterochromatin in human chromosomes. *Cytogenetic and Genome Research*, 10(2), 81–86.

Baker, R.E., Fitzgerald-Hayes, M., O'Brien, T.C. (1989). Purification of the yeast centromere binding protein CP1 and a mutational analysis of its binding site. *The Journal of Biological Chemistry*, 264(18), 10843–10850.

Bernard, P., Maure, J.F., Partridge, J.F., Genier, S., Javerzat, J.P., Allshire, R.C. (2001). Requirement of heterochromatin for cohesion at centromeres. *Science*, 294(5551), 2539–2542.

Blower, M.D., Sullivan, B.A., Karpen, G.H. (2002). Conserved organization of centromeric chromatin in flies and humans. *Developmental Cell*, 2(3), 319–330.

Boveri, T. (2018). *Zellen-Studien Vol. 2: Die Befruchtung und Teilung des Eies von Ascaris megalocephala*. Forgotten Books, London.

Boveri, T. (2019). *Zellen-Studien Vol. 4: Ueber die Natur der Centrosomen*. Forgotten Books, London.

Bram, R.J. and Kornberg, R.D. (1987). Isolation of a *Saccharomyces cerevisiae* centromere DNA-binding protein, its human homolog, and its possible role as a transcription factor. *Molecular and Cellular Biology*, 7(1), 403–409.

Buchwitz, B.J., Ahmad, K., Moore, L.L., Roth, M.B., Henikoff, S. (1999). A histone-H3-like protein in *C. elegans*. *Nature*, 401(6753), 547–548.

Cai, M. and Davis, R.W. (1990). Yeast centromere binding protein CBF1, of the helix-loop-helix protein family, is required for chromosome stability and methionine prototrophy. *Cell*, 61(3), 437–446.

Caspersson, T., Zech, L., Johansson, C., Modest, E.J. (1970). Identification of human chromosomes by DNA-binding fluorescent agents. *Chromosoma*, 30(2), 215–227.

Chang, C.-H., Chavan, A., Palladino, J., Wei, X., Martins, N.M.C., Santinello, B., Chen, C.-C., Erceg, J., Beliveau, B.J., Wu, C.-T. et al. (2019). Islands of retroelements are major components of Drosophila centromeres. *PLoS Biology*, 17(5), e3000241.

Chatterjee, G., Sankaranarayanan, S.R., Guin, K., Thattikota, Y., Padmanabhan, S., Siddharthan, R., Sanyal, K. (2016). Repeat-associated fission yeast-like regional centromeres in the ascomycetous budding yeast *Candida tropicalis*. *PLoS Genetics*, 12(2), e1005839.

Cheeseman, I.M. (2014). The kinetochore. *Cold Spring Harbor Perspectives in Biology*, 6(7), a015826.

Cheng, Z., Dong, F., Langdon, T., Ouyang, S., Buell, C.R., Gu, M., Blattner, F.R., Jiang, J. (2002). Functional rice centromeres are marked by a satellite repeat and a centromere-specific retrotransposon. *The Plant Cell*, 14(8), 1691–1704.

Chinault, A.C. and Carbon, J. (1979). Overlap hybridization screening: Isolation and characterization of overlapping DNA fragments surrounding the leu2 gene on yeast chromosome III. *Gene*, 5(2), 111–126.

Choo, K.H., Vissel, B., Nagy, A., Earle, E., Kalitsis, P. (1991). A survey of the genomic distribution of alpha satellite DNA on all the human chromosomes, and derivation of a new consensus sequence. *Nucleic Acids Research*, 19(6), 1179–1182.

Clarke, L. (1998). Centromeres: Proteins, protein complexes, and repeated domains at centromeres of simple eukaryotes. *Current Opinion in Genetics & Development*, 8(2), 212–218.

Clarke, L. and Carbon, J. (1980a). Isolation of a yeast centromere and construction of functional small circular chromosomes. *Nature*, 287(5782), 504–509.

Clarke, L. and Carbon, J. (1980b). Isolation of the centromere-linked CDC10 gene by complementation in yeast. *Proceedings of the National Academy of Sciences of the United States of America*, 77(4), 2173–2177.

Clarke, L., Baum, M., Marschall, L.G., Ngan, V.K., Steiner, N.C. (1993). Structure and function of *Schizosaccharomyces pombe* centromeres. *Cold Spring Harbor Symposia on Quantitative Biology*, 58, 687–695.

Copenhaver, G.P., Nickel, K., Kuromori, T., Benito, M.I., Kaul, S., Lin, X., Bevan, M., Murphy, G., Harris, B., Parnell, L.D. et al. (1999). Genetic definition and sequence analysis of Arabidopsis centromeres. *Science*, 286(5449), 2468–2474.

Cortes-Silva, N., Ulmer, J., Kiuchi, T., Hsieh, E., Cornilleau, G., Ladid, I., Dingli, F., Loew, D., Katsuma, S., Drinnenberg, I.A. (2020). CenH3-independent kinetochore assembly in Lepidoptera requires CCAN, including CENP-T. *Current Biology: CB*, 30(4), 561-572.e10.

Coughlan, A.Y. and Wolfe, K.H. (2019). The reported point centromeres of *Scheffersomyces stipitis* are retrotransposon long terminal repeats. *Yeast*, 36(5), 275–283.

Darlington, C.D. (1936a). The external mechanics of the chromosomes III-Meiosis in triploids. *Proceedings of the Royal Society of London. Series B – Biological Sciences*, 121(823), 290–300.

Darlington, C.D. (1936b). The external mechanics of the chromosomes I – The scope of enquiry. *Proceedings of the Royal Society of London. Series B – Biological Sciences*, 121(823), 264–273.

Dong, F., Miller, J.T., Jackson, S.A., Wang, G.L., Ronald, P.C., Jiang, J. (1998). Rice (*Oryza sativa*) centromeric regions consist of complex DNA. *Proceedings of the National Academy of Sciences of the United States of America*, 95(14), 8135–8140.

Douglass, A.P., Offei, B., Braun-Galleani, S., Coughlan, A.Y., Martos, A.A.R., Ortiz-Merino, R.A., Byrne, K.P., Wolfe, K.H. (2018). Population genomics shows no distinction between pathogenic *Candida krusei* and environmental *Pichia kudriavzevii*: One species, four names. *PLoS Pathogens*, 14(7), e1007138.

Drets, M.E. and Shaw, M.W. (1971). Specific banding patterns of human chromosomes. *Proceedings of the National Academy of Sciences of the United States of America*, 68(9), 2073–2077.

Drinnenberg, I.A., de Young, D., Henikoff, S., Malik, H.S. (2014). Recurrent loss of CenH3 is associated with independent transitions to holocentricity in insects. *eLife*, 3, e0367.

Dujon, B. (2010). Yeast evolutionary genomics. *Nature Reviews Genetics*, 11(7), 512–524.

Earnshaw, W.C. and Rothfield, N. (1985). Identification of a family of human centromere proteins using autoimmune sera from patients with scleroderma. *Chromosoma*, 91(3–4), 313–321.

Fitzgerald-Hayes, M., Clarke, L., Carbon, J. (1982). Nucleotide sequence comparisons and functional analysis of yeast centromere DNAs. *Cell*, 29(1), 235–244.

Flemming, W. (1882). *Zellsubstanz, Kern und Zelltheilung (in German)*. F.C. W. Vogel.

Fukagawa, T. and Earnshaw, W.C. (2014). The centromere: Chromatin foundation for the kinetochore machinery. *Developmental Cell*, 30(5), 496–508.

Garavís, M., Méndez-Lago, M., Gabelica, V., Whitehead, S.L., González, C., Villasante, A. (2015). The structure of an endogenous Drosophila centromere reveals the prevalence of tandemly repeated sequences able to form i-motifs. *Scientific Reports*, 5, 13307.

Gassmann, R., Rechtsteiner, A., Yuen, K.W., Muroyama, A., Egelhofer, T., Gaydos, L., Barron, F., Maddox, P., Essex, A., Monen, J. et al. (2012). An inverse relationship to germline transcription defines centromeric chromatin in *C. elegans*. *Nature*, 484(7395), 534–537.

Gladyshev, E. and Kleckner, N. (2016). Recombination-independent recognition of DNA homology for repeat-induced point mutation (RIP) is modulated by the underlying nucleotide sequence. *PLoS Genetics*, 12(5), e1006015.

Gordon, J.L., Byrne, K.P., Wolfe, K.H. (2011). Mechanisms of chromosome number evolution in yeast. *PLoS Genetics*, 7(7), e1002190.

Heckmann, S., Schroeder-Reiter, E., Kumke, K., Ma, L., Nagaki, K., Murata, M., Wanner, G., Houben, A. (2011). Holocentric chromosomes of *Luzula elegans* are characterized by a longitudinal centromere groove, chromosome bending, and a terminal nucleolus organizer region. *Cytogenetic and Genome Research*, 134(3), 220–228.

Hooff, J.J., Tromer, E., Wijk, L.M., Snel, B., Kops, G.J. (2017). Evolutionary dynamics of the kinetochore network in eukaryotes as revealed by comparative genomics. *EMBO Reports*, 18(9), 1559–1571.

Hughes-Schrader, S. (1944). A primitive coccid chromosome cycle in *Puto* sp. *The Biological Bulletin*, 87(3), 167–176.

Hughes-Schrader, S. and Ris, H. (1941). The diffuse spindle attachment of coccids, verified by the mitotic behavior of induced chromosome fragments. *Journal of Experimental Zoology*, 87(3), 429–456.

Jain, M., Olsen, H.E., Turner, D.J., Stoddart, D., Bulazel, K.V., Paten, B., Haussler, D., Willard, H.F., Akeson, M., Miga, K.H. (2018). Linear assembly of a human centromere on the Y chromosome. *Nature Biotechnology*, 36(4), 321–323.

Janbon, G., Ormerod, K.L., Paulet, D., Byrnes III, E.J., Yadav, V., Chatterjee, G., Mullapudi, N., Hon, C.-C., Billmyre, R.B., Brunel, F. et al. (2014). Analysis of the genome and transcriptome of *Cryptococcus neoformans* var. *grubii* reveals complex RNA expression and microevolution leading to virulence attenuation. *PLoS Genetics*, 10(4), e1004261.

Jones, K.W. and Corneo, G. (1971). Location of satellite and homogeneous DNA sequences on human chromosomes. *Nature New Biology*, 233(43), 268–271.

Jones, K.W., Purdom, I.F., Prosser, J., Corneo, G. (1974). The chromosomal localisation of human satellite DNA I. *Chromosoma*, 49(2), 161–171.

Kapoor, S., Zhu, L., Froyd, C., Liu, T., Rusche, L.N. (2015). Regional centromeres in the yeast Candida lusitaniae lack pericentromeric heterochromatin. *Proceedings of the National Academy of Sciences of the United States of America*, 112(39), 12139–12144.

Kasinathan, S. and Henikoff, S. (2018). Non-B-Form DNA is enriched at centromeres. *Molecular Biology and Evolution*, 35(4), 949–962.

Kobayashi, N., Suzuki, Y., Schoenfeld, L.W., Müller, C.A., Nieduszynski, C., Wolfe, K.H., Tanaka, T.U. (2015). Discovery of an unconventional centromere in budding yeast redefines evolution of point centromeres. *Current Biology: CB*, 25(15), 2026–2033.

Kursel, L.E. and Malik, H.S. (2016). Centromeres. *Current Biology: CB*, 26(12), R487–R490.

Le, M.H., Duricka, D., Karpen, G.H. (1995). Islands of complex DNA are widespread in Drosophila centric heterochromatin. *Genetics*, 141(1), 283–303.

Lechner, J. and Carbon, J. (1991). A 240 kd multisubunit protein complex, CBF3, is a major component of the budding yeast centromere. *Cell*, 64(4), 717–725.

Locke, D.P., Hillier, L.W., Warren, W.C., Worley, K.C., Nazareth, L.V., Muzny, D.M., Yang, S.-P., Wang, Z., Chinwalla, A.T., Minx, P. et al. (2011). Comparative and demographic analysis of orangutan genomes. *Nature*, 469(7331), 529–533.

Loftus, B.J., Fung, E., Roncaglia, P., Rowley, D., Amedeo, P., Bruno, D., Vamathevan, J., Miranda, M., Anderson, I.J., Fraser, J.A. et al. (2005). The genome of the basidiomycetous yeast and human pathogen *Cryptococcus neoformans*. *Science*, 307(5713), 1321–1324.

Lynch, D.B., Logue, M.E., Butler, G., Wolfe, K.H. (2010). Chromosomal G + C content evolution in yeasts: Systematic interspecies differences, and GC-poor troughs at centromeres. *Genome Biology and Evolution*, 2, 572–583.

Ma, J. and Jackson, S.A. (2006). Retrotransposon accumulation and satellite amplification mediated by segmental duplication facilitate centromere expansion in rice. *Genome Research*, 16(2), 251–259.

Malik, H.S. and Henikoff, S. (2009). Major evolutionary transitions in centromere complexity. *Cell*, 138(6), 1067–1082.

Manuelidis, L. (1978a). Chromosomal localization of complex and simple repeated human DNAs. *Chromosoma*, 66(1), 23–32.

Manuelidis, L. (1978b). Complex and simple sequences in human repeated DNAs. *Chromosoma*, 66(1), 1–21.

Manuelidis, L. and Wu, J.C. (1978). Homology between human and simian repeated DNA. *Nature*, 276(5683), 92–94.

Marie-Nelly, H., Marbouty, M., Cournac, A., Liti, G., Fischer, G., Zimmer, C., Koszul, R. (2014). Filling annotation gaps in yeast genomes using genome-wide contact maps. *Bioinformatics*, 30(15), 2105–2113.

Marques, A., Ribeiro, T., Neumann, P., Macas, J., Novák, P., Schubert, V., Pellino, M., Fuchs, J., Ma, W., Kuhlmann, M. et al. (2015). Holocentromeres in Rhynchospora are associated with genome-wide centromere-specific repeat arrays interspersed among euchromatin. *Proceedings of the National Academy of Sciences*, 112(44), 13633–13638.

McKay, R.D. (1973). The mechanism of G and C banding in mammalian metaphase chromosomes. *Chromosoma*, 44(1), 1–14.

Mellor, J., Jiang, W., Funk, M., Rathjen, J., Barnes, C.A., Hinz, T., Hegemann, J.H., Philippsen, P. (1990). CPF1, a yeast protein which functions in centromeres and promoters. *The EMBO Journal*, 9(12), 4017–4026.

Melters, D.P., Paliulis, L.V., Korf, I.F., Chan, S.W.L. (2012). Holocentric chromosomes: Convergent evolution, meiotic adaptations, and genomic analysis. *Chromosome Research*, 20(5), 579–593.

Miga, K.H., Newton, Y., Jain, M., Altemose, N., Willard, H.F., Kent, W.J. (2014). Centromere reference models for human chromosomes X and Y satellite arrays. *Genome Research*, 24(4), 697–707.

Mishra, P.K., Baum, M., Carbon, J. (2007). Centromere size and position in *Candida albicans* are evolutionarily conserved independent of DNA sequence heterogeneity. *Molecular Genetics and Genomics: MGG*, 278(4), 455–465.

Morales, L., Noel, B., Porcel, B., Marcet-Houben, M., Hullo, M.-F., Sacerdot, C., Tekaia, F., Leh-Louis, V., Despons, L., Khanna, V. et al. (2013). Complete DNA sequence of *Kuraishia capsulata* illustrates novel genomic features among budding yeasts (Saccharomycotina). *Genome Biology and Evolution*, 5(12), 2524–2539.

Musacchio, A. and Desai, A. (2017). A molecular view of kinetochore assembly and function. *Biology*, 6(1), 5.

Nagaki, K. and Murata, M. (2005). Characterization of CENH3 and centromere-associated DNA sequences in sugarcane. *Chromosome Research: An International Journal on the Molecular, Supramolecular and Evolutionary Aspects of Chromosome Biology*, 13(2), 195–203.

Nagaki, K., Tsujimoto, H., Sasakuma, T. (1998). A novel repetitive sequence of sugar cane, SCEN family, locating on centromeric regions. *Chromosome Research: An International Journal on the Molecular, Supramolecular and Evolutionary Aspects of Chromosome Biology*, 6(4), 295–302.

Nagaki, K., Talbert, P.B., Zhong, C.X., Dawe, R.K., Henikoff, S., Jiang, J. (2003). Chromatin immunoprecipitation reveals that the 180-bp satellite repeat is the key functional DNA element of *Arabidopsis thaliana* centromeres. *Genetics*, 163(3), 1221–1225.

Nagaki, K., Cheng, Z., Ouyang, S., Talbert, P.B., Kim, M., Jones, K.M., Henikoff, S., Buell, C.R., Jiang, J. (2004). Sequencing of a rice centromere uncovers active genes. *Nature Genetics*, 36(2), 138–145.

Nagaki, K., Kashihara, K., Murata, M. (2005). Visualization of diffuse centromeres with centromere-specific histone H3 in the holocentric plant *Luzula nivea*. *The Plant Cell*, 17(7), 1886–1893.

Nagaki, K., Walling, J., Hirsch, C., Jiang, J., Murata, M. (2009). Structure and evolution of plant centromeres. *Progress in Molecular and Subcellular Biology*, 48, 153–179.

Navarro-Mendoza, M.I., Pérez-Arques, C., Panchal, S., Nicolás, F.E., Mondo, S.J., Ganguly, P., Pangilinan, J., Grigoriev, I.V., Heitman, J., Sanyal, K., Garre, V. (2019). Early diverging fungus *Mucor circinelloides* lacks centromeric histone CENP-A and displays a mosaic of point and regional centromeres. *Current Biology*, 29(22), 3791-3802.e6.

Nergadze, S.G., Piras, F.M., Gamba, R., Corbo, M., Cerutti, F., McCarter, J.G.W., Cappelletti, E., Gozzo, F., Harman, R.M., Antczak, D.F. et al. (2018). Birth, evolution, and transmission of satellite-free mammalian centromeric domains. *Genome Research*, 28(6), 789–799.

Nonomura, K.I. and Kurata, N. (1999). Organization of the 1.9-kb repeat unit RCE1 in the centromeric region of rice chromosomes. *Molecular & General Genetics: MGG*, 261(1), 1–10.

Nusbaum, C., Mikkelsen, T.S., Zody, M.C., Asakawa, S., Taudien, S., Garber, M., Kodira, C.D., Schueler, M.G., Shimizu, A., Whittaker, C.A. et al. (2006). DNA sequence and analysis of human chromosome 8. *Nature*, 439(7074), 331–335.

Ohama, T., Suzuki, T., Mori, M., Osawa, S., Ueda, T., Watanabe, K., Nakase, T. (1993). Non-universal decoding of the leucine codon CUG in several Candida species. *Nucleic Acids Research*, 21(17), 4039–4045.

Ola, M., O'Brien, C.E., Coughlan, A.Y., Ma, Q., Donovan, P.D., Wolfe, K.H., Butler, G. (2020). Polymorphic centromere locations in the pathogenic yeast *Candida parapsilosis*. *Genome Research*, 30(5), 684–696.

Padmanabhan, S., Thakur, J., Siddharthan, R., Sanyal, K. (2008). Rapid evolution of Cse4p-rich centromeric DNA sequences in closely related pathogenic yeasts, *Candida albicans* and *Candida dubliniensis*. *Proceedings of the National Academy of Sciences of the United States of America*, 105(50), 19797–19802.

Paweletz, N. (2001). Walther Flemming: Pioneer of mitosis research. *Nature Reviews Molecular Cell Biology*, 2(1), 72–75.

Peterson, J.B. and Ris, H. (1976). Electron-microscopic study of the spindle and chromosome movement in the yeast Saccharomyces cerevisiae. *Journal of Cell Science*, 22(2), 219–242.

Piras, F.M., Nergadze, S.G., Magnani, E., Bertoni, L., Attolini, C., Khoriauli, L., Raimondi, E., Giulotto, E. (2010). Uncoupling of satellite DNA and centromeric function in the genus *Equus*. *PLoS Genetics*, 6(2), e1000845.

Pluta, A.F., Mackay, A.M., Ainsztein, A.M., Goldberg, I.G., Earnshaw, W.C. (1995). The centromere: Hub of chromosomal activities. *Science*, 270(5242), 1591–1594.

Polizzi, C. and Clarke, L. (1991). The chromatin structure of centromeres from fission yeast: Differentiation of the central core that correlates with function. *The Journal of Cell Biology*, 112(2), 191–201.

Rhind, N., Chen, Z., Yassour, M., Thompson, D.A., Haas, B.J., Habib, N., Wapinski, I., Roy, S., Lin, M.F., Heiman, D.I. et al. (2011). Comparative functional genomics of the fission yeasts. *Science*, 332(6032), 930–936.

Sankaranarayanan, S.R., Ianiri, G., Coelho, M.A., Reza, M.H., Thimmappa, B.C., Ganguly, P., Vadnala, R.N., Sun, S., Siddharthan, R., Tellgren-Roth, C. et al. (2020). Loss of centromere function drives karyotype evolution in closely related Malassezia species. *eLife*, 9, e53944.

Sanyal, K., Baum, M., Carbon, J. (2004). Centromeric DNA sequences in the pathogenic yeast Candida albicans are all different and unique. *Proceedings of the National Academy of Sciences*, 101(31), 11374–11379.

Schotanus, K., Soyer, J.L., Connolly, L.R., Grandaubert, J., Happel, P., Smith, K.M., Freitag, M., Stukenbrock, E.H. (2015). Histone modifications rather than the novel regional centromeres of *Zymoseptoria tritici* distinguish core and accessory chromosomes. *Epigenetics & Chromatin*, 8, 41.

Schrader, F. (1935). Notes an the mitotic behavior of long chromosomes. *Cytologia*, 6(4), 422–430.

Schueler, M.G., Higgins, A.W., Rudd, M.K., Gustashaw, K., Willard, H.F. (2001). Genomic and genetic definition of a functional human centromere. *Science*, 294(5540), 109–115.

Seabright, M. (1972). The use of proteolytic enzymes for the mapping of structural rearrangements in the chromosomes of man. *Chromosoma*, 36(2), 204–210.

Selker, E.U. (1990). Premeiotic instability of repeated sequences in *Neurospora crassa*. *Annual Review of Genetics*, 24, 579–613.

Senaratne, A.P., Muller, H., Fryer, K.A., Kawamoto, M., Katsuma, S., Drinnenberg, I.A. (2020). Formation of the CenH3-Deficient holocentromere in Lepidoptera avoids active chromatin. *Current Biology: CB*, 31, 173–181.e7.

Shang, W.-H., Hori, T., Toyoda, A., Kato, J., Popendorf, K., Sakakibara, Y., Fujiyama, A., Fukagawa, T. (2010). Chickens possess centromeres with both extended tandem repeats and short non-tandem-repetitive sequences. *Genome Research*, 20(9), 1219–1228.

Smith, K.M., Phatale, P.A., Sullivan, C.M., Pomraning, K.R., Freitag, M. (2011). Heterochromatin is required for normal distribution of *Neurospora crassa* CenH3. *Molecular and Cellular Biology*, 31(12), 2528–2542.

Steiner, F.A. and Henikoff, S. (2014). Holocentromeres are dispersed point centromeres localized at transcription factor hotspots. *eLife*, 3, e02025.

Sun, X. (2003). Sequence analysis of a functional drosophila centromere. *Genome Research*, 13(2), 182–194.

Sun, X., Wahlstrom, J., Karpen, G. (1997). Molecular structure of a functional drosophila centromere. *Cell*, 91(7), 1007–1019.

Talbert, P.B., Kasinathan, S., Henikoff, S. (2018). Simple and complex centromeric satellites in drosophila sibling species. *Genetics*, 208(3), 977–990.

Thakur, J., Talbert, P.B., Henikoff, S. (2015). Inner kinetochore protein interactions with regional centromeres of fission yeast. *Genetics*, 201(2), 543–561.

Thon, M.R., Pan, H., Diener, S., Papalas, J., Taro, A., Mitchell, T.K., Dean, R.A. (2006). The role of transposable element clusters in genome evolution and loss of synteny in the rice blast fungus *Magnaporthe oryzae*. *Genome Biology*, 7(2), R16.

Waldeyer, W. (1888). Ueber Karyokinese und ihre Beziehungen zu den Befruchtungsvorgängen. *Archiv für Mikroskopische Anatomie*, 32(1), 1–122.

Waye, J.S. and Willard, H.F. (1989). Human beta satellite DNA: Genomic organization and sequence definition of a class of highly repetitive tandem DNA. *Proceedings of the National Academy of Sciences of the United States of America*, 86(16), 6250–6254.

Willard, H.F. (1985). Chromosome-specific organization of human alpha satellite DNA. *American Journal of Human Genetics*, 37(3), 524–532.

Willard, H.F. and Waye, J.S. (1987). Chromosome-specific subsets of human alpha satellite DNA: Analysis of sequence divergence within and between chromosomal subsets and evidence for an ancestral pentameric repeat. *Journal of Molecular Evolution*, 25(3), 207–214.

Yadav, V., Sun, S., Billmyre, R.B., Thimmappa, B.C., Shea, T., Lintner, R., Bakkeren, G., Cuomo, C.A., Heitman, J., Sanyal, K. (2018). RNAi is a critical determinant of centromere evolution in closely related fungi. *Proceedings of the National Academy of Sciences of the United States of America*, 115(12), 3108–3113.

Yadav, V., Yang, F., Reza, M.H., Liu, S., Valent, B., Sanyal, K., Naqvi, N.I. (2019). Cellular dynamics and genomic identity of centromeres in cereal blast fungus. *mBio*, 10(4), e01581-19.

Yamane, T., Ogawa, T., Matsuoka, M. (2008). Derivation of consensus sequence for protein binding site in *Yarrowia lipolytica* centromere. *Journal of Bioscience and Bioengineering*, 105(6), 671–674.

Zhong, C.X., Marshall, J.B., Topp, C., Mroczek, R., Kato, A., Nagaki, K., Birchler, J.A., Jiang, J., Dawe, R.K. (2002). Centromeric retroelements and satellites interact with maize kinetochore protein CENH3. *The Plant Cell*, 14(11), 2825–2836.

5

Evolution and Functions of Telomeres

Arturo LONDOÑO-VALLEJO
Télomères et cancer, Institut Curie, CNRS UMR3347, Paris, France

Telomeres are the nucleoprotein structures that cap the extremities of linear chromosomes in all eukaryotes. These structures are absolutely essential to protect chromosome extremities from aberrant repair reactions that would lead to chromosome end-to-end fusions and from degradation reactions that may cause the loss of essential genetic material. Telomeres were initially conceptualized around 1930 based on their protective function observed in experimental systems that included plants and insects (McClintock 1938; Muller 1938). Currently, it is the single chromosome structure that has been more tightly associated with human aging and disease. Indeed, telomere shortening is both a major mechanism to prevent tumor development but also a major obstacle to tissue renewal and regeneration. In this chapter, we will discuss the origin and evolution of telomeres, the basics of telomere metabolism and the way such structures impact cell function and organismal lifespan.

5.1. Primary structure of telomeres

Telomeres are typically constituted of short tandemly repeated sequences. In all vertebrates, which shared a common ancestor more than 400 million years ago, the canonical telomere sequence is the hexamer 5'-TTAGGG-3', repeated several hundreds to several thousands of times depending on the species (see Figure 5.1) (Lejnine et al. 1995; Moyzis et al. 1988).

For a color version of all figures in this chapter, see http://www.iste.co.uk/richard/function.zip.

Function and Evolution of Repeated DNA Sequences,
coordinated by Guy-Franck RICHARD.

5′**NNNNNN**TTAGGG**(TTAGGG)**$_n$ TTAGGGTTAG***GGTTAGGGTTAG*** 3′
3′**NNNNNN**AATCCC**(AATCCC)**$_n$ AATCCCAATC 5′

3 1 2

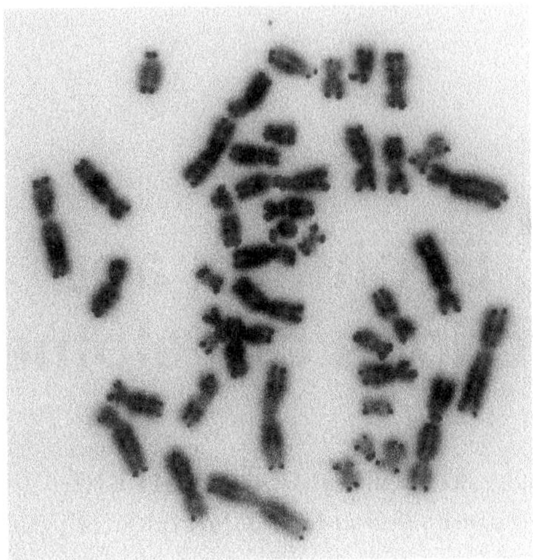

Figure 5.1. *On top, canonical nucleotide sequence of vertebrate telomeres. The hexamer TTAGGG (1) found on the top strand (5'>3') is repeated several hundreds to several thousand times (depending on the species, telomerase status, replication history). All telomeres end up with a G-rich single stranded extension (2), which is absolutely required for telomere protection and is the substrate for elongation reactions. In human chromosomes, telomeres are either directly connected to chromosome specific sequences (3) or through juxtatelomeric sequences composed of degenerate telomere repeats or more complex types of repeated sequences found at several extremities. At the bottom, we can see metaphase chromosome spreading, prepared from a normal diploid human cell and visualized under the microscope after hybridization with a telomere specific fluorescent probe (in red). Chromosomes are counterstained with DAPI (gray). The size (intensity) of the telomeric signals reflects the number of the telomere repeats. The differences in telomere lengths between chromosome extremities reflect allelic, inherited differences*

The same motif is found in other distantly related metazoans (including the Choanozoa, the unicellular sister group of the clade Animalia) (Traut et al. 2007). This extraordinary conservation suggests an almost perfect fit for telomere function. In humans, telomeres are 8–15 kb long in somatic cells and tend to shorten with age for reasons explained later. In order to be functional, that is, to efficiently protect chromosome ends, telomeres must have a minimum number of telomeric repeats (Capper et al. 2007). This is true for all eukaryotes, including unicellular organisms and plants, where telomere repeated motifs show variability. For instance, among budding yeasts, important differences in terms of repeat length, base composition and tract heterogeneity are found between closely related species (Cohn et al. 1998). Nevertheless, all telomeric repeats remain conspicuously rich in G and C with a strong strand bias (Teixeira and Gilson 2005): Gs are mostly or exclusively present on the strand running 5' to 3' toward the end. The reasons explaining this bias (or explaining the divergence between organisms) are not clear but may be related to very fundamental metabolic aspects, such as replication, and the possibility that G-rich compositions favor the formation of secondary structures, such as G-quadruplexes, that would help telomeres achieve their protective role (Bryan 2020). Also, as discussed later in this chapter, these G-rich motifs are always specifically recognized by dedicated proteins, thus imposing further constraints in (or, perhaps in some cases, favoring) their evolution potential.

Another critical feature of telomere structures is that they invariably end with a 3' single stranded extension, the G-rich overhang (Wright et al. 1997), bound by a telomere-specific protein (Baumann and Cech 2001), thus conferring to chromosome ends a highly distinctive property that is crucial to differentiate them from other double strand breaks (Wu et al. 2006). These 3' overhangs also constitute the substrates of specialized reverse transcriptases, called telomerases (Lingner et al. 1997), which use a specific RNA as a template to add more repeats to the 3'-end, thus resulting in telomere lengthening (Collins 2006; Morin 1989). In particular situations (specifically in the absence of telomerase), other strategies, mainly based on homologous recombination-based activities, may be used by the cell to add telomere repeats de novo and therefore to lengthen telomeres (Lundblad and Blackburn 1993; Murnane et al. 1994; Reddel et al. 1997). How telomeric synthesis by telomerases, and therefore how telomere length is controlled, will be discussed in another section.

It is important to note that in nature there are noticeable exceptions to the G-rich, short tandem repeat-based structures of telomeres. Indeed, a special type of telomeres is observed in drosophila and other insects, in which the telomere sequence is constituted of long, repeated stretches that span coding sequences (Biessmann et al. 1997; Casacuberta and Pardue 2003; Kamnert et al. 1997). Genes contained in these long repeats code for transposases that are able to insert new repeats at the end of chromosomes, thus maintaining the telomere structure and length. It is not known how such structures evolved instead of the short TTAGGG motif, which is considered by many to be the ancestor sequence (Fulneckova et al. 2013). Nevertheless, they may display common features, such as G-rich sequences on the 3' overhang (Abad and Villasante 1999), and, as we will see in the next section, transposon-mediated chromosome end healing has been proposed as the original mechanism for telomere formation after linearization of the circular chromosome ancestor.

5.1.1. *Origin and evolution of telomeres*

In the prokaryote world, circular chromosomes are the rule and are considered as the primitive structure of all genetic material (Volff and Altenbuchner 2000). The occurrence of a double strand break disrupting the circle will trigger an immediate stress response that will lead to repair by direct ligation of the extremities, thus preserving chromosome integrity while shutting down the stress response. The emergence of linearized chromosomes can be visualized as the result of an alternative mechanism of DNA repair in which repeated retrotransposon insertion at the exposed extremities occurred, allowing for the first time the modification of these extremities by the presence of repeats that acted both as a deterrent for re-ligation and as a buffer of dispensable genetic material (see Figure 5.2) (Garavis et al. 2013). The DNA composition of such repeats could then have been under further selection so that only those sequences most apt to prevent/control repair reactions and to serve as target sites for proteins able to modify the immediate environment would persist throughout evolution, leading to the emergence of telomere-specific nucleoprotein complexes.

On the other hand, the transition from a prokaryote to a eukaryote world did not only concern the linearization of chromosomes, but also the development of a nuclear membrane and the complexification of genomes that led to the fragmentation of genomes in multiple linear chromosomes, requiring the development of a performant tubulin-mediated chromosome segregation apparatus. As proposed by Villasante et al. (2007), retrotransposons may have been involved in both processes. Firstly, transposon insertion within coding sequences, which gave rise to introns, forced the selection of splicing mechanisms and the need to separate transcription and translation, thus leading to nuclear compartmentalization. On the other hand, repeated retrotransposon insertion at the end of linear chromosomes may have created the opportunity for a strong tubulin nucleation center which would become the centromere. Such structures eventually competed out the original prokaryotic partition systems and likely contributed to the fragmentation of chromosomes as a consequence of double strand breaks introduced to resolve conflicts between two newly created and oppositely located centromeres. The newly exposed extremities would then become new telomeres or would be fused to other extremities. Further down in evolution, the presence of both telomeres and centromeres and the stabilization of multiple linear chromosomes together allowed for the emergence of the meiotic process, thus encouraging genetic diversity and fostering speciation (Villasante et al. 2007).

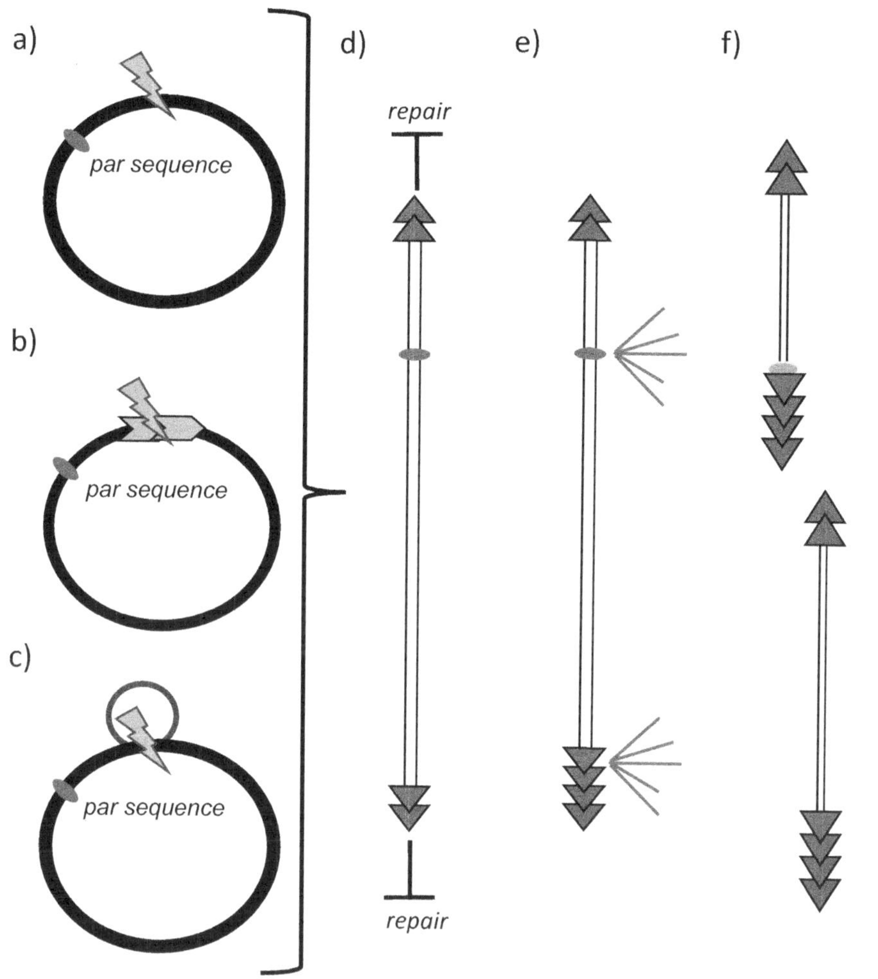

Figure 5.2. *Emergence and evolution of linear chromosomes (simplified from Villasante et al. 2007)*

NOTES ON FIGURE 5.2.– *The circular chromosome carried by the ancestor prokaryote was frequently subjected to challenges affecting its integrity, such as double strands breaks arising spontaneously during replication (a), or created during the recombining activities of indigenous transposable elements (b) or foreign plasmids (c). These double strand breaks would immediately lead to a stress response that mobilizes all cell resources to reestablish circularity (DNA repair by ligation). However, the reiterated insertion of repeated sequences at the extremities created by breaks (and perhaps most likely belonging to a transposable element) may have created the conditions for the suppression of such stress response, thus leading to the persistence of exposed extremities that were the target of further insertions/ recombination events that contributed to telomere expansions (blue triangles in d). During these expansions, repeated sequences may have acquired microtubule-nucleation properties (neocentromeres, green triangles in e), which could compete with sequences driving the actin-based mechanisms for chromosome partitioning. The co-existence of neocentromeres with par sequences would lead to chromosome segregation conflicts during mitosis due to the binding of microtubules (green bundle) to two different regions on the same chromosome. These conflicts should be resolved through chromosome fragmentation (thus creating double strand breaks), thus exposing new extremities which will be new targets for repeat insertion, telomere and neocentromere formation (f).*

5.1.2. *Nucleoprotein structure of telomeres*

In all organisms studied so far, telomeres are the target of specific sets of proteins, some of which specifically recognize telomeric sequences (de Lange 2009). In humans, for instance, the proteins TRF1 and TRF2 bind to the double stranded part of telomeric repeats, whereas POT1 binds to the G-rich telomeric overhang. These three proteins directly interact with three other telomere-specific proteins to form the Shelterin complex: TRF2 binds to RAP1 and TIN2, TRF1 binds to TIN2 and TPP1 and POT1 binds to TPP1 and likely to TRF2 (see Figure 5.3) (Palm and de Lange 2008; Kendellen et al. 2009). These multiple interactions make it possible for the Shelterin complex to exist in a variety of stoichiometry states all along the telomere sequence, although the functional consequences of local differences in Shelterin composition remain to be explored. In any case, Shelterin is absolutely required for telomere protection in mammals since the absence of key members such as TRF1/2, POT1 or TIN2 leads to telomere "uncapping" and telomere dysfunction, revealed by local DNA damage responses and end-to-end fusions, even in the presence of long stretches of telomere repeats (Sfeir and de Lange 2012). In lower eukaryotes, including fission yeast and ciliates, similar complexes are found although the conservation of individual members throughout evolution is limited to structural motifs rather than defined consensus sequences (Linger and Price 2009; Giraud-Panis et al. 2018). In fact, fast evolution in such complexes appear to be required in order to adapt not only to telomere sequence variation but also to other sources of variation such as changes in interacting partners that connect telomere biology and metabolism to other functions in the genome and the cell (Saint-Leandre et al. 2020).

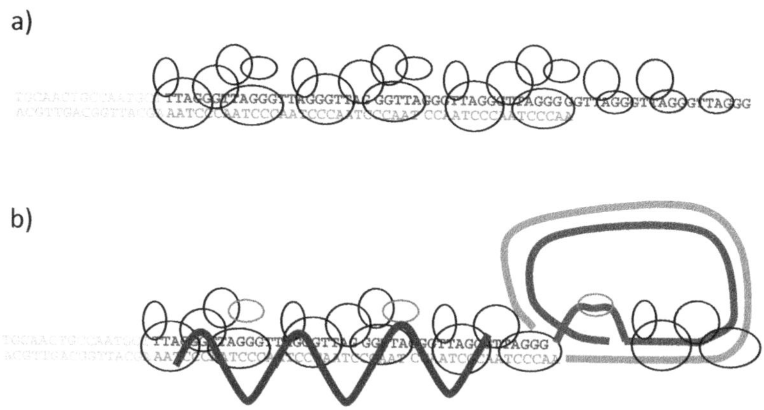

Figure 5.3. *The nucleoprotein structure of human telomeres. The telomere-specific protein complex Shelterin directly recognize the telomeres through the TRF1 (red circles) and TRF2 (dark blue ovals) subunits, which bind the double stranded portion of the TTAGGG repeated sequence, and through POT1 (green ovals) which binds the G-rich single stranded overhang (a). TIN2 (light yellow circles) binds both TRF1 and TRF2, while RAP1 (light blue ovals) binds TRF2. On the other hand, TPP1 (orange circles) binds both TIN2 and POT1. The six-member complex is absolutely required for correct telomere function (protection of chromosome extremities, chromosome-end replication and telomere length homeostasis). The presence of the Shelterin complex (mainly TRF2) promotes the formation of a T-loop (b) by the invasion of the double stranded part of the telomere by the single stranded overhang, thus displacing the G-strand. The presence of TERRA, the G-rich RNA transcribed from telomere repeats (thick blue wavy line) contributes to the telomere chromatin structure*

In higher eukaryotes (that is, mice, humans but also plants) and very much like other chromosome structures, double stranded telomere sequences are wrapped around histones (Blasco 2004; Vaquero-Sedas and Vega-Palas 2011). The nucleosome organization of telomeres appears to be only a little more compact than the rest of the genome (Lejnine et al. 1995). On the other hand, telomeres have been described as a heterochromatic structure due to the presence of post-transcriptional histone modifications usually associated with chromatin compaction (Blasco 2004). Even if these histone modifications are not more enriched at telomeres than at other parts of the genome (Cubiles et al. 2018), or are even physiologically absent in particular contexts (Gauchier et al. 2019), their experimental abrogation in other contexts has profound consequences in the way telomeres are maintained, allowing abnormal elongation and de-repression of recombination reactions (Benetti et al. 2007). In spite of their relative compact state, telomeres are actively transcribed to produce TERRA, a long non-coding RNA that remains in part associated with the chromosome ends (Azzalin et al. 2007). TERRA transcription initiates in the juxtatelomeric region of several chromosome ends where transcription control elements are sometimes found (Nergadze et al. 2009). However, most of the TERRA detected in mouse and human cells, for instance, originates from one or two extremities (Viceconte et al. 2021). It is therefore likely that TERRA molecules associate in trans with other chromosome extremities, thus introducing a trafficking aspect to TERRA biogenesis and function. The presence of TERRA at telomeres, which has been demonstrated in many organisms, very likely impacts the telomeric structure (Azzalin and Lingner 2008). Upon transcription, TERRA may remain bound to the transcribed strand, thus creating a so-called R-loop in which the G-rich telomeric strand is displaced (Pfeiffer et al. 2013; Balk et al. 2014). Such R-loops are expected to locally modify the nucleoprotein composition of the telomere since not only nucleosomes will not be allowed to reform but TRF1 and TRF2 will no longer bind to the unwound telomere repeats (Nie et al. 2021). On the other hand, POT1 will be able to bind the displaced G-rich strand. The presence of R-loops at telomeres have been demonstrated in many lower and higher eukaryotes but their accumulation is always related to a dysfunctional state where the mechanisms to remove such structures become deficient or when modifications of the chromatin allow a higher level of transcription of telomeric repeats (Balk et al. 2014; Graf et al. 2017). On the other hand, the accumulation of R-loops at telomeres has consequences at different levels including the perturbation of the normal progression of the replication fork or the accumulation of recombinogenic structures (Petti et al. 2019). Therefore, under normal circumstances, it is believed that a variety of mechanisms exist to prevent R-loop accumulation at telomeres. Nevertheless, and as noted above, many TERRA molecules remain associated with telomeres most likely through interactions with the many RNA-binding proteins normally found there such as members of the HNRNP family of proteins (Lopez de Silanes et al. 2010; Scheibe et al. 2013). The presence of TERRA at telomeres plays a fundamental role as suggested by the severe consequences of the elimination of TERRA on telomere stability. In lower eukaryotes, such as the budding yeast, TERRA seems required for telomere length homeostasis by helping address telomerase to the same chromosome end from which it originated (Luke et al. 2008; Cusanelli et al. 2013). Such mechanism has not been described in higher eukaryotes but further supports the view that trafficking may be an important component of TERRA biogenesis (Viceconte et al. 2021). The intimate mechanisms explaining how TERRA impact telomere stability and length homeostasis in human cells remain to be described.

5.2. A telomere specific higher order structure: the T-loop

In 1999, the Griffith's and the de Lange's labs demonstrated, using electron microscopy, that mammalian telomeres can form loops in mammalian cells (Griffith et al. 1999). The proposed mechanism of formation involves the invasion by the 3' overhang extension of the more centromeric double strand part of the telomere, displacing the G-rich strand to hybridize to the C-rich strand, thus forming a D-loop (Griffith et al. 1999). The resulting structure, named a T-loop (see Figure 5.3), will therefore hide the extremity of the chromosome, which will no longer be seen as a double strand break by the cell. Theoretically, such an arrangement will be sufficient to protect the extremity from repair reactions and degradation. Importantly, the presence of sheltering components, such as TRF2, promotes the folding of the telomeric DNA and the strand invasion reaction, thus by providing another explanation for why when TRF2 is removed, end-to-end fusions occur even in the presence of telomeric repeats (Timashev and De Lange 2020). This elegant mode of chromosome end protection has probably been used for very long time during evolution and could even predate the emergence of the ancestor TTAGGG vertebrate sequence (de Lange 2004). However, up until today it has not been totally clear whether every chromosome end in a cell should adopt this conformation in order to be fully protected or whether this event is only favored when telomeres are sufficiently long. Indeed, recent studies have suggested that when telomeres shorten, some sort of protection against end-to-end fusion may be exerted by RAP1, even in the absence of TRF2 (and therefore of T-loop) (Lototska et al. 2020). T-loops have been seen in other vertebrates but they do not seem to exist in lower eukaryotes, where telomeres tend to be much shorter. In the latter, though, other types of higher order structures may exist (folding back structures, for instance), which may also play a role in telomere functions (de Bruin et al. 2000).

5.2.1. *Telomere replication, a fundamental mechanism for telomere maintenance*

Telomere replication is a critical process that differs in many ways from the replication events that occur in the rest of the genome during the S-phase. A first major difference stems from the fact that replication is unidirectional as only one fork progresses toward the end (see Figure 5.4(a)). As a result, the possibility of rescuing any fork blockade by a convergent replication fork, as it occurs in the rest of the genome, does not exist at telomeres. It is then reasonable to think that every mechanistic aspect during telomere replication has been perfected throughout evolution to ensure that this unique fork progresses unimpeded until it reaches the very end of the extremity. One intriguing aspect is the repeated sequence itself.

Indeed, sequences that are G-rich and that display a strong G/C bias are usually considered difficult to replicate when they are found in interstitial places (Miller et al. 2006; Belotserkovskii et al. 2010). Therefore, replication of telomere sequences should be particularly challenging. And they are, as demonstrated by their exquisite fragility when replication mechanisms are perturbed (Rizzo et al. 2009; Zimmermann et al. 2014; Sun et al. 2015; Majerska et al. 2018; Porreca et al. 2020).

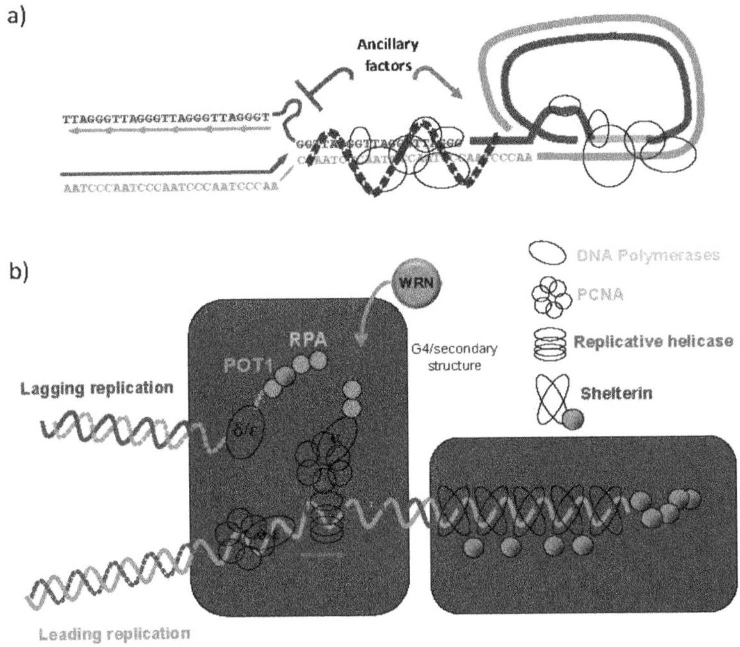

Figure 5.4. *Mechanisms of telomere replication*

NOTES ON FIGURE 5.4.– *Telomeres are replicated unidirectionally: the C-rich strand is replicated by a continuous mechanism (red long arrow) while the G-rich strand replication is discontinuous (lagging mechanism, green short arrows). The recruitment of ancillary factors such as helicases, toposiomerases, endonucleases and chromatin modifiers by the Shelterin complex during replication help in the progression of the replication fork toward the end, including through the base of the T-loop. TERRA (dotted blue line) must also be dealt with to allow normal progression of the fork. Many of these mechanisms are yet to be fully characterized. In (b), an example of a G-quadruplex (G4) being formed on the lagging strand is shown. Under normal circumstances, the helicase WRN is recruited through its interactions with the Shelterin complex and unwinds the G4, thus preventing the blockage of the replication on the lagging strand and the further accumulation of single strands. In the absence of WRN, the amount of exposed single strand increases, but contrary to what happens elsewhere in the genome where RPA binds to this single strand, the exposed G-rich strand is rather bound by POT1, thus preventing the triggering of the arrest of the replicative helicase, which then will be able to reach the end of the chromosome. This mechanism of replication fork uncoupling ensures that the telomere associated with the sister chromatid replicated by continuous is fully replicated.*

This fragility is readily demonstrated using fluorescent probes that target telomere repeats (Sfeir et al. 2009). Telomeres usually appear as condensed dots at the end of metaphasic chromosomes, each sister chromatid displaying a single dot (see Figure 5.1). Under replication stress conditions these dots are doubled or acquire an abnormal shape. Such phenotypes are detected upon treatment of cells with genotoxic drugs that prevent normal replication fork progression or when telomeric factors are targeted for downregulation or are invalidated by knockout (Rizzo et al. 2009; Sfeir et al. 2009; Zimmermann et al. 2014; Sun et al. 2015; Majerska et al. 2018; Porreca et al. 2020). One specific difficulty that may appear during the progression of the replication fork is the possibility for the G-rich strand, which is being replicated by lagging mechanisms, to form secondary structures, in particular G-quadruplexes (Bryan 2020) (see Figure 5.4). The strong potential of telomeric repeats, specially TTAGGG-based sequences, to form G4s has been demonstrated in vitro (Chen 1992; Wang and Patel 1992; Smith and Feigon 1993). In vivo, telomeric G4s have been shown to be highly prevalent in some organisms such as Stylonychia (Schaffitzel et al. 2001); however, they are not easily detected in higher eukaryotes, including human cells, although evidence does exist about their existence (Granotier et al. 2005; Fernando et al. 2008). Nevertheless, the use of small molecules able to specifically target and stabilize G4s has been instrumental both to demonstrate that G4s form at telomeres during replication and to reveal that the formation of stable G4 is detrimental for telomere replication (Mergny et al. 2001; Riou et al. 2002; Gomez et al. 2003). It is clear then that mechanisms normally exist to prevent such structures to accumulate on the lagging strand during telomere replication. One such mechanism relies on the activity of Q-type helicases (Mendoza et al. 2016; Lansdorp and van Wietmarschen 2019), such as BLM, WRN and, in lower organisms, PIF1, which are recruited through direct interactions with Shelterin components in response to stress or in response to chromatin-dependent cues during replication (Ferrarelli et al. 2013; Zimmermann et al. 2014). In budding yeast, it has been experimentally demonstrated that the absence of Pif1 leads to a replication-mediated instability of G4-forming sequences at interstitial loci (Ribeyre et al. 2009; Piazza et al. 2010). Interestingly, this instability only occurs when the G-rich sequence is on the leading strand, not when it is replicated by lagging mechanisms (Lopes et al. 2011). Similar results were observed in contexts where yeast cells were treated with G4 ligands, thus indicating that G4 formation on the leading strand is the reason for replication fork blockade and genome instability (Piazza et al. 2010). This extraordinary observation would suggest that, contrary to what has been proposed for many years, G-rich strands during replication do not, under physiological conditions, form G4s when they are replicated by lagging mechanisms, which is the case of telomeres. On the other hand, telomeric G-rich single strands exposed during replication will be immediately targeted, perhaps in competition with RPA, the replicative single strand-binding complex, by their cognate protein POT1 (see Figure 5.4(b)). It has been shown that binding of POT1 to telomere single strands counteract the formation of secondary structures such as G4s (Zaug et al. 2005; Chaires et al. 2020), very similarly to RPA (Salas et al. 2006; Audry et al. 2015), thus further preventing G4 formation while protecting the exposed single strand.

Another unique phenomenon that occurs at telomeres is the uncoupling of leading and lagging replication mechanisms. Elsewhere in the genome, when the replicative polymerases encounter an obstacle, single strands accumulate due to the continuous activity of the replicative helicase MCM. As mentioned above, RPA binds to single stranded DNA and, when replication is blocked, its accumulation above a defined threshold triggers the activation of signaling pathways that command the MCM to stop, thus interrupting replication on both strands. At telomeres, it has been demonstrated that the replication of the C-rich strand can proceed even if the replication of the G-rich strand has stopped, a phenomenon seen in the absence of the helicase WRN (see Figure 5.4(b)) (Crabbe et al. 2004; Arnoult et al. 2009). This helicase would act by preventing the formation of secondary structures on the lagging strand, thus allowing replication to proceed on that strand. But why when this lagging replication is interrupted, the accumulation of G-rich single strands fails to signal the MCM helicase to stop (Arnoult et al. 2009). In fact, the binding of POT1 to the exposed G-rich strand limits the accumulation of RPA, thus tampering with the RPA-dependent signaling activities and allowing MCM to progress, so that the replication can continue on the leading strand up to the end (see Figure 5.4(b)). Indeed, in situations where both WRN and POT1 are deficient, RPA accumulates and the replication of the C-strand is also affected, thus leading to the loss of telomere sequences on both sister chromatids (Arnoult et al. 2009).

Finally, the replication machinery at telomeres must face a particularly crowded nucleoprotein environment with telomere-specific proteins tightly bound to the telomere repeats, the presence of particular nucleic acid structures, including R-loops and T-loops, and the existence of topological constraints (see Figure 5.4(b)). Concerning the Shelterin complex, strong experimental evidence in vivo and in cellulo supports the idea that instead of constituting an obstacle to the replication fork as initially suggested by in vitro experiments, the presence of Shelterin is required for smoothing fork progression (Sfeir and de Lange 2012; Pinzaru et al. 2020; Porreca et al. 2020; Nie et al. 2021). The Shelterin complex interact with a myriad of replication ancillary factors (which together form the so-called "telosome"; Xin et al. 2008) that help in all aspects of DNA replication at telomeres by resolving all sorts of nucleic acid- or protein-dependent obstacles (reviewed in Gilson and Geli 2007; Cicconi and Chang 2020). Major actors in these interventions are topoisomerase II, various helicases, Fanconi anemia factors and endo-/exonucleases. Whenever an obstacle cannot be removed, a permanent replication blockade arises that leads to the permanent loss of downstream repeated sequences. These losses are without immediate consequence if enough telomere repeats have been already replicated so that the capping function will be preserved in the daughter cell. If not enough telomere sequences have been replicated at the end of S-phase, and in the absence of an elongating mechanism, a DNA damage response is triggered with the risk of end-to-end fusions to arise, most likely between sister chromatids.

As just seen, telomere replication is an extremely important mechanism to maintain telomere integrity. However, under physiological conditions, it also remains a major and unavoidable source of telomere shortening. Indeed, as replication reaches the end of the chromosome, two major events, exclusively seen at telomeres, occur. First, on the lagging strand, the 5' RNA moiety of the last Okazaki fragment is degraded, thus leaving the last portion of the parental G-rich strand unreplicated on that sister chromatid (see Figure 5.5).

This unreplicated portion reproduces the G-rich overhang found in the parental chromosome, therefore creating a fully functional telomere on that sister chromatid. Secondly, on the leading strand, the replication machinery will fully copy the C-rich strand thus creating a blunt end (see Figure 5.5). In order to convert this blunt end into a functional telomere, the parental C-rich strand must be partially degraded, thus creating the G-rich overhang. This "controlled" degradation of the parental strand on the leading sister chromatid constitutes a net loss of telomeric material that will be, in the absence of a lengthening mechanism, transmitted by that sister chromatid to one of the daughter cells. This "loss" is what has been called the "end-replication problem" to explain how telomeres shorten with every cell cycle. This name is nevertheless inappropriate insofar the loss of telomeric material is mostly due to a controlled degradation of the C-strand and not, as stated in the initial paradigm, to the inability of polymerases working on the lagging strand to completely replicate the G-rich strand (Watson 1972; Olovnikov 1973). Instead, the "end-replication problem" turns out to be a "solution" to reconstitute the sister telomere replicated by lagging mechanisms (see Figure 5.5) (Lingner et al. 1995).

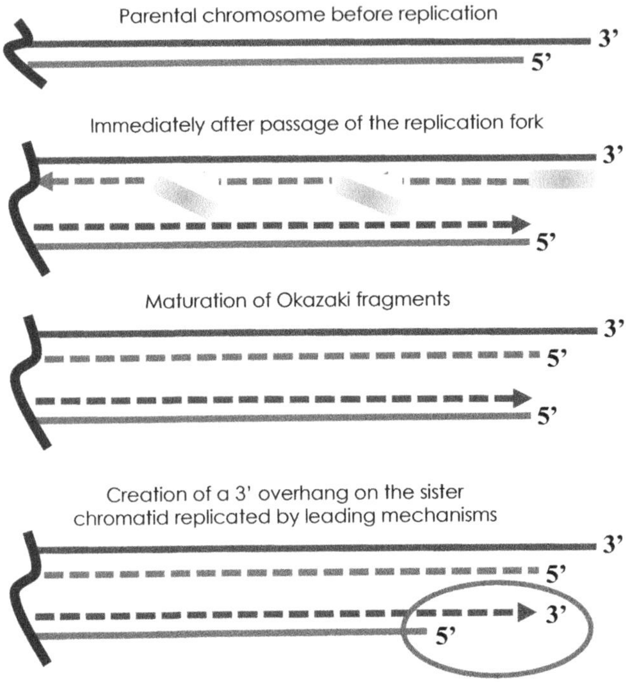

Figure 5.5. *The end-replication problem at telomeres is a leading mechanism problem. The end-replication problem contends that strands replicated by lagging mechanisms in a linear template will be replicated incompletely due to the 5' > 3' directionality of DNA polymerases and their need to be primed by RNA (pink rectangle). At telomeres, the degradation of the RNA moiety that primes the last Okazaki fragment on the lagging strand reproduces the 5' gap that already existed in the parental template without net loss of telomere material. On the leading strand, on the other hand, the replication mechanism allows replication to reach the 5' end thus creating a blunt end, which is incompatible with a telomeric function. In order to produce a functional telomere, the C-rich template is partially degraded in a controlled way to create a 3' overhang on that sister chromatid, with a net loss of telomere material*

5.3. Telomere lengthening mechanisms

All telomeres possess a dedicated mechanism for elongation. In the vast majority of cases, it is the telomerase. As already mentioned, it is a reverse transcriptase that uses a specific RNA to add nucleotides to the 3' end of telomeres (Lingner et al. 1997). While the catalytic subunits of all telomerases have been rather well conserved throughout evolution, the telomerase RNAs are much more divergent, even though some structural motifs can be recognized and serve as binding sites to the catalytic subunit and other ancillary factors that control RNA biogenesis, trafficking and enzymatic activity (Theimer and Feigon 2006). The reverse transcriptase catalytic subunit and the RNA are all the components that are needed to add telomeric repeats to 3' substrates in vitro (Autexier and Greider 1994; Weinrich et al. 1997; Beattie et al. 1998). In live cells, however, these two components interact with many other factors that play important roles in the biogenesis and trafficking of the ribonucleoprotein complex, as well as in telomere elongation. Depending on the organism, the telomerase RNA transits through specific locations in the nucleus (for instance, nucleolus and Cajal bodies in human cells, or through the cytoplasm in the case of the budding yeast) before joining the catalytic subunit (Teixeira et al. 2002; Tomlinson et al. 2006; Gallardo et al. 2008). How the telomerase holoenzyme is recruited to telomeres to exert its activity and how this activity is locally regulated has been the focus of intense research during the past decades, although many fundamental questions remain open. Following a current view, telomerase would be active only during the S-phase and will bind to individual substrates after replication is completed (Schmidt et al. 2016). The recruitment of human telomerase to telomeres involves the direct interaction of the catalytic subunit with the Shelterin complex component TPP1 (Nandakumar et al. 2012; Zhong et al. 2012; Zhang et al. 2013). However, a productive interaction (i.e. resulting in repeat addition) will only occur when the template on the RNA moiety is aligned with the last nucleotides on the 3' end so that a full repeat can be synthesized (Laprade et al. 2020). Once this synthesis is done, the enzyme should translocate to reposition the RNA template so that another repeat is added. The efficiency with which these nucleotides are added and the speed at which the holoenzyme translocates are key determinants of the enzyme processivity. The number of repeats that a telomerase can add to a single telomere is highly variable as judged by the time the enzyme spends in association with single telomeres in live cells and it is likely that local factors will influence this association (Schmidt et al. 2016; Laprade et al. 2020). For instance, it is known that telomerase-dependent elongation preferentially occurs on shorter telomeres. It has then been proposed that a repeat counting mechanism exists based on the presence of the number of TRF1 molecules bound to telomere repeats (van Steensel and de Lange 1997). TRF1, through its interaction with POT1, will then control the accessibility of telomerase to its substrate because the binding of POT1 to the 3' end can have an obliterating effect, thus excluding this extremity from telomerase activity (Loayza and De Lange 2003). On the other hand, binding of POT1 to the G-rich overhang may also help present the substrate to the enzyme by preventing the formation of G4s (Zaug et al. 2005). Indeed, it has been shown that in contexts when G-rich overhangs are very long, elongation by telomerase requires, in addition to POT1, the activity of helicases, such as RTEL1 (Porreca et al. 2018). Lastly, it should be noted that an experimentally induced excess of telomerase activity in a human cell will lead to hyper-elongated telomeres, thus upsetting the counting mode that controls telomerase activity at single telomere ends (Cristofari and Lingner 2006). Thus, control of expression of telomerase is an important aspect of telomere biology that determines the telomere length homeostasis in the whole organism.

Other telomerase-independent mechanisms to elongate telomeres exist, mainly based on homologous recombination, collectively known as alternative lengthening of telomeres (ALT) (Claude and Decottignies 2020). They are, however, entirely suppressed under physiological conditions perhaps because such a mode of elongation is not compatible with length regulation and is prone to genome instability. Indeed, recombination-based telomere elongation involves the invasion by a telomeric 3' end of another telomere in order to create a D-loop that can be extended to the end of the invaded chromosome by a mechanism called break-induced replication (BIR) (Roumelioti et al. 2016). This reaction will result in the net elongation of the invading telomere. In order to be efficient in maintaining telomere length in the cell as a whole, the capacity to invade should only be acquired by very short telomeres and only long telomeres should be targeted for invasion. However, such control mechanisms do not appear to exist in ALT cells. Instead, telomeres are randomly affected by gains and losses and invasion of telomere-like sequences found at interstitial places in human cells frequently lead to chromosome translocation and instability (Murnane et al. 1994; Draskovic et al. 2009; Marzec et al. 2015). As a result, ALT is almost exclusively observed under conditions where telomerase is absent and as a result of strong selection pressure. In humans, it is only observed in a limited proportion of tumor cells (Claude and Decottignies 2020).

5.4. Telomere length homeostasis

As mentioned previously, telomere length (the number of telomeric repeats) is a critical parameter of telomere function as critically short telomeres are not able to exert their protective function. In spite of this, there seems to be an imposed upper limit to telomere lengths in humans, even though telomere length appears to be highly variable between individuals, tissues, cells within the same tissue and even between chromosome ends in the same cell (Lansdorp et al. 1996; Londoño-Vallejo et al. 2001; Demanelis et al. 2020). Indeed, there is no other structure in the (human or not) genome with such dramatic dynamics. This dynamics, or length homeostasis, is the result of an equilibrium between lengthening reactions and shortening events, such as the ones described in the previous section, mostly associated with the S-phase. As noted above, a major mechanism for telomere lengthening is telomerase activity, which adds telomere repeats to the overhangs of telomeres. Several factors control telomerase activity in cells. In lower eukaryotes, such as budding yeast, telomerase is always expressed but its availability during replication is regulated to compensate for the loss of telomere repeats (Londoño-Vallejo and Wellinger 2012). Indeed, telomerase activity is regulated in cis by telomere-specific factors that bind to the double stranded part of telomeres or that act in trans to directly control either the accessibility to the substrate or the active status of the telomerase complex through post-translational modifications (Vasianovich and Wellinger 2017). In this way, telomerase is allowed to work more efficiently on relatively short telomeres than on relatively long telomeres to maintain a telomere length of around 300 bp.

The situation in human cells is totally different. Telomerase expression, and in particular that of the catalytic subunit TERT, is transcriptionally repressed in somatic cells after embryogenesis (Cong et al. 2002). Thus, telomere shortening events dominate the telomere length homeostasis landscape for the rest of the human's life. Indeed, the most remarkable observation is the progressive shortening of telomeres with age (Harley et al. 1990; Hastie et al. 1990; Chen et al. 2011) (see Figure 5.6).

Interestingly, there are marked differences in telomere length among healthy individuals belonging to the same age class (Chiang et al. 2010). Most of these differences are explained by genetic variations found in a few genes, including genes coding for TERT, the telomerase RNA, as well as factors involved in telomerase biogenesis and in telomere replication, thus underscoring the role of well-known lengthening and shortening mechanisms in telomere homeostasis in physiological contexts (Codd et al. 2013). However, endogenous factors (such as estrogen production) and life styles (such as cigarette smoking) do have an impact on telomere length (longer telomeres are found in women than in men and shorter telomeres are found in smokers in comparison to non-smokers) (Vyas et al. 2020). Interestingly, the extent of variation in telomere lengths observed in most age classes tend to be remarkably reduced in the very old, otherwise healthy, individuals (>90 years old) arguing for the existence of a minimal telomere length compatible with a healthy human (Halaschek-Wiener et al. 2008) (see Figure 5.6). However, it is not clear whether these individuals initially carried long telomeres that progressively shortened at the same pace than the rest of the healthy population or had relatively short telomeres that were better maintained by replication mechanisms and therefore shortened at a lower rate. Interestingly, chromosome ends display telomere length differences within the same cell, constituting true allelic polymorphisms as they are transmitted from parents to offspring, and these differences are maintained throughout life (Graakjaer et al. 2006). This observation supports the view following which the efficiency of replication mechanisms, and therefore the impact of shortening events at individual chromosome extremities, are strongly influenced by cis-acting, genetically or epigenetically determined, elements.

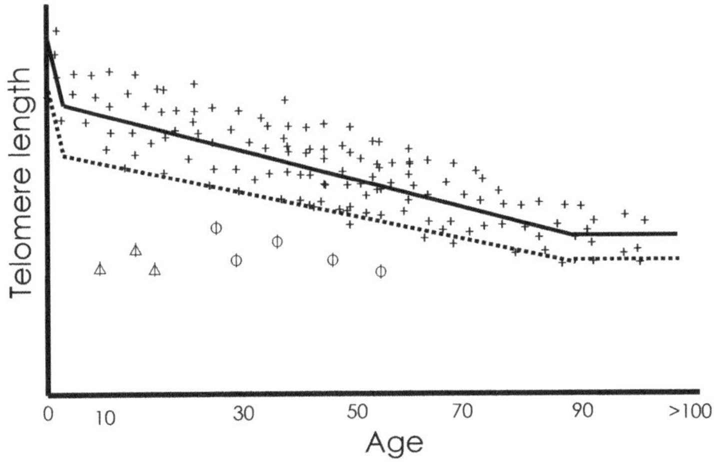

Figure 5.6. *Telomere length shortens with age and exaggerated shortening is associated with premature aging. The graph represents a hypothetical population study using either flow-FISH- or PCR-based methodologies to measure telomere length in peripheral white blood cells. Telomeres are long at birth (15–18 kb) and shorten rapidly in the first years of life because of intense cell proliferation and in spite of the presence of telomerase activity in bone marrow cells progenitors. After that, telomere lengths then shorten very progressively with age (23–47 bp/year) (Chen et al. 2011) to reach a lower level around 6–8 kb. The important dispersion of measurements within age classes mainly reflects the genetic heterogeneity among individuals and the fact that telomere length and (likely) the strength of telomerase activity are genetically determined. However, sex and exogenous factors (such as smoking) have a significant impact on telomere lengths. Note that the dispersion of values at an advanced age tends to decrease, suggesting a minimal level of telomere length to ensure proper organismal function (Halaschek-Wiener et al. 2008). Individuals with telomeres shorter than the 10th percentile (indicated by the dotted line) and clinical manifestations are probably affected by a short telomere syndrome. Φ indicates examples of adults affected by mutations in any of the telomerase components or in Shelterin subunits. ⚠ indicates children having inherited a dominant mutation from the affected parent (for instance, one Φ). These children tend to be affected earlier in life and have shorter telomeres than (or as short telomeres as) their parents in spite of their young age (anticipation)*

5.5. Telomeres and genome organization and function

Telomeres not only protect chromosome ends but they also impact the way the rest of the chromosome's organization and function. In budding yeast, telomeres are permanently attached to the nuclear envelope, thus limiting chromosome arm movements, already constrained by the fact that centromeres in the yeast are also constrained as they are found in a Rabl-like conformation in which centromeres remain attached to the spindle pole body after mitosis (Londoño-Vallejo and Wellinger 2012). In human cells, some specific telomeres are found to be almost permanently associated with the nuclear envelope, although a much broader association of up to 50% of all telomeres with the nuclear envelope is transiently observed during nuclear envelope reformation in telophase, right after chromosome segregation (Crabbe et al. 2012). Interestingly, human telomeres associated with the nuclear envelope are replicated during the last part of the S-phase, similar to what is seen in budding yeast, where telomeres are all replicated late when compared to the rest of the genome (Arnoult et al. 2010). Telomeres also play a role in sister chromatid cohesion as they are the last region to be separated during chromosome segregation (Canudas and Smith 2009; Hsiao and Smith 2009). Non-separation of telomeres, due to the absence of tankyrase, will impact chromosome migration during anaphase leading to mis-segregation (Hsiao and Smith 2009).

Telomeres also affect the way genes found in subtelomeric regions are expressed, through a phenomenon called "telomere position effect" (TPE). In the context of TPE, first described in the budding yeast (Gottschling et al. 1990; Evans et al. 1998), genes located closely to long telomeres tend to be repressed, and their expression will be reactivated upon telomere shortening. In human cells, TPE has been experimentally documented using juxtatelomeric reporter transgenes (Baur et al. 2001), although it appears that TPE in human cells can be exerted at much greater distances, arguing for the existence of long range physical contacts in cis between the telomeric structure and interstitial chromosome regions (Robin et al. 2014). The frequency and extent of such phenomena remain unknown but it is expected to play an important physiological role during embryonic development and cell differentiation.

Lastly, during meiosis, all telomeres associate with the nuclear envelope and migrate to the same place to form a "bouquet" (Scherthan et al. 1996). Bouquet formation is deemed to be important for the alignment of homologues and subsequent chromosome dynamics during spermatogenesis.

5.6. Cell senescence, aging and disease

The detection by the cell of a critically short telomere will trigger a local DNA damage response (DDR), followed by a repair reaction (fusion to another available extremity). However, the presence of several dysfunctional telomeres and the consequent increase in the level of DDR will lead to the activation of a general stress response in which activation of the p53 pathway is the leading event (Roake and Artandi 2017). This activation will have a broad impact on the transcriptional program of the cell, propping the cell into senescence (state of permanent cell cycle arrest) or into apoptosis (induced programmed cell dead). This telomere-driven cell senescence/apoptosis mechanism, together with the telomere shortening associated with age and the accumulation of senescent cells in the tissues of old individuals, supports the central role of telomeres during the aging process (Tuttle et al. 2021). Indeed, aging is characterized by the increasing difficulty to renew lost or damaged cells in all organs, thus impacting their regeneration capacity and function. A clear demonstration of the link between telomere maintenance and aging came with the discovery that mutations in telomerase components or in Shelterin were responsible for premature aging syndromes such as dyskeratosis congenita (Mitchell et al. 1999). On the other hand, mutations in genes involved in telomere maintenance are associated with an extended list of aging-related diseases (such as bone marrow dysfunctions, liver and lung fibrosis, etc.) (Armanios 2009). Lastly, very often in familial telomere-related diseases, clinical manifestations tend to occur at earlier ages with every generation (Vulliamy et al. 2004; Armanios et al. 2005). This phenomenon, called anticipation, indicates that the offspring that received the causal mutation also received shorter telomeres from the affected parent, thus compounding the telomere maintenance defect (Goldman et al. 2005).

Finally, of all aging-related diseases, the most closely related to telomeres is cancer (DePinho 2000). There are two major reasons for this. Firstly, most tumor cells reactivate a telomere maintenance mechanism (most of the time telomerase), which allows them to proliferate indefinitely (Harley et al. 1994). The mechanisms of reactivation of telomerase (or of ALT mechanisms) are diverse but will not be discussed here. The other reason comes from observations following which patients affected by cancer tend to have shorter telomeres than healthy people of the same age (Wu et al. 2003). Short telomeres have since been considered a risk factor for cancer, likely because they increase the chance for the precancerous cell to accumulate chromosome aberrations, a characteristic of most cancers (Der-Sarkissian et al. 2004; Londoño-Vallejo 2008). However, as shown in prospective studies where individuals with short telomeres were followed for at least 20 years, while these individuals showed a higher risk of mortality as well as of suffering cardiovascular and inflammatory diseases, they did not show an increase in cancer incidence (Cawthon et al. 2003). This conundrum has been recently solved by other epidemiological studies where mendelian randomization approaches have been used; that is, instead of measuring telomere lengths, individuals were genotyped for genetic variations associated with long or short telomeres (Rode et al. 2015). These studies showed that individuals with a genetic composition compatible with short telomeres were significantly protected against cancers (Rode et al. 2015). This important observation lines up very nicely with all the experimental evidence which has been cumulated over the last 30 years, supporting the role of telomere shortening as a strong barrier to uncontrolled proliferation and therefore as a major antitumor mechanism (Gunes et al. 2018).

5.7. Conclusion

Telomeres are remarkable structures not only because of their relatively simple primary sequences which contrast with the highly complex functions they fulfill, but also for the extreme dynamics they experience both at the single cell and the organismal level. Much still remains to be understood to fully comprehend the biological roles of telomeres in genome organization and genomic stress as well as the broad impact of their metabolism in virtually all aspects of eukaryote life.

5.8. Acknowledgments

Work in the telomeres and cancer laboratory has been supported along the years with multiple grants from ANR, INCa, Fondation ARC, La Ligue contre le Cancer, La Fondation de France, La Fondation pour les Maladies Rares, as well as internal grant calls from the Curie Institute. I thank all former and present members of the lab, as well as all collaborators. I apologize to all colleagues in the telomere field whose original works were not cited due to space limitations.

5.9. References

Abad, J.P. and Villasante, A. (1999). The 3' non-coding region of the *Drosophila melanogaster* HeT-A telomeric retrotransposon contains sequences with propensity to form G-quadruplex DNA. *FEBS Lett.*, 453, 59–62.

Armanios, M. (2009). Syndromes of telomere shortening. *Annu. Rev. Genomics Hum. Genet.*, 10, 45–61.

Armanios, M., Chen, J.L., Chang, Y.P., Brodsky, R.A., Hawkins, A., Griffin, C.A., Eshleman, J.R., Cohen, A.R., Chakravarti, A., Hamosh, A. et al. (2005). Haploinsufficiency of telomerase reverse transcriptase leads to anticipation in autosomal dominant dyskeratosis congenita. *Proc. Natl. Acad. Sci. USA*, 102, 15960–15964.

Arnoult, N., Saintome, C., Ourliac-Garnier, I., Riou, J.F., Londoño-Vallejo, A. (2009). Human POT1 is required for efficient telomere C-rich strand replication in the absence of WRN. *Genes Dev.*, 23, 2915–2924.

Arnoult, N., Schluth-Bolard, C., Letessier, A., Drascovic, I., Bouarich-Bourimi, R., Campisi, J., Kim, S.H., Boussouar, A., Ottaviani, A., Magdinier, F. et al. (2010). Replication timing of human telomeres is chromosome arm-specific, influenced by subtelomeric structures and connected to nuclear localization. *PLoS Genet.*, 6, e1000920.

Audry, J., Maestroni, L., Delagoutte, E., Gauthier, T., Nakamura, T.M., Gachet, Y., Saintome, C., Geli, V., Coulon, S. (2015). RPA prevents G-rich structure formation at lagging-strand telomeres to allow maintenance of chromosome ends. *EMBO J.*, 34, 1942–1958.

Autexier, C. and Greider, C.W. (1994). Functional reconstitution of wild-type and mutant tetrahymena telomerase. *Genes Dev.*, 8, 563–575.

Azzalin, C.M. and Lingner, J. (2008). Telomeres: The silence is broken. *Cell Cycle*, 7, 1161–1165.

Azzalin, C.M., Reichenbach, P., Khoriauli, L., Giulotto, E., Lingner, J. (2007). Telomeric repeat containing RNA and RNA surveillance factors at mammalian chromosome ends. *Science*, 318, 798–801.

Balk, B., Dees, M., Bender, K., Luke, B. (2014). The differential processing of telomeres in response to increased telomeric transcription and RNA-DNA hybrid accumulation. *RNA Biol.*, 11, 95–100.

Baumann, P. and Cech, T.R. (2001). Pot1, the putative telomere end-binding protein in fission yeast and humans. *Science*, 292, 1171–1175.

Baur, J.A., Zou, Y., Shay, J.W., Wright, W.E. (2001). Telomere position effect in human cells. *Science*, 292, 2075–2077.

Beattie, T.L., Zhou, W., Robinson, M.O., Harrington, L. (1998). Reconstitution of human telomerase activity in vitro. *Curr. Biol.*, 8, 177–180.

Belotserkovskii, B.P., Liu, R., Tornaletti, S., Krasilnikova, M.M., Mirkin, S.M., Hanawalt, P.C. (2010). Mechanisms and implications of transcription blockage by guanine-rich DNA sequences. *Proc. Natl. Acad. Sci. USA*, 107, 12816–12821.

Benetti, R., Gonzalo, S., Jaco, I., Schotta, G., Klatt, P., Jenuwein, T., Blasco, M.A. (2007). Suv4-20h deficiency results in telomere elongation and derepression of telomere recombination. *J. Cell. Biol.*, 178, 925–936.

Biessmann, H., Walter, M.F., Mason, J.M. (1997). Drosophila telomere elongation. *Ciba. Found. Symp.*, 211, 53–67.

Blasco, M.A. (2004). Telomere epigenetics: A higher-order control of telomere length in mammalian cells. *Carcinogenesis*, 6, 6.

Bryan, T.M. (2020). G-quadruplexes at telomeres: Friend or foe? *Molecules*, 25(16), 3686.

Canudas, S. and Smith, S. (2009). Differential regulation of telomere and centromere cohesion by the Scc3 homologues SA1 and SA2, respectively, in human cells. *J. Cell Biol.*, 187, 165–173.

Capper, R., Britt-Compton, B., Tankimanova, M., Rowson, J., Letsolo, B., Man, S., Haughton, M., Baird, D.M. (2007). The nature of telomere fusion and a definition of the critical telomere length in human cells. *Genes Dev.*, 21, 2495–2508.

Casacuberta, E. and Pardue, M.L. (2003). Transposon telomeres are widely distributed in the Drosophila genus: TART elements in the virilis group. *Proc. Natl. Acad. Sci. USA*, 7, 7.

Cawthon, R.M., Smith, K.R., O'Brien, E., Sivatchenko, A., Kerber, R.A. (2003). Association between telomere length in blood and mortality in people aged 60 years or older. *Lancet*, 361, 393–395.

Chaires, J.B., Gray, R.D., Dean, W.L., Monsen, R., Deleeuw, L.W., Stribinskis, V., Trent, J.O. (2020). Human POT1 unfolds G-quadruplexes by conformational selection. *Nucleic Acids Res.*, 48, 4976–4991.

Chen, F.M. (1992). Sr^{2+} facilitates intermolecular G-quadruplex formation of telomeric sequences. *Biochemistry*, 31, 3769–3776.

Chen, W., Kimura, M., Kim, S., Cao, X., Srinivasan, S.R., Berenson, G.S., Kark, J.D., Aviv, A. (2011). Longitudinal versus cross-sectional evaluations of leukocyte telomere length dynamics: Age-dependent telomere shortening is the rule. *J. Gerontol. A Biol. Sci. Med. Sci.*, 66, 312–319.

Chiang, Y.J., Calado, R.T., Hathcock, K.S., Lansdorp, P.M., Young, N.S., Hodes, R.J. (2010). Telomere length is inherited with resetting of the telomere set-point. *Proc. Natl. Acad. Sci. USA*, 107, 10148–10153.

Cicconi, A. and Chang, S. (2020). Shelterin and the replisome: At the intersection of telomere repair and replication. *Curr. Opin. Genet Dev.*, 60, 77–84.

Claude, E. and Decottignies, A. (2020). Telomere maintenance mechanisms in cancer: Telomerase, ALT or lack thereof. *Curr. Opin. Genet. Dev.*, 60, 1–8.

Codd, V., Nelson, C.P., Albrecht, E., Mangino, M., Deelen, J., Buxton, J.L., Hottenga, J.J., Fischer, K., Esko, T., Surakka, I. et al. (2013). Identification of seven loci affecting mean telomere length and their association with disease. *Nat. Genet.*, 45, 422–427, 427e1-2.

Cohn, M., Mceachern, M.J., Blackburn, E.H. (1998). Telomeric sequence diversity within the genus Saccharomyces. *Curr. Genet.*, 33, 83–91.

Collins, K. (2006). The biogenesis and regulation of telomerase holoenzymes. *Nat. Rev. Mol. Cell. Biol.*, 7, 484–494.

Cong, Y.S., Wright, W.E., Shay, J.W. (2002). Human telomerase and its regulation. *Microbiol. Mol. Biol. Rev.*, 66, 407–425.

Crabbe, L., Verdun, R.E., Haggblom, C.I., Karlseder, J. (2004). Defective telomere lagging strand synthesis in cells lacking WRN helicase activity. *Science*, 306, 1951–1953.

Crabbe, L., Cesare, A.J., Kasuboski, J.M., Fitzpatrick, J.A., Karlseder, J. (2012). Human telomeres are tethered to the nuclear envelope during postmitotic nuclear assembly. *Cell Rep*, 2, 1521–1529.

Cristofari, G. and Lingner, J. (2006). Telomere length homeostasis requires that telomerase levels are limiting. *EMBO J*, 25, 565–574.

Cubiles, M.D., Barroso, S., Vaquero-Sedas, M.I., Enguix, A., Aguilera, A., Vega-Palas, M.A. (2018). Epigenetic features of human telomeres. *Nucleic Acids Res.*, 46, 2347–2355.

Cusanelli, E., Romero, C.A., Chartrand, P. (2013). Telomeric noncoding RNA TERRA is induced by telomere shortening to nucleate telomerase molecules at short telomeres. *Mol. Cell.*, 51, 780–791.

De Bruin, D., Kantrow, S.M., Liberatore, R.A., Zakian, V.A. (2000). Telomere folding is required for the stable maintenance of telomere position effects in yeast. *Mol. Cell. Biol.*, 20, 7991–8000.

De Lange, T. (2004). T-loops and the origin of telomeres. *Nat. Rev. Mol. Cell. Biol.*, 5, 323–329.

De Lange, T. (2009). How telomeres solve the end-protection problem. *Science*, 326, 948–952.

Demanelis, K., Jasmine, F., Chen, L.S., Chernoff, M., Tong, L., Delgado, D., Zhang, C., Shinkle, J., Sabarinathan, M., Lin, H. et al. (2020). Determinants of telomere length across human tissues. *Science*, 369, eaaz6876.

Depinho, R.A. (2000). The age of cancer. *Nature*, 408, 248–254.

Der-Sarkissian, H., Bacchetti, S., Cazes, L., Londoño-Vallejo, J.A. (2004). The shortest telomeres drive karyotype evolution in transformed cells. *Oncogene*, 23, 1221–1228.

Draskovic, I., Arnoult, N., Steiner, V., Bacchetti, S., Lomonte, P., Londoño-Vallejo, A. (2009). Probing PML body function in ALT cells reveals spatiotemporal requirements for telomere recombination. *Proc. Natl. Acad. Sci. USA*, 106, 15726–15731.

Evans, S.K., Sistrunk, M.L., Nugent, C.I., Lundblad, V. (1998). Telomerase, ku, and telomeric silencing in *Saccharomyces cerevisiae*. *Chromosoma*, 107, 352–358.

Fernando, H., Rodriguez, R., Balasubramanian, S. (2008). Selective recognition of a DNA G-quadruplex by an engineered antibody. *Biochemistry*, 47, 9365–9371.

Ferrarelli, L.K., Popuri, V., Ghosh, A.K., Tadokoro, T., Canugovi, C., Hsu, J.K., Croteau, D.L., Bohr, V.A. (2013). The RECQL4 protein, deficient in Rothmund-Thomson syndrome is active on telomeric D-loops containing DNA metabolism blocking lesions. *DNA Repair (Amst)*, 12, 518–528.

Fulneckova, J., Sevcikova, T., Fajkus, J., Lukesova, A., Lukes, M., Vlcek, C., Lang, B.F., Kim, E., Elias, M., Sykorova, E. (2013). A broad phylogenetic survey unveils the diversity and evolution of telomeres in eukaryotes. *Genome Biol. Evol.*, 5, 468–483.

Gallardo, F., Olivier, C., Dandjinou, A.T., Wellinger, R.J., Chartrand, P. (2008). TLC1 RNA nucleo-cytoplasmic trafficking links telomerase biogenesis to its recruitment to telomeres. *EMBO J.*, 27, 748–757.

Garavis, M., Gonzalez, C., Villasante, A. (2013). On the origin of the eukaryotic chromosome: The role of noncanonical DNA structures in telomere evolution. *Genome Biol. Evol.*, 5, 1142–1150.

Gauchier, M., Kan, S., Barral, A., Sauzet, S., Agirre, E., Bonnell, E., Saksouk, N., Barth, T.K., Ide, S., Urbach, S. et al. (2019). SETDB1-dependent heterochromatin stimulates alternative lengthening of telomeres. *Sci Adv.*, 5, eaav3673.

Gilson, E. and Geli, V. (2007). How telomeres are replicated. *Nat. Rev. Mol. Cell. Biol.*, 8, 825–838.

Giraud-Panis, M.J., Ye, J., Gilson, E. (2018). TRFH domain: At the root of telomere protein evolution? *Cell Res.*, 28, 7–8.

Goldman, F., Bouarich, R., Kulkarni, S., Freeman, S., Du, H.Y., Harrington, L., Mason, P.J., Londoño-Vallejo, A., Bessler, M. (2005). The effect of TERC haploinsufficiency on the inheritance of telomere length. *Proc. Natl. Acad. Sci. USA*, 102, 17119–17124.

Gomez, D., Aouali, N., Londoño-Vallejo, A., Lacroix, L., Megnin-Chanet, F., Lemarteleur, T., Douarre, C., Shin-Ya, K., Mailliet, P., Trentesaux, C. et al. (2003). Resistance to the short term antiproliferative activity of the G-quadruplex ligand 12459 is associated with telomerase overexpression and telomere capping alteration. *J. Biol. Chem.*, 278, 50554–50562.

Gottschling, D.E., Aparicio, O.M., Billington, B.L., Zakian, V.A. (1990). Position effect at *S. cerevisiae* telomeres: Reversible repression of Pol II transcription. *Cell*, 63, 751–762.

Graakjaer, J., Der-Sarkissian, H., Schmitz, A., Bayer, J., Thomas, G., Kolvraa, S., Londoño-Vallejo, J.A. (2006). Allele-specific relative telomere lengths are inherited. *Hum. Genet.*, 119, 344–350.

Graf, M., Bonetti, D., Lockhart, A., Serhal, K., Kellner, V., Maicher, A., Jolivet, P., Teixeira, M.T., Luke, B. (2017). Telomere length determines TERRA and R-loop regulation through the cell cycle. *Cell*, 170, 72–85, e14.

Granotier, C., Pennarun, G., Riou, L., Hoffschir, F., Gauthier, L.R., De Cian, A., Gomez, D., Mandine, E., Riou, J.F., Mergny, J.L. et al. (2005). Preferential binding of a G-quadruplex ligand to human chromosome ends. *Nucleic Acids Res.*, 33, 4182–4190.

Griffith, J.D., Comeau, L., Rosenfield, S., Stansel, R.M., Bianchi, A., Moss, H., De Lange, T. (1999). Mammalian telomeres end in a large duplex loop. *Cell*, 97, 503–514.

Gunes, C., Avila, A.I., Rudolph, K.L. (2018). Telomeres in cancer. *Differentiation*, 99, 41–50.

Halaschek-Wiener, J., Vulto, I., Fornika, D., Collins, J., Connors, J.M., Le, N.D., Lansdorp, P.M., Brooks-Wilson, A. (2008). Reduced telomere length variation in healthy oldest old. *Mech. Ageing Dev.*, 129, 638–641.

Harley, C.B., Futcher, A.B., Greider, C.W. (1990). Telomeres shorten during ageing of human fibroblasts. *Nature*, 345, 458–460.

Harley, C.B., Kim, N.W., Prowse, K.R., Weinrich, S.L., Hirsch, K.S., West, M.D., Bacchetti, S., Hirte, H.W., Counter, C.M., Greider, C.W. et al. (1994). Telomerase, cell immortality, and cancer. *Cold Spring Harb. Symp. Quant. Biol.*, 59, 307–315.

Hastie, N.D., Dempster, M., Dunlop, M.G., Thompson, A.M., Green, D.K., Allshire, R.C. (1990). Telomere reduction in human colorectal carcinoma and with ageing. *Nature*, 346, 866–868.

Hsiao, S.J. and Smith, S. (2009). Sister telomeres rendered dysfunctional by persistent cohesion are fused by NHEJ. *J. Cell Biol.*, 184, 515–526.

Kamnert, I., Lopez, C.C., Rosen, M., Edstrom, J.E. (1997). Telomeres terminating with long complex tandem repeats. *Hereditas*, 127, 175–180.

Kendellen, M.F., Barrientos, K.S., Counter, C.M. (2009). POT1 association with TRF2 regulates telomere length. *Mol. Cell. Biol.*, 29(20), 5611–5619.

Lansdorp, P.M. and Van Wietmarschen, N. (2019). Helicases FANCJ, RTEL1 and BLM act on guanine quadruplex DNA in vivo. *Genes (Basel)*, 10(11), 870.

Lansdorp, P.M., Verwoerd, N.P., Van De Rijke, F.M., Dragowska, V., Little, M.T., Dirks, R.W., Raap, A.K., Tanke, H.J. (1996). Heterogeneity in telomere length of human chromosomes. *Hum. Mol. Genet.*, 5, 685–691.

Laprade, H., Querido, E., Smith, M.J., Guerit, D., Crimmins, H., Conomos, D., Pourret, E., Chartrand, P., Sfeir, A. (2020). Single-molecule imaging of telomerase RNA reveals a recruitment-retention model for telomere elongation. *Mol. Cell.*, 79, 115–126, e6.

Lejnine, S., Makarov, V.L., Langmore, J.P. (1995). Conserved nucleoprotein structure at the ends of vertebrate and invertebrate chromosomes. *Proc. Natl. Acad. Sci. USA*, 92, 2393–2397.

Linger, B.R. and Price, C.M. (2009). Conservation of telomere protein complexes: Shuffling through evolution. *Crit. Rev. Biochem. Mol. Biol.*, 44, 434–446.

Lingner, J., Cooper, J.P., Cech, T.R. (1995). Telomerase and DNA end replication: No longer a lagging strand problem? *Science*, 269, 1533–1534.

Lingner, J., Hughes, T.R., Shevchenko, A., Mann, M., Lundblad, V., Cech, T.R. (1997). Reverse transcriptase motifs in the catalytic subunit of telomerase. *Science*, 276, 561–567.

Loayza, D. and De Lange, T. (2003). POT1 as a terminal transducer of TRF1 telomere length control. *Nature*, 423, 1013–1018.

Londoño-Vallejo, J.A. (2008). Telomere instability and cancer. *Biochimie*, 90, 73–82.

Londoño-Vallejo, J.A. and Wellinger, R.J. (2012). Telomeres and telomerase dance to the rhythm of the cell cycle. *Trends Biochem. Sci.*, 37, 391–399.

Londoño-Vallejo, J.A., Dersarkissian, H., Cazes, L., Thomas, G. (2001). Differences in telomere length between homologous chromosomes in humans. *Nucleic Acids Res.*, 29, 3164–3171.

Lopes, J., Piazza, A., Bermejo, R., Kriegsman, B., Colosio, A., Teulade-Fichou, M.P., Foiani, M., Nicolas, A. (2011). G-quadruplex-induced instability during leading-strand replication. *EMBO J.*, 30, 4033–4046.

Lopez de Silanes, I., Stagno d'Alcontres, M., Blasco, M.A. (2010). TERRA transcripts are bound by a complex array of RNA-binding proteins. *Nat. Commun.*, 1, 33.

Lototska, L., Yue, J.X., Li, J., Giraud-Panis, M.J., Songyang, Z., Royle, N.J., Liti, G., Ye, J., Gilson, E., Mendez-Bermudez, A. (2020). Human RAP1 specifically protects telomeres of senescent cells from DNA damage. *EMBO Rep.*, 21, e49076.

Luke, B., Panza, A., Redon, S., Iglesias, N., Li, Z., Lingner, J. (2008). The Rat1p 5' to 3' exonuclease degrades telomeric repeat-containing RNA and promotes telomere elongation in Saccharomyces cerevisiae. *Mol. Cell*, 32, 465–477.

Lundblad, V. and Blackburn, E.H. (1993). An alternative pathway for yeast telomere maintenance rescues est1-senescence. *Cell*, 73, 347–360.

Majerska, J., Feretzaki, M., Glousker, G., Lingner, J. (2018). Transformation-induced stress at telomeres is counteracted through changes in the telomeric proteome including SAMHD1. *Life Sci. Alliance*, 1, e201800121.

Marzec, P., Armenise, C., Perot, G., Roumelioti, F.M., Basyuk, E., Gagos, S., Chibon, F., Dejardin, J. (2015). Nuclear-receptor-mediated telomere insertion leads to genome instability in ALT cancers. *Cell*, 160, 913–927.

McClintock, B. (1938). The production of homozygous deficient tissues with mutant characteristics by means of the aberrant mitotic behavior of ring-shaped chromosomes. *Genetics*, 23, 315–376.

Mendoza, O., Bourdoncle, A., Boule, J.B., Brosh Jr., R.M., Mergny, J.L. (2016). G-quadruplexes and helicases. *Nucleic Acids Res.*, 44, 1989–2006.

Mergny, J.L., Lacroix, L., Teulade-Fichou, M.P., Hounsou, C., Guittat, L., Hoarau, M., Arimondo, P.B., Vigneron, J.P., Lehn, J.M., Riou, J.F. et al. (2001). Telomerase inhibitors based on quadruplex ligands selected by a fluorescence assay. *Proc. Natl. Acad. Sci. USA*, 98, 3062–3067.

Miller, K.M., Rog, O., Cooper, J.P. (2006). Semi-conservative DNA replication through telomeres requires Taz1. *Nature*, 440, 824–828.

Mitchell, J.R., Wood, E., Collins, K. (1999). A telomerase component is defective in the human disease dyskeratosis congenita. *Nature*, 402, 551–555.

Morin, G.B. (1989). The human telomere terminal transferase enzyme is a ribonucleoprotein that synthesizes TTAGGG repeats. *Cell*, 59, 521–529.

Moyzis, R.K., Buckingham, J.M., Cram, L.S., Dani, M., Deaven, L.L., Jones, M.D., Meyne, J., Ratliff, R.L., Wu, J.R. (1988). A highly conserved repetitive DNA sequence, (TTAGGG)n, present at the telomeres of human chromosomes. *Proc. Natl. Acad. Sci. USA*, 85, 6622–6626.

Muller, H. (1938). The remaking of chromosomes. *Collecting Net.*, 13, 181–198.

Murnane, J.P., Sabatier, L., Marder, B.A., Morgan, W.F. (1994). Telomere dynamics in an immortal human cell line. *EMBO J.*, 13, 4953–4962.

Nandakumar, J., Bell, C.F., Weidenfeld, I., Zaug, A.J., Leinwand, L.A., Cech, T.R. (2012). The TEL patch of telomere protein TPP1 mediates telomerase recruitment and processivity. *Nature*, 492, 285–289.

Nergadze, S.G., Farnung, B.O., Wischnewski, H., Khoriauli, L., Vitelli, V., Chawla, R., Giulotto, E., Azzalin, C.M. (2009). CpG-island promoters drive transcription of human telomeres. *RNA*, 15, 2186–2194.

Nie, X., Xiao, D., Ge, Y., Xie, Y., Zhou, H., Zheng, T., Li, X., Liu, H., Huang, H., Zhao, Y. (2021). TRF2 recruits nucleolar protein TCOF1 to coordinate telomere transcription and replication. *Cell Death Differ*, 28, 1062–1075.

Olovnikov, A.M. (1973). A theory of marginotomy. The incomplete copying of template margin in enzymic synthesis of polynucleotides and biological significance of the phenomenon. *J. Theor. Biol.*, 41, 181–190.

Palm, W. and De Lange, T. (2008). How shelterin protects mammalian telomeres. *Annu. Rev. Genet.*, 42, 301–334.

Petti, E., Buemi, V., Zappone, A., Schillaci, O., Broccia, P.V., Dinami, R., Matteoni, S., Benetti, R., Schoeftner, S. (2019). SFPQ and NONO suppress RNA:DNA-hybrid-related telomere instability. *Nat. Commun.*, 10, 1001.

Pfeiffer, V., Crittin, J., Grolimund, L., Lingner, J. (2013). The THO complex component Thp2 counteracts telomeric R-loops and telomere shortening. *EMBO J.*, 32, 2861–2871.

Piazza, A., Boule, J.B., Lopes, J., Mingo, K., Largy, E., Teulade-Fichou, M.P., Nicolas, A. (2010). Genetic instability triggered by G-quadruplex interacting Phen-DC compounds in *Saccharomyces cerevisiae*. *Nucleic Acids Res.*, 38, 4337–4348.

Pinzaru, A.M., Kareh, M., Lamm, N., Lazzerini-Denchi, E., Cesare, A.J., Sfeir, A. (2020). Replication stress conferred by POT1 dysfunction promotes telomere relocalization to the nuclear pore. *Genes Dev.*, 34, 1619–1636.

Porreca, R.M., Glousker, G., Awad, A., Matilla Fernandez, M.I., Gibaud, A., Naucke, C., Cohen, S.B., Bryan, T.M., Tzfati, Y., Draskovic, I. et al. (2018). Human RTEL1 stabilizes long G-overhangs allowing telomerase-dependent over-extension. *Nucleic Acids Res.*, 46, 4533–4545.

Porreca, R.M., Herrera-Moyano, E., Skourti, E., Law, P.P., Gonzalez Franco, R., Montoya, A., Faull, P., Kramer, H., Vannier, J.B. (2020). TRF1 averts chromatin remodelling, recombination and replication dependent-break induced replication at mouse telomeres. *Elife*, 9, e49817.

Reddel, R.R., Bryan, T.M., Murnane, J.P. (1997). Immortalized cells with no detectable telomerase activity. A review. *Biochemistry (Mosc.)*, 62, 1254–1262.

Ribeyre, C., Lopes, J., Boule, J.B., Piazza, A., Guedin, A., Zakian, V.A., Mergny, J.L., Nicolas, A. (2009). The yeast Pif1 helicase prevents genomic instability caused by G-quadruplex-forming CEB1 sequences in vivo. *PLoS Genet.*, 5, e1000475.

Riou, J.F., Guittat, L., Mailliet, P., Laoui, A., Renou, E., Petitgenet, O., Megnin-Chanet, F., Helene, C., Mergny, J.L. (2002). Cell senescence and telomere shortening induced by a new series of specific G-quadruplex DNA ligands. *Proc. Natl. Acad. Sci. USA*, 99, 2672–2677.

Rizzo, A., Salvati, E., Porru, M., d'Angelo, C., Stevens, M.F., d'Incalci, M., Leonetti, C., Gilson, E., Zupi, G., Biroccio, A. (2009). Stabilization of quadruplex DNA perturbs telomere replication leading to the activation of an ATR-dependent ATM signaling pathway. *Nucleic Acids Res.*, 37, 5353–5364.

Roake, C.M. and Artandi, S.E. (2017). Control of cellular aging, tissue function, and cancer by p53 downstream of telomeres. *Cold Spring Harb. Perspect. Med.*, 7, a026088.

Robin, J.D., Ludlow, A.T., Batten, K., Magdinier, F., Stadler, G., Wagner, K.R., Shay, J.W., Wright, W.E. (2014). Telomere position effect: Regulation of gene expression with progressive telomere shortening over long distances. *Genes Dev.*, 28, 2464–2476.

Rode, L., Nordestgaard, B.G., Bojesen, S.E. (2015). Peripheral blood leukocyte telomere length and mortality among 64,637 individuals from the general population. *J. Natl. Cancer Inst.*, 107, djv074.

Roumelioti, F.M., Sotiriou, S.K., Katsini, V., Chiourea, M., Halazonetis, T.D., Gagos, S. (2016). Alternative lengthening of human telomeres is a conservative DNA replication process with features of break-induced replication. *EMBO Rep.*, 17, 1731–1737.

Saint-Leandre, B., Christopher, C., Levine, M.T. (2020). Adaptive evolution of an essential telomere protein restricts telomeric retrotransposons. *Elife*, 9, e60987.

Salas, T.R., Petruseva, I., Lavrik, O., Bourdoncle, A., Mergny, J.L., Favre, A., Saintomé, C. (2006). Human replication protein A unfolds telomeric G-quadruplexes. *Nucleic Acids Res.*, 34, 4857–4865.

Schaffitzel, C., Berger, I., Postberg, J., Hanes, J., Lipps, H.J., Pluckthun, A. (2001). In vitro generated antibodies specific for telomeric guanine-quadruplex DNA react with Stylonychia lemnae macronuclei. *Proc. Natl. Acad. Sci. USA*, 98, 8572–8577.

Scheibe, M., Arnoult, N., Kappei, D., Buchholz, F., Decottignies, A., Butter, F., Mann, M. (2013). Quantitative interaction screen of telomeric repeat-containing RNA reveals novel TERRA regulators. *Genome Res.*, 23, 2149–2157.

Scherthan, H., Weich, S., Schwegler, H., Heyting, C., Harle, M., Cremer, T. (1996). Centromere and telomere movements during early meiotic prophase of mouse and man are associated with the onset of chromosome pairing. *J. Cell Biol.*, 134, 1109–1125.

Schmidt, J.C., Zaug, A.J., Cech, T.R. (2016). Live cell imaging reveals the dynamics of telomerase recruitment to telomeres. *Cell*, 166, 1188–1197, e9.

Sfeir, A. and De Lange, T. (2012). Removal of shelterin reveals the telomere end-protection problem. *Science*, 336, 593–597.

Sfeir, A., Kosiyatrakul, S.T., Hockemeyer, D., Macrae, S.L., Karlseder, J., Schildkraut, C.L., De Lange, T. (2009). Mammalian telomeres resemble fragile sites and require TRF1 for efficient replication. *Cell*, 138, 90–103.

Smith, F.W. and Feigon, J. (1993). Strand orientation in the DNA quadruplex formed from the Oxytricha telomere repeat oligonucleotide d(G4T4G4) in solution. *Biochemistry*, 32, 8682–8692.

Sun, L., Tan, R., Xu, J., Laface, J., Gao, Y., Xiao, Y., Attar, M., Neumann, C., Li, G.M., Su, B. et al. (2015). Targeted DNA damage at individual telomeres disrupts their integrity and triggers cell death. *Nucleic Acids Res.*, 43, 6334–6347.

Teixeira, M.T. and Gilson, E. (2005). Telomere maintenance, function and evolution: The yeast paradigm. *Chromosome Res.*, 13, 535–548.

Teixeira, M.T., Forstemann, K., Gasser, S.M., Lingner, J. (2002). Intracellular trafficking of yeast telomerase components. *EMBO Rep.*, 3, 652–659.

Theimer, C.A. and Feigon, J. (2006). Structure and function of telomerase RNA. *Curr. Opin. Struct. Biol.*, 16, 307–318.

Timashev, L.A. and De Lange, T. (2020). Characterization of t-loop formation by TRF2. *Nucleus*, 11, 164–177.

Tomlinson, R.L., Ziegler, T.D., Supakorndej, T., Terns, R.M., Terns, M.P. (2006). Cell cycle-regulated trafficking of human telomerase to telomeres. *Mol. Biol. Cell*, 17, 955–965.

Traut, W., Szczepanowski, M., Vitkova, M., Opitz, C., Marec, F., Zrzavy, J. (2007). The telomere repeat motif of basal Metazoa. *Chromosome Res.*, 15, 371–382.

Tuttle, C.S.L., Luesken, S.W.M., Waaijer, M.E.C., Maier, A.B. (2021). Senescence in tissue samples of humans with age-related diseases: A systematic review. *Ageing Research Reviews*, 68, 101334.

Van Steensel, B. and De Lange, T. (1997). Control of telomere length by the human telomeric protein TRF1. *Nature*, 385, 740–743.

Vaquero-Sedas, M.I. and Vega-Palas, M.A. (2011). On the chromatin structure of eukaryotic telomeres. *Epigenetics*, 6, 1055–1058.

Vasianovich, Y. and Wellinger, R.J. (2017). Life and death of yeast telomerase RNA. *J. Mol. Biol.*, 429, 3242–3254.

Viceconte, N., Loriot, A., Lona Abreu, P., Scheibe, M., Fradera Sola, A., Butter, F., De Smet, C., Azzalin, C.M., Arnoult, N., Decottignies, A. (2021). PAR-TERRA is the main contributor to telomeric repeat-containing RNA transcripts in normal and cancer mouse cells. *RNA*, 27, 106–121.

Villasante, A., Abad, J.P., Mendez-Lago, M. (2007). Centromeres were derived from telomeres during the evolution of the eukaryotic chromosome. *Proc. Natl. Acad. Sci. USA*, 104, 10542–10547.

Volff, J. and Altenbuchner, J. (2000). A new beginning with new ends: Linearisation of circular chromosomes during bacterial evolution. *FEMS Microbiol. Lett.*, 186, 143–150.

Vulliamy, T., Marrone, A., Szydlo, R., Walne, A., Mason, P.J., Dokal, I. (2004). Disease anticipation is associated with progressive telomere shortening in families with dyskeratosis congenita due to mutations in TERC. *Nat Genet.*, 36, 447–449.

Vyas, C.M., Ogata, S., Reynolds, C.F., Mischoulon, D., Chang, G., Cook, N.R., Manson, J.E., Crous-Bou, M., De Vivo, I., Okereke, O.I. (2020). Telomere length and its relationships with lifestyle and behavioural factors: Variations by sex and race/ethnicity. *Age and Ageing*, 50, 838–846.

Wang, Y. and Patel, D.J. (1992). Guanine residues in d(T2AG3) and d(T2G4) form parallel-stranded potassium cation stabilized G-quadruplexes with anti glycosidic torsion angles in solution. *Biochemistry*, 31, 8112–8119.

Watson, J.D. (1972). Origin of concatemeric T7 DNA. *Nat. New Biol.*, 239, 197–201.

Weinrich, S.L., Pruzan, R., Ma, L., Ouellette, M., Tesmer, V.M., Holt, S.E., Bodnar, A.G., Lichtsteiner, S., Kim, N.W., Trager, J.B. et al. (1997). Reconstitution of human telomerase with the template RNA component hTR and the catalytic protein subunit hTRT. *Nat Genet.*, 17, 498–502.

Wright, W.E., Tesmer, V.M., Huffman, K.E., Levene, S.D., Shay, J.W. (1997). Normal human chromosomes have long G-rich telomeric overhangs at one end. *Genes Dev.*, 11, 2801–2809.

Wu, X., Amos, C.I., Zhu, Y., Zhao, H., Grossman, B.H., Shay, J.W., Luo, S., Hong, W.K., Spitz, M.R. (2003). Telomere dysfunction: A potential cancer predisposition factor. *J. Natl. Cancer Inst.*, 95, 1211–1218.

Wu, L., Multani, A.S., He, H., Cosme-Blanco, W., Deng, Y., Deng, J.M., Bachilo, O., Pathak, S., Tahara, H., Bailey, S.M. et al. (2006). Pot1 deficiency initiates DNA damage checkpoint activation and aberrant homologous recombination at telomeres. *Cell*, 126, 49–62.

Xin, H., Liu, D., Songyang, Z. (2008). The telosome/shelterin complex and its functions. *Genome Biol.*, 9, 232.

Zaug, A.J., Podell, E.R., Cech, T.R. (2005). Human POT1 disrupts telomeric G-quadruplexes allowing telomerase extension in vitro. *Proc. Natl. Acad. Sci. USA*, 102, 10864–10869.

Zhang, Y., Chen, L.Y., Han, X., Xie, W., Kim, H., Yang, D., Liu, D., Songyang, Z. (2013). Phosphorylation of TPP1 regulates cell cycle-dependent telomerase recruitment. *Proc. Natl. Acad. Sci. USA*, 110, 5457–5462.

Zhong, F.L., Batista, L.F., Freund, A., Pech, M.F., Venteicher, A.S., Artandi, S.E. (2012). TPP1 OB-fold domain controls telomere maintenance by recruiting telomerase to chromosome ends. *Cell*, 150, 481–494.

Zimmermann, M., Kibe, T., Kabir, S., De Lange, T. (2014). TRF1 negotiates TTAGGG repeat-associated replication problems by recruiting the BLM helicase and the TPP1/POT1 repressor of ATR signaling. *Genes Dev.*, 28, 2477–2491.

6

G-quadruplexes: Structure, Detection and Functions

Emilia Puig LOMBARDI
SFR Necker, CNRS UAR3633, Inserm US24, Paris, France

6.1. From guanine-guanine base-pairing to a secondary structure

6.1.1. *G-quartets*

In most settings, when describing the structure of DNA, we think of the base pairing model described by Watson and Crick over 60 years ago: adenine (A) pairs with thymine (T) and guanine (G) pairs with cytosine (C). However, guanines can interact with each other via the so-called Hoogsteen non-canonical base pairing (Gellert et al. 1962) and thus form planar structures when four guanines are engaged: the resulting guanosine tetraplex is called a G-quartet (see Figure 6.1). While canonical Watson–Crick G-C base pairs involve three hydrogen bonds (see the red arrows for possible sites; Figure 6.1), each guanine residue in a G-quartet acts as both acceptor and donor of two hydrogen bonds on its Hoogsteen side (see the blue arrows for possible sites; Figure 6.1).

For a color version of all figures in this chapter, see http://www.iste.co.uk/richard/function.zip.

Function and Evolution of Repeated DNA Sequences,
coordinated by Guy-Franck RICHARD.

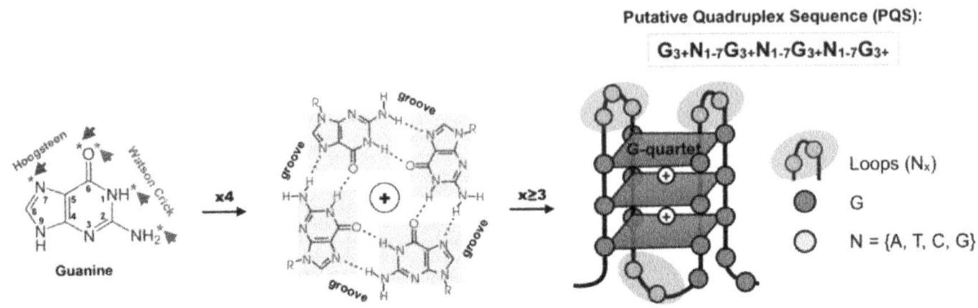

Figure 6.1. *From guanines to the G-quadruplex structure. From left to right: a guanine residue; four guanines (highlighted in blue) forming a planar G-quartet (R: DNA sugar-phosphate backbone); the staking of multiple G-quartets forming a G-quadruplex. Details on the putative quadruplex sequence (PQS) definition and usage can be found in section 6.3*

Furthermore, the planar G-quartets are stabilized by the presence of monovalent cations, mainly K^+ and Na^+ (Mergny et al. 1998), meaning that the interaction with such particles having a single charge is necessary for the formation of the G-quartet. Let us note that the physiologically most relevant interacting cation is K^+ due to its high concentration in cells, but other monovalent cations – and even some divalent (i.e. two positive charges) cations – have been shown to also play a role in the formation of G-quartets, either in physiological or experimental conditions (extensively reviewed in Bhattacharyya et al. (2016)). The cations interact with the oxygen atoms (see Figure 6.1) of the four guanines and are essential to the self-stacking of G-quartets (Williamson et al. 1989): in other terms, the assembly of multiple quartets does not require any energy input. Ultimately, the stacking of at least two G-quartets results in a three-dimensional structure: the G-quadruplex (G4) structure (Sen and Gilbert 1988; see Figure 6.1). Four guanosine nucleosides within a tetrad can exist in either *anti* or *syn* conformations with respect to the glycosidic bond angle of the DNA strands involved (see Figure 6.2). In addition, the cavities limited by the DNA sugar-phosphate backbones form grooves of varying dimensions, depending on the overall topology of the quadruplex (see Figure 6.2). Importantly, another specificity of the G4 structure is that all quadruplexes have four grooves, unlike the canonical double helix which has only two grooves.

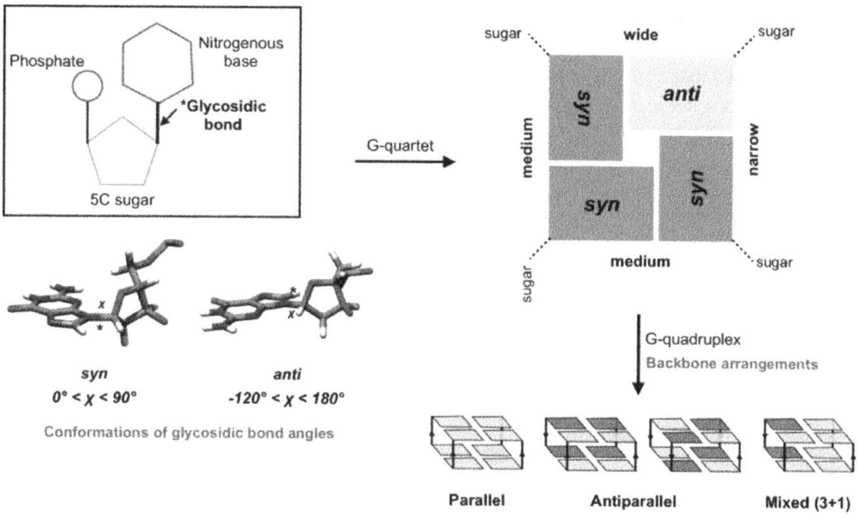

Figure 6.2. *Strand orientation and G-quadruplex topologies. The glycosidic bonds of the guanines involved in a G-quartet can adopt syn or anti conformations depending on the glycosidic torsion angles (noted χ in the schematics). An example of groove dimensions is given with the representation of a G-quartet constituted of three syn (purple) and one anti (gray) guanine orientations. Ultimately, depending on the geometry of the guanines involved in the quartets, a G-quadruplex structure can adopt different topologies (parallel, antiparallel, mixed)*

6.1.2. *Folding into a G-quadruplex structure*

The piled G-quartets linked by the nucleic acid phosphate backbone constitute the invariant fundamental structural unit of all G-quadruplex structures. In addition, all G4s have a central ion channel and, as aforementioned, four grooves of variable dimensions: these features are common to all G-quadruplexes (see Figure 6.3 and Box 6.1). However, important structural particularities exist.

The Protein Data Bank at rcsb.org is a landmark resource for structural biology, in the form of a freely accessible web repository containing hundreds of thousands of curated macromolecular structures of "biological interest" (such as proteins, but also nucleic acids, G-quadruplexes, triplex DNA and i-motifs). PDB also refers to the associated file format type of the entries in the database: the PDB file for a protein contains detailed information about the structure, including notably the position of each atom in a three-dimensional space. Such data are obtained by diverse structural biology approaches such as X-ray crystallography, NMR spectroscopy and cryo-electron microscopy. Importantly, each database entry has a unique PDB identifier in the form of a case-insensitive four-character code: for instance, the 139D entry corresponds to the NMR spectroscopy structure of a parallel-stranded DNA G-quadruplex published in 1993. Finally, the database offers learning modules (starting, of course, with a guide to PDB data types and a dedicated educational portal) and visualization tools (3D representations of molecules powered by MolView). The PDB files can be downloaded and processed for visualization using external tools, such as PyMOL and UCSF Chimera (among others), as depicted in the figures of this chapter.

Box 6.1. *PDB – Protein Data Bank*

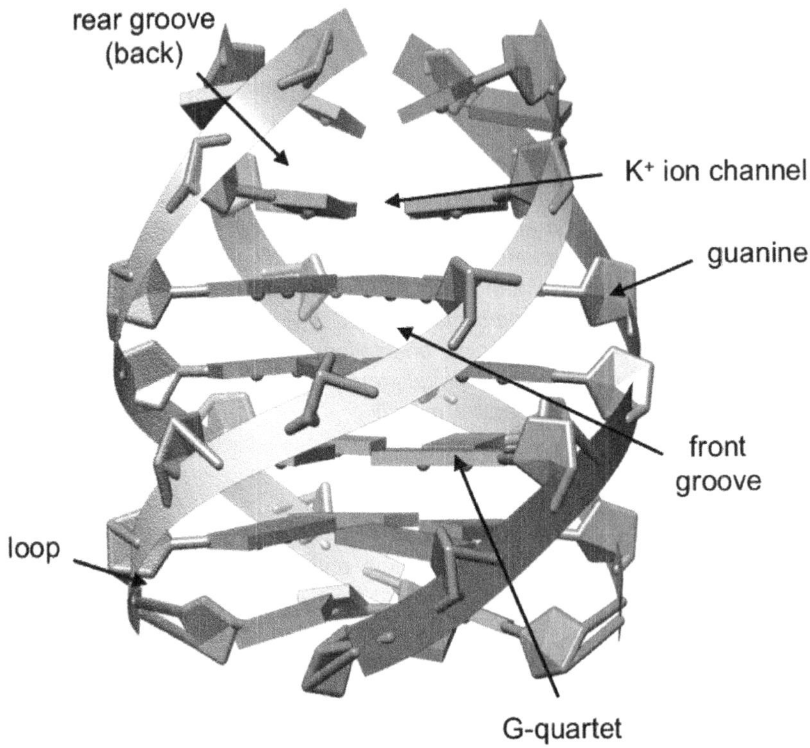

Figure 6.3. *Parallel-stranded DNA G-quadruplex model reconstruction. The UCSF Chimera software was used for the reconstruction of the model based on data accessed from the PDB, entry 139D. G-quartets are in light blue, loop regions are represented in green (regardless of sequence content, A/T/C or G) and the DNA sugar-phosphate backbone is light gray. Invariable features of a G4 are labeled*

6.2. The G4 structure: variations on a theme

To go even further, there is a large number of additional structural variables which lead to a rich topological and structural variety of quadruplexes: there is not an invariant G-quadruplex structure, but rather several possible variant structures depending mainly on the number of stacked G-quartets (at least 2), the number of G-rich strands involved in the fold (unimolecular, bimolecular or tetramolecular G4s – see below) and the sequence, length and orientation of nucleotide loops separating the G-rich quartets (see Figures 6.1 and 6.2).

First, G4 structures can be assembled from a single DNA strand containing several runs of guanines separated by a variable number of nucleotides (loops) and folded back three times, constituting a monomolecular G4. G-quadruplexes can also be formed from two nucleic acid strands, typically identical, each folded back once, thus constituting a bimolecular G4. Finally, they can involve the association of three or four strands, forming tetramolecular quadruplexes (see Burge et al. (2006) for additional details). While monomolecular G4s are folded from only one DNA strand – they are called "intrastrand" – bimolecular or tetramolecular quadruplexes are necessarily "interstrand" topological configurations as they engage multiple nucleic acid strands (Víglaský et al. 2010).

The directionality of the sugar-phosphate backbone defines the orientation of each nucleic acid strand within a G4. As aforementioned, depending on the orientation of the base toward the sugar, the guanines can adopt two different conformations within the G-quartet, *syn* and *anti* (see Figure 6.2). A parallel G4 structure is obtained when all the guanines have the same *anti* conformation and the strands possess the same directionality (Sen and Gilbert 1988). Conversely, G-quartets can contain both *syn*- and *anti*-guanines, resulting in 16 possible orientations that lead to four categories of G4 topologies (reviewed in König et al. (2010)): (i) a parallel conformation containing four strands oriented in the same direction, (ii) a mixed conformation (3+1) where three strands are oriented in one direction and the fourth is oriented in the opposite direction, and two types of anti-parallel conformations, (iii) one with two adjacent stands oriented in the same direction and the others in the opposite direction with *syn–syn–anti–anti* glycosidic angles, (iv) and one where two adjacent stands are oriented in the opposite direction and the glycosidic angles are *syn–anti–syn–anti* (see Figure 6.2).

The loops in G4 structures are the nucleotides connecting runs of consecutive guanines that constitute the G-quartets core, and add a further level of diversity to G4 topologies already described. Indeed, loops can adopt diverse conformations, but their sequence content (i.e. which nucleotides A/T/C/G are in the loops) and length also contribute to structural complexity. In terms of conformation, the loops can be grouped into different categories (see Figure 6.4), for instance: (i) diagonal loops that connect two opposing anti-parallel strands (Wang and Patel 1993), (ii) lateral (also called "edge-wise") loops that connect two adjacent anti-parallel strands (Haider et al. 2002) and (iii) v-shaped loops connecting two corners of a G-quartet core in which a support column is missing (Kuryavyi and Patel 2010).

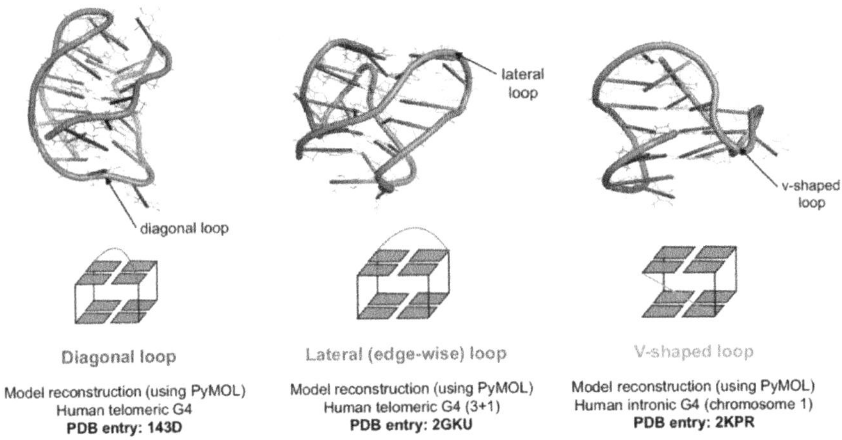

Figure 6.4. *Examples of loop conformations from published G-quadruplex structures. The PyMOL software was used for the visualization of the models based on data accessed from the PDB (accession codes are indicated below each schematic model). The DNA sugar-phosphate backbone is represented in orange in all models*

The sequence identity of the loops not only differentiates quadruplexes in their primary sequence (described in section 6.3 and Figure 6.1), but also plays an important role in determining the folding pattern and stability of G-quadruplexes. Indeed, to a large extent, the conformation that a loop adopts depends on the length of its sequence (Crnugelj et al. 2003). Furthermore, the presence of short loops (and especially single nucleotide loops) favors folding into very stable parallel structures, whereas longer loops of more than three nucleotides tend to favor anti-parallel folds (Rachwal et al. 2007; Guédin et al. 2010). Initially, possible loop length limits have been set based on biophysical studies carried out on G4 structures folded in laboratory settings (with up to seven nucleotides in the loop regions), and also postulated that longer loops (over seven nucleotides) would have a destabilizing effect (Hazel et al. 2004; Bugaut and Balasubramanian 2008).

6.2.1. *RNA G-quadruplexes (rG4)*

Studies focusing on RNA G-quadruplexes (also noted rG4s) were performed less intensively than on DNA G4s. However, both coding and non-coding RNA sequences containing several blocks of consecutive guanines could potentially also fold into quadruplexes, and it is reasonable to think that the single-stranded nature of RNA molecules in the cell makes them more likely to form intrastrand, unimolecular G4s. Moreover, it has been shown that many rG4s are remarkably stable structures, which can be linked to their preferred adopted conformation (Saccà et al. 2005). Indeed, in contrast with DNA, all known naturally occurring rG4s adopt parallel folds, due to the strong propensity of guanosines present in RNA molecules for the *anti* conformation specifically (Malgowska et al. 2016). To date, there is only one reported exception: a non-canonical G-quadruplex motif found in an artificial selected 97-nt RNA fluorogenic oligonucleotides named *Spinach* (Warner et al. 2014), which folds into a two-layer quadruplex.

6.2.2. *Exceptions to the rule(s): non-canonical G-quadruplexes*

Over the past few years, the definition describing assumed G-quadruplex-forming sequences has been expanded. First, sequences with a loop of 10 and even up to 15 nucleotides were found capable of folding into stable G4s, provided that the other two loops are sufficiently short (Agrawal et al. 2014). Furthermore, recent structural studies have unveiled exceptions to the consensus, which can be termed "non-canonical" G-quadruplexes. Notably, bulges (i.e. projections of bases outside of the G-quartet core) have been described in G4 structures and are commonly responsible for interactions with other nucleic acids or proteins (Mukundan and Phan 2013). Another peculiarity are structures with incomplete tetrads forming "G-*triads*" instead of "G-*quartets*", which result from sequences lacking four tracts of three consecutive guanines in their primary sequence and thus drift from the consensus quadruplex sequence (Heddi et al. 2016). Finally, recent studies have revealed quadruplex-forming sequences with unusually long loops: for instance, a 26-nucleotide loop stabilized as a hairpin (see Box 6.2), that forms within the G-quadruplex encoded in the promoter sequence of the hTERT gene (Palumbo et al. 2009).

Several non-canonical nucleic acid (DNA or RNA) structures, which compete with Watson–Crick pairing forms (typically, the B-DNA double helix), have been identified in the past few years. This includes G-quadruplexes, as discussed in this chapter, but other structures have also been described and constitute over 10% of the human genome. Briefly, three additional such examples are hairpins, cruciforms and triplex DNA (also known as H-DNA). Palindromic DNA sequences or inverted sequence repeats are sequences characterized by an internal symmetry so that bases involved can switch between interstrand and intrastrand base pairing. A hairpin structure is formed from single-stranded DNA with ends complementary to each other and in particular, a cruciform structure forms when both strands are involved. Triplex DNA is folded from nucleotide sequences where three nitrogenous bases are linked by both Watson–Crick and Hoogsteen hydrogen bonds: in this setting, a third DNA strand is folded into the major groove of the duplex DNA (Watson—Crick), an interaction that is mediated by the non-canonical Hoogsteen base pairing.

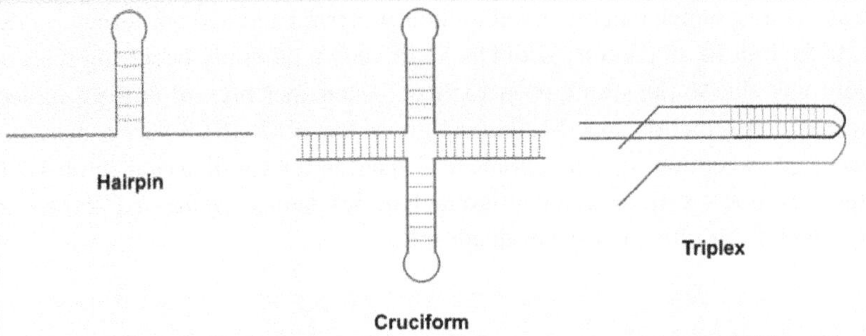

Box 6.2. *Other non-canonical nucleic acid structures*

6.3. Finding G-quadruplexes in a genome

There are a number of largely described experimental techniques that have been used to validate the quadruplex-forming capacity of particular nucleotide sequences, including methods that provide very detailed structural information, such as nuclear magnetic resonance (NMR) spectroscopy (Webba da Silva 2007) and X-ray crystallography (Campbell and Parkinson 2007). Other approaches will indicate the presence of a folded G-quadruplex in an experimental setting, namely UV melting (Mergny et al. 1998; Rachwal and Fox 2007), circular dichroism (CD) spectroscopy (Paramasivan et al. 2007) and the use of fluorescent tags for visualization in microscopy settings (Mergny and Maurizot 2001). However, none of these techniques is suitable nor sufficiently high-throughput to scan and identify new G-quadruplexes on a whole-genome scale. This limitation has made the use of different forms of predictive algorithms essential when pursuing the identification of potential G-quadruplexes at the level of an entire genome. Interestingly, most studies aiming to characterize G-quadruplex structures have so far combined computational (or in silico) predictions with experimental biophysical evidence (in vitro laboratory work) of G4 folding. As a matter of fact, further showing the connection between the two approaches, the first algorithms for in silico G4 detection were developed based on rules that were obtained from the results of a variety of biophysical and biochemical experiments carried out over the past two decades.

Here, we will first briefly present some of the experimental approaches for G4 detection and then discuss in more detail the computational techniques currently used to identify sequences with the potential of forming G-quadruplexes in genomes.

6.3.1. *Experimental methods for G-quadruplex detection*

6.3.1.1. *Biophysical and biochemical methods*

Currently, and even when using computational approaches relying on solid experimental data validations for their predictions, primary sequence information alone is not sufficient to determine whether a given region of a nucleic acid may or may not fold into a quadruplex in a living cell. Several experimental approaches can be used to provide evidence of G4 formation, specific folding topologies and particularly, to confirm the folding of putative quadruplex sequences that were predicted computationally. The existence of a stable quadruplex can be validated by the combined use of several techniques (summarized in Table 6.1), notably UV fluorescence and circular dichroism, ideally together with structural information provided by X-ray crystallography or NMR analyses. The description of these complex techniques is beyond the scope of this chapter. However, the references indicated in Table 6.1 should provide sufficient details for an interested reader to obtain further information.

Method	Information	Nucleic acid	Reference
Circular dichroism	Topology	DNA and RNA	Paramasivan et al. (2007)
UV melting	Thermostability	DNA and RNA	Mergny et al. (1998)
FRET melting	Thermostability	DNA and RNA	Mergny et al. (2001)
NMR	3D structure	DNA and RNA	Webba da Silva (2007)
X-ray crystallography	3D structure	DNA and RNA	Parkinson et al. (2002)
Fluorescent probes	G4 presence or absence	DNA and RNA	Kreig et al. (2015)
Polymerase stop assay	Starting location at nucleotide resolution	DNA	Han et al. (1999)

DMS and piperidine cleavage assay	Location and structural reactivity at guanine nucleotide resolution	DNA	Williamson et al. (1989)
RNase T1	Location and structural reactivity at guanine nucleotide resolution	RNA	Morris and Basu (2009)
In-line probing	Location and structural reactivity at single nucleotide resolution	RNA	Beaudoin et al. (2013)
RTS	Starting location at nucleotide resolution	RNA	Kwok et al. (2015)
SHALiPE (and DMSLiPE)	Location and structural reactivity at single nucleotide resolution	RNA	Kwok et al. (2016)
FOLDeR	Location and structural reactivity at single nucleotide resolution	RNA	Weldon et al. (2017)

Table 6.1. *Biophysical and biochemical methods to study G-quadruplexes (for additional comments on these techniques, see Kwok and Merrick (2017))*

6.3.1.2. *Visualization in living cells*

The early development of probes capable of binding G-quadruplexes provided one method to visualize these particular nucleic acid structures in live cells. For instance, antibodies are specific in their recognition of molecular structures and have been widely used to detect proteins-binding sites in both DNA and RNA. The first biologically relevant visualization of quadruplex formation was reported two decades ago in a protozoan organism, the ciliate *Stylonychia lemnae*. This experiment used a G4-selective single chain variable fragment of an antibody known as "Sty49" to show that G4s form at telomeres in the macronuclei of *S. lemnae* (Schaffitzel et al. 2001). This ciliate was chosen for imagery as it contains very large amounts of telomeric DNA, thus offering an abundance of putative G-quadruplex targets to visualize. More recently, two structure-specific antibodies were independently developed and are now commercially available to perform G4 detection by immunofluorescence in a variety of cells. These antibodies, named BG4 (Biffi et al. 2013) and 1H6 (Henderson et al. 2014), are suitable for whole-cell immunofluorescence on fixed mammalian cells, although it has not been demonstrated whether they are sensitive/resolutive enough to the level of a single

G-quadruplex structure or can only detect G-quadruplex clusters (i.e. high local densities of multiple G4s that are in close physical proximity). Interestingly, RNA G-quadruplexes are detectable using the BG4 antibody (Biffi et al. 2014), but 1H6 is apparently not able to do so. To note, the same laboratory that first described the 1H6 antibody later showed that it can also recognize single stranded T-rich DNA sequences (Kazemier et al. 2017); thus, this cross-reactivity needs to be taken into account in the interpretation of experiments using the 1H6 antibody.

6.3.1.3. *G-quadruplex-specific next-generation sequencing methods*

Finally, the development of next-generation sequencing (NGS) techniques has provided the opportunity of designing experimental, high-throughput, G4-specific genome-wide detection methods (see Box 6.3). As an example, in order to map G4 structures genome-wide in specific cell lines (namely, normal human epidermal keratinocytes NHEKs and spontaneously immortalized precancerous keratinocytes HaCaT), a recent study used the aforementioned BG4 antibody in combination with the sequencing technique termed ChIP-seq (Hänsel-Hertsch et al. 2016). ChIP-seq, or Chromatin ImmunoPrecipitation followed by DNA sequencing, is an experimental method relying on NGS that allows the genome-wide detection of DNA-binding sites for transcription factors and other proteins of interest. It requires the DNA-bound protein (or binding element of interest) to be immunoprecipitated using a specific antibody, followed by purification and sequencing. Here, BG4-bound G4 DNA was subjected to the ChIP-seq protocol, which allowed us to identify approximately 10,000 G4s in HaCaT cells, but only about 1,000 in NHEK cells. In both cases, the detected G4s were located principally in regulatory and nucleosome-depleted regions of the genome, thus suggesting that an open chromatin (for instance, in regions of active transcription) is likely to favor the folding of G4 structures in live cells.

Other recent sequencing-based G4 detection approaches include G4-seq (Chambers et al. 2015) and rG4-seq for RNA quadruplexes (Kwok et al. 2018). In the human genome, detection using the G4-seq technique identified over 700,000 widespread canonical and non-canonical G-quadruplexes; about 70-fold more than with BG4 ChIP-seq mentioned before and about twofold more than the number of G-quadruplexes reported by the first generation of computational G4 detection methods (see section 6.3.2.1). Interestingly, many detected quadruplexes were found to contain bulges or unusually long loops (>7 nucleotides), as were the majority of G4 found by ChIP (79%; Hansel-Hertsch et al. 2016). Furthermore, there was a significant degree of overlap between the ChIP- and G4-seq experiments: 87% of the ChIP G4 DNA fragments contained sequences also identified by G4-seq, pointing to the reliability of both methods. To date, the rG4-seq method has only been reported for the human genome, whereas DNA G4 maps were recently made available for 11 additional species (Marsico et al. 2019): *Mus musculus* (mouse), *Danio rerio* (zebrafish), *Drosophila melanogaster* (fruit fly), *Arabidopsis thaliana* (thale cress), the nematode *Caenorhabditis elegans*, the budding yeast *Saccharomyces cerevisiae*, the parasites *Leishmania major*, *Trypanosoma brucei* and *Plasmodium falciparum*, as well as two bacteria, *Rhodobacter sphaeroides* and *Escherichia coli*.

> Next-generation sequencing, also simply known as high-throughput sequencing, is a term used to describe widespread, large-scale and automated DNA or RNA sequencing techniques. These techniques rely on three common steps, involving (i) sequencing library preparation (based on the fragmentation of the nucleic acids into small pieces followed by the random sampling of fragments), (ii) the amplification (repeatedly replicating an oligonucleotide) of the libraries, and (iii) the sequencing of the amplified libraries (chemistry depending on the chosen technology, the most common one being Illumina). NGS can be applied to sequence entire genomes (whole genome sequencing, WGS) or targeted regions of a genome only (for instance, only gene exons, an approach known as whole exome sequencing, WES); to measure the presence and quantity of RNA in a cell and thus analyze its transcriptome (RNA-seq); to determine regions of open chromatin accessibility and identify active promoters, enhancers and other regulatory elements (ATAC-seq); to perform a genome-wide detection of DNA-binding sites for transcription factors and other proteins (ChIP-seq); among many other applications.

Box 6.3. *Next-generation sequencing*

6.3.2. *Computational methods*

The algorithmic development in G-quadruplex detection initially used relatively simple regular expression matching approaches, which were further enriched through the use of sliding windows, score calculation and, most recently, machine learning approaches.

6.3.2.1. *Mining for G-quadruplex sequence motifs*

A regular expression (*regex*) is a discrete sequence of characters that defines a search pattern. From an algorithmic point of view, regular expressions are particularly useful in extracting information from any type of text by searching for one or more matches of a specific search pattern (i.e. a specific character sequence): this approach is called regular expression matching. As previously mentioned, biophysical data led to the definition of a motif sequence for intramolecular G4s comprising stretches of guanines (also referred to as G runs or G tracts) separated by gaps or loop sequences of limited length, which were predicted to fold into stable G4s (Hazel et al. 2004). The first analyses of potential G4-forming sequences in the human genome used the regular expression $G_{3-5}N_{1-7}G_{3-5}N_{1-7}G_{3-5}N_{1-7}G_{3-5}$ to match G4 sequences, which led to the identification of 376,000 putative G4s in the reference genome assembly (Huppert and Balasubramanian 2005; Todd et al. 2005). The chosen regular expression required a matching sequence to rigorously satisfy several criteria: (i) the motif should contain four guanine runs, (ii) each of the four guanine runs should have a length of three to five nucleotides (G_{3-5}), (iii) the sequence must contain three loop regions between the guanine runs and (iv) the lengths of the three loop regions are comprised between 1 and 7 nucleotides (N_{1-7}), where N is any of the four nucleotides {A,T,C,G}. Many scripting languages provide frameworks for regular expression matching implementation, for instance the following syntax is used in Python: '([gG]{3,5}\w{1,7}){3}[gG]{3,5}' (or '([cC]{3,5}\w{1,7}){3}[cC]{3,5}' in the C-rich strand, as both G- and C-rich patterns can be taken into account). This type of search produces a binary output (i.e. a motif sequence was found or not) without any judgment as to the folding likelihood in a living cell. Since 2005, the $G_{3-5}N_{1-7}G_{3-5}N_{1-7}G_{3-5}N_{1-7}G_{3-5}$ motif (or slight variants of the same underlying expression using different limits on the length of the loops or the number of guanines in the G runs) has been used in most analytical studies. Although this may seem like a simple approach, let us note that the global pattern $G_{x1}N_{L1}G_{x2}N_{L2}G_{x3}N_{L3}G_{x4}$ has some important features which can be used to distinguish and categorize sequences for complex analyses, namely the individual loop sequences (L) and the length of the guanine runs (x). A loop with a length between 1 and 7 nucleotides can give rise to up to 21,844 possible sequences. The possible combinations of three loops gives 1.04×10^{13} outcomes, which can partly explain the limited number of studies focusing on the loop nucleotide identity of the different G4s found in a genome (and just focus on the absolute numbers and sizes of sequences found).

Furthermore, the quadruplex-forming G-rich sequence (QGRS) mapper algorithm takes a slightly different approach from the simple regular expression matching technique, by introducing the concept of scoring each of the possible sequences in order to rank them and hence predicting the most likely sequence when there are several alternatives (Kikin et al. 2006). Its implementation uses a more flexible motif definition: $G_{x1}N_{L1}G_{x2}N_{L2}G_{x3}N_{L3}G_{x4}$, where the length of the G tracts is defined as $x \geq 2$ nucleotides (as a reminder, originally the consensus quadruplex sequence stipulated G tracts of at least three consecutive guanines) and allows for at most one of the loops to be of "zero" length. The scoring method (termed "G-score" by the authors of this algorithm) is based on three considerations derived from experimental studies: (i) shorter loops seem to be more common than longer loops, (ii) loop sizes tend to be of similar length in the same sequence and (iii) the greater the number of G-quartets, the more stable the G4 is predicted to be.

Nevertheless, the G-score method is far from ideal, as the experimental data supporting the score's significance are limited and verification of some of the attributes considered in the scoring process is lacking in the published versions of the QGRS mapper guidelines.

> Oligonucleotides – or simply 'oligos' – are short, usually single-stranded, nucleic acid molecules commonly manufactured in the laboratory for specific purposes such as creating primers for sequencing, to be used as molecular probes or to conduct routine experiments such as, for instance, polymerase chain reaction (PCR). The formation of non-canonical DNA structures can complicate oligo design, as the folding of such structures can change the properties of the synthesized nucleic acid. Indeed, G-quadruplexes have been shown to be more hydrophobic than the canonical B-DNA double-helix, and this change in physicochemical properties can result in lower yields and purities during the manufacturing process. In addition, the formation of a quadruplex within the oligo could cause clumping and interfere with its purity assessment. Thus, it appears essential to dispose of reliable G4 prediction tools when designing oligos prior to manufacturing, in order to avoid – if possible – the folding of a potential quadruplex within DNA and/or RNA oligos, notably by selecting sequences with a limited number of G repeats.

Box 6.4. *G-quadruplexes, a challenge for synthetic nucleic acid design*

6.3.2.2. *State-of-the-art algorithms*

Alternatively, algorithms based on sliding window approaches have also been developed and used to detect potential G4s within a genome. A sliding window is a problem-solving technique relying on an algorithm that works exactly as its name states: a window is formed over some part of the data of interest (e.g. the primary sequence of a given chromosome), and this window can slide over the data to capture different portions of it. This detection method does not define strict individual G4 sequence boundaries nor a strict sequence composition as the regular expression matching approach and therefore makes it possible to identify non-canonical G4s (if any are present in the genome of interest). An example of the application of this technique to the identification of G-quadruplexes is the G4Hunter algorithm, which also introduces the concept of validation stages using large-scale experimental data. This program takes into account G-richness (reflecting the fraction of guanines in the sequence) and G-skewness (reflecting G/C asymmetry between the complementary nucleic acid strands) of a given sequence and provides a G-quadruplex propensity (or likelihood) score (Bedrat et al. 2016). The G-skewness parameter was introduced to consider the observed experimental destabilization effect caused by nearby cytosine (C) residues in the vicinity of a G-quadruplex: indeed, as C can base pair with G, it could ultimately obstruct G-quartet formation by impeding the formation of non-canonical G-G base pairs (Beaudoin et al. 2014).

The G4Hunter method was tested and validated using 392 G4 sequences validated in vitro and by an in-depth analysis of all possible G4-forming sequences in the GC-rich human mitochondrial genome. The implementation of this algorithm is slightly more intricate than the motif matching method, as it requires the setting of additional parameters: the window size and a score threshold for G4 detection. The user can choose default values for these parameters, which are based on the outcomes of validation analyses carried by the authors of the G4Hunter method. A window size of 25 nucleotides was used for most of the validation analyses described and seems to be a reasonable choice given that it corresponds to the actual size of many experimentally characterized and validated G-quadruplexes. As for the score threshold, a value of 1.25 results in a good discrimination of G4 versus non-G4 sequences that are picked-up during the search (Bedrat et al. 2016) and represents a good compromise between sensitivity and sensibility. Indeed, and this applies to all such algorithmic approaches, setting a high score threshold considerably minimizes the number of false positives (i.e. a sequence called a G4 when in reality it is not one), but results in a high number of false negatives (i.e a dismissed sequence that in reality is a true G4); therefore, to perform an extensive exploration of G4 potential within a genome or a set of target sequences, lower thresholds should be privileged. The main limitation of this method still remains the empirical choice by the user of a score threshold for detection when running the program on genomes that have not been benchmarked in the reported validation studies (notably, anything other than model organisms).

Overall, the previous approaches mentioned were mostly developed on the basis of expert knowledge of the G-quadruplex field, involving the integration of biophysical and biochemical data, as well as insight from resolved G4 structures. Furthermore, these detection approaches considered a limited number of observed G4 structures that could actually depict an incomplete picture of the wide variety of possible – whether known or still not described to date – quadruplex conformations (as discussed in section 6.2). Such strategies would not be necessarily suitable if the goal was to predict new conformations or sequences purely by computational investigation, especially when focusing on non-model organisms. Therefore, the newest approaches that are being developed in the field are increasingly centered on the use of machine learning algorithms (fundamental methods applied in computational biology are reviewed in detail in Angermueller et al. (2016)), which let the data drive the predictions without prior assumptions. These approaches avoid predefined motif definitions (for instance, the use of a consensus quadruplex sequence or regular expression) and minimize G4 structure stability assumptions to improve the analytical accuracy on non-canonical or unanticipated quadruplex sequences, but are commonly thought to be relatively obscure when it comes to providing further insight into the predictive features determining G4 formation (so-called "black-boxes", see Box 6.5). One of the first G4 detection methods relying on a machine learning algorithm is the G4RNA screener tool, which implements a minimal machine learning model (a single-layer artificial neural network) that trains itself in the recognition of G4-prone sequences based on the experimentally validated G4s available in the G4RNA database (149 G4 and 179 non-G4), and reports a score illustrating the similarity of a given input sequence to known rG4s (Garant et al. 2015, 2017). This detection method demonstrated an excellent predictive power for RNA G4s and was especially efficient in discarding randomly selected sequences (Garant et al. 2017). More recently, this tool was tested on a larger panel of in vitro validated rG4s (nearly 4,000 structures; Kwok et al. 2018) and gave comparable – if not better – prediction outcomes than the widely used G4Hunter algorithm for classification performance. In a similar fashion, the Quadron algorithm (Sahakyan et al. 2017) uses tree-based gradient boosting machines (GBMs, a regression and classification method) as its model's central framework and was trained on a large experimental G4-formation dataset (over 700,000 sequences) obtained for the human genome using the aforementioned G4-seq methodology. Interestingly, this program includes a graphical interface for non-expert users, which facilitates access to the tool, giving a simple output file containing the sequence and location (start position, length, strand) of the detected G4 sequences, as well as a prediction score based on comparisons with the experimental G4-seq method. The authors state that score values above 19 indicate that the corresponding PQS is predicted to fold into a highly stable G-quadruplex (Sahakyan et al. 2017).

As opposed to classical linear models, machine learning is a data analytics technique that relies on algorithms using computational methods to "learn" information directly from observed data and without basing predictions on predetermined model equations. Machine learning models are often referred as "black boxes" to illustrate the apparent difficulty in interpreting these models (tracing back how predictions were made or which model features were important to output prediction results). However, widely used machine learning algorithms (such as the random forest approach) have few statistical assumptions and are particularly well suited to handle "real world" data. In addition, sometimes fitting "simple" linear models (e.g. linear regression) can be tricky, as these models have constraining statistical assumptions (typically, data normality). With intricate biological data, it is very rarely the case that data follow such assumptions, and applying data transformations or processing that would allow it to meet these requirements can be a difficult task. Genome-wide G-quadruplex identification is a good example to apply machine learning approaches that would allow the integrateion of several search criteria for sequence prediction (that would be too complicated to integrate into simple algorithms, such a regular expression matching or simple scoring): sequence content, length, flanking sequence G/C content, even unknown sequence characteristics coming from biophysical studies (selecting for a sequence without apparent particularities to the human eye). Many algorithms have been recently published, such as G4RNA screener (feedforward, single-layer artificial neural network; Garant et al. 2017), PENGUINN (model based on convolutional neural networks; Klimentova et al. 2020), DeepG4 (deep learning approach; Rocher et al. 2020) or Quadron (straightforward tree-based gradient boosting machines approach, a regression and classification method; Sahakyan et al. 2017). Quadron is a good example of an application of a machine learning model with good interpretability, as the authors of this method provide a clear list of over 50 sequence-based features ranked by relative importance for the achieved G4 prediction quality of the model.

Box 6.5. *Machine learning approaches: not always a "black box"*

This program assesses formation propensity for canonical sequence motifs only, reducing the false positive and false discovery rates of the classical pattern matching method. However, to date, its methodology has not been extended to account for non-canonical sequences.

6.3.2.3. *G-quadruplex sequence content in different genomes*

All computational G-quadruplex prediction approaches have their drawbacks and limitations despite the recent advances in the field and the introduction of validation steps based on experimental data. For the first group of algorithms, accounting for variability is heavily restricted, meaning that only the same kind of structures can be looked for, which generally excludes non-canonical quadruplexes (see section 6.2.2). For machine learning methods, the current limitations mostly rely on the quality and quantity of the available training datasets: the collections of experimentally validated G4 structures and their sequences remain limited in size (in particular, for RNA G-quadruplexes) and most of the experimental data is only available for the human genome (although NGS-based G4-seq data have recently been published for 12 species, as aforementioned). Another difficulty could be associated with the quality of the reference genome used for G4-potential evaluation: the reference genome assemblies (i.e. digital DNA sequence databases or the primary genome sequences obtained by high-throughput sequencing) present in large databases such as Ensembl or UCSC Genomes are carefully curated. However, in most references, long runs of repetitive sequences (minisatellites, CpG islands, long stretches of mono- or dinucleotide repeats) are missing or arbitrarily truncated in the references, as they are particularly tricky to assemble, which may lead to an underestimation of the genome-wide G4 motif content.

To provide the reader with concrete examples of G4 quadruplex prediction by different computational tools, Table 6.2 includes the outcomes of three different detection approaches (regular expression matching, the sliding window-based

G4Hunter and the machine learning-based Quadron) tested on 12 reference genome assemblies (including the human genome).

Genome	Size (Mb)	%GC	Regular expression ($G_{3\text{-}5}N_{1\text{-}7}G_{3\text{-}5}N_{1\text{-}7}G_{3\text{-}5}N_{1\text{-}7}G_{3\text{-}5}$)	G4Hunter	Quadron (score ≥19)
Human	3,095.69	37.8	369,647	2,890,423	376,129
Mouse	2,730.87	42.6	491,803	2,724,011	542,232
Zebrafish	1,371.72	36.8	52,215	263,185	58,430
D. melanogaster	143.73	42.1	10,250	110,024	11,795
C. elegans	100.29	35.4	2,227	36,136	2,358
S. cerevisiae	12.16	38.4	38	2,701	67
L. major	32.86	59.6	11,462	100,569	13,577
T. brucei	35.83	46.8	1,647	29,219	1,852
P. falciparum	23.33	19.6	306	4,341	109
A. thaliana	119.67	36.1	1,232	25,786	811
R. sphaeroides	4.64	68.8	633	10,107	497
E. coli	4.60	50.8	52	1,701	21

Table 6.2. *Putative G-quadruplex sequences in 12 genomes (for a recent and detailed review on these computational methods, see Puig Lombardi and Londoño-Vallejo (2020))*

Provided that a reasonable G4Hscore threshold is chosen (1.25 in this case), the number of hits found at the genome-wide level is systematically higher using G4Hunter than the amount found using the classical regular expression matching approach, which can be explained, at least partly, by the detection of non-canonical quadruplex sequences by G4Hunter. Results obtained with Quadron and the regular expression matching search are very similar for every species; however, the machine learning algorithm provides an additional layer of information by outputting a folding potential score for all canonical (with four tracts of $G_{3\text{-}5}$) G4 sequences found. The amount of potential G4s found by Quadron significantly drops when filtering the results by such a score ($Q \geq 19$; numbers not shown), with 20%–84% of the raw hits being filtered out. This observation perfectly illustrates the impact that false positives can have in G4 estimation when using classical regular expression matching approaches. Overall, the data briefly described here support the view according to which the number of G4-folding potential sites within the human genome needs to be significantly revised upward compared to the widely cited figure of ~376,000 putative quadruplex sequences (still commonly cited since its first estimation in 2005), and points to the importance of further including and studying non-canonical sequences.

To finish, let us note that there is still relatively limited information on the evolutionary conservation and distribution characteristics of G-quadruplexes in eukaryotic genomes. A comprehensive study has been carried out in yeast genomes (Capra et al. 2010), pointing to that fact that DNA G4 sequences are more conserved than expected by chance (and even among other related fungi) in the yeast model organism *S. cerevisiae*, thus suggesting that many of these sequence motifs could fold into quadruplexes in live cells. Large studies in mammalian (human in particular) genomes are more recent, and largely dependent on the development of genome-wide G4 sequence detection methods during the past decade. With the publication of recent G4-seq data in 12 species (including mammals), it has been shown that stable G4s occur at regulatory non-coding regions (predominantly gene promoters) and that these stable quadruplexes are conserved across several organisms (Marsico et al. 2019). Furthermore, bioinformatic analyses have established that particularly stable G4 motifs (namely putative quadruplex sequences containing single-nucleotide loops, of the type $G_3N_1G_3N_1G_3N_1G_3$, where N = {T,C} mainly) tend to be suppressed in mammals (Puig Lombardi et al. 2019). Moreover, this study showed that elevated genetic instability (e.g. accumulation of DNA mutations) in gene coding regions is directly associated with predicted high G4 stability. Very recently, a comprehensive exploration of genome-wide G4 predicted motifs was performed in 37 species at 14 representative evolutionary nodes (considered critical in the eukaryotic clade evolutionary tree), ranging from single-celled fungi to higher mammals, aiming to uncover the possible evolutionary patterns and associated functional implications of G4 sequences in eukaryotic genomes (Wu et al. 2021). This descriptive study suggests that higher order organisms may have developed and conserved G4s as key instruments of complex and reversible transcriptional regulatory mechanisms involved in diverse cellular and physiological activities.

6.4. Biological roles of G-quadruplexes

6.4.1. *First role attributed to quadruplexes: their formation in telomeres*

The first telomeric motif was characterized in the ciliate *Tetrahymena thermophila* (Blackburn and Gall 1978). As mentioned above for *S. lemnae* (see section 6.3.1.2), ciliates have the advantage of having a very large number of chromosomes in their nucleus. Interestingly, the chromosome ends of this ciliate consist of the highly repeated motif [GGGGTT] (or [AACCCC] in the complementary C-rich strand): this type of G-rich motif has been largely conserved during evolution and has, given its composition, G4-forming potential in many eukaryotes, including the human telomeric repeat $[TTAGGG]_n$. Indeed, it was later shown that human telomeric repeats do form G-quadruplex structures (Parkinson et al. 2002).

Direct in vivo evidence at the macronuclear telomeres of *S. lemnae* shows that the G-quadruplex structures are resolved during DNA replication and suggests that G-quadruplexes may act as a telomeric capping structure (Schaffitzel et al. 2001; Paeschke et al. 2008): the capping process ensures that telomeres are protected from degradation and fusion, protecting these chromosome ends from degradation (in particular, subsequent to being recognized as damaged DNA). The observation that G-quadruplex unfolding is regulated by a cell cycle dependent phosphorylation of a telomere-end binding protein (termed TEBPβ) and a telomerase associated helicase (RecQ-like) provided an important mechanistic insight into the spatial and temporal regulation of dynamic G-quadruplex folding and unfolding by helicases (Paeschke et al. 2008). Helicase proteins use the energy of nucleoside triphosphate hydrolysis in order to alter the DNA or RNA structure, and in particular, several helicases have been shown to play roles in the dynamic remodeling of G-quadruplexes (for a detailed list, see Koch Lerner and Sale (2019)). Indeed, as G4s may form spontaneously when DNA single strands are exposed during replication or transcription, specific helicases recognize and unwind G4 DNA to maintain genetic stability (Maizels and Gray 2013). For human telomeres, the first indication that G-quadruplexes may be present came from the observation that treating cells with small molecules that bind to G4s impedes their unwinding by helicases (these drugs, or G4 ligands, are said to stabilize the G4 structure), impaired telomere metabolism and led to telomere shortening (Burger et al. 2005). Approaches to identify G-quadruplexes at human telomeres have included the binding of a radio-labeled G-quadruplex ligands to metaphase chromosomes (Granotier et al. 2005) and more recently, the use of engineered, structure-specific antibodies against human G-quadruplexes (Biffi et al. 2013; Henderson et al. 2014). In all cases, signals were detected at the ends of chromosomes, albeit not all ends. The resolution of light microscopy is insufficiently high to decipher whether binding occurred at the very end of the chromosome or at subtelomeric regions, also known to adopt G-quadruplex structures in vitro (Piazza et al. 2012). However, the finding that a number of helicases that are known to unwind G-quadruplexes in vitro, localize to telomeres and are required for telomere integrity in vivo, provide strong (even if circumstantial) evidence for the existence of such structures at mammalian telomeres (Paeschke et al. 2010).

6.4.2. *Predictions based on bioinformatic analyses*

Technical limitations left aside (see section 6.3.2), computational analyses of the human reference genome – using the consensus $G_{3\text{-}5}N_{1\text{–}7}G_{3\text{-}5}N_{1\text{–}7}G_{3\text{-}5}N_{1\text{–}7}G_{3\text{-}5}$ motif – revealed that the localization of putative quadruplex sequences is non-random and that these sequences colocalize with functional regions of the genome. Indeed, there is a high abundance of such sequences in three repetitive functional domains: in telomeres (Huppert and Balasubramanian 2005), which consist of 5–15 kb (depending on the cell type) of $[TTAGGG]_n$ repeats; within ribosomal DNA (rDNA) (Drygin et al. 2009); and in the so-called immunoglobulin heavy-chain switch regions (Yu et al. 2003), which are all long, repetitive and G-rich. They are also significantly enriched in gene promoters (Huppert and Balasubramanian 2007), at the 5′-end of first introns (Eddy and Maizels 2008), but conversely, significantly depleted in the coding regions of most genes (Eddy and Maizels 2006). More recently, it has also been suggested that 90% of potential human DNA replication origins contain putative quadruplex sequences, and in higher densities near frequently fired (active or used within a given cell cycle) origins (Besnard et al. 2012). All of these associations found in silico point to the fact that G-quadruplex structures potentially play diverse regulatory roles in important genomic regions and provide interesting evidence for complementary functional validations in live cells.

6.4.2.1. *Quadruplex motifs in transcription and translation*

Several bioinformatic studies conducted over the past decades have extensively characterized the distribution of G4 motifs within human genes (Eddy and Maizels 2006, 2009; Eddy et al. 2011; Gray et al. 2014). Overall, G4s are enriched at the transcription start site (TSS), the 5′-UTR (untranslated region), and the 5-end of the first intron of genes, and conversely, depleted in exons. In particular, G4 motifs located downstream of the TSS and in the non-template DNA strand (also termed sense or positive strand) will be transcribed into the pre-messenger RNA (pre-mRNA), meaning they could fold into either DNA or RNA quadruplex structures, which makes functions of these quadruplexes particularly interesting to study (Eddy and Maizels 2008). Further analyses show that G4 motifs occur in 10%–15% of human genes, specifically in the 5′-UTR region of their encoded mRNA, suggesting that G4 RNA structures may promote translational repression (Bugaut and Balasubramanian 2012). Furthermore, at many genes, the RNA polymerase Pol2 effectively initiates transcription but pauses near the 5′-end, and this pausing correlates with enrichment of G4 motifs (Eddy et al. 2011).

In silico predictions have also revealed that particular RNA G4 motifs are present within the 5′-UTR of many genes, and the involvement of G-quadruplexes in the regulation of translation has been demonstrated by both in vitro and in vivo studies (Bugaut and Balasubramanian 2012). An RNA helicase is a class of enzyme that can rearrange RNA secondary structures and RNA-protein interactions (ATP-dependent process). A recent investigation into the role of a specific RNA helicase, eIF4A, has revealed that the hallmark of eIF4A-dependent transcripts is a 12-nucleotide signature $[CGG]_4$ that is able to form RNA G-quadruplex structures in an experimental setting (Wolfe et al. 2014).

6.4.2.2. *Quadruplex motifs in DNA replication*

Replication origin identification by high-throughput sequencing of short nascent DNA strands (a technique called SNS-seq) in the human genome revealed that the majority of the ~250,000 origins found overlap with canonical G4 sequence motifs (Besnard et al. 2012). This study identified over half (67%) of the putative quadruplex sequences in the genome as replication origins, and the fraction drops only slightly using more relaxed criteria (notably, loop sizes >7 nucleotides) for G4 motif identification (Maizels and Gray 2013), which was later confirmed by other sequencing studies (Mesner et al. 2013; Picard et al. 2014). Nevertheless, a more recent report in which highly purified Okazaki fragments from two cell types (HeLa and GM06990) were sequenced (resulting in a better origin identification resolution) identifies only about 9,000 large replication initiation zones (Petryk et al. 2016) and only a weak enrichment for G-quadruplexes within these regions. This suggests that G4s may not overlap the origins themselves, but rather their immediate vicinity, perhaps being necessary although not sufficient for origin function. Putative roles for quadruplexes in the vicinity of replication origins include assisting the positioning nucleosomes flanking the origin (Cayrou et al. 2015), interaction with protein factors required for the formation of the replication origin (Keller et al. 2014) or cooperation with adjacent regulatory sequence modules (Valton et al. 2014).

DNA replication presents an optimal context for DNA G4s to be formed, given that DNA becomes unwound and transiently single-stranded for this process to occur. In theory, the occurrence of secondary structure should present a physical obstacle to replication, but in practice such impediments should be effectively resolved by accessory enzymes, in particular helicases, as aforementioned. However, when helicases involved in replication and repair are faulty (e.g. as a result of mutations, as is often the case in cancer cells), then replication may become impaired or delayed (Nguyen et al. 2014). To note, G-quadruplex-related replication arrest is enhanced by G4-binding small molecules that can stabilize the G4 structure, as the ligand-bound quadruplexes are inaccessible to helicase unwinding compared to the native ones (Biffi et al. 2013).

6.4.2.3. *Quadruplex motifs in minisatellites*

Eukaryotic genomes contain tandem repeat sequences, constituting up to 10% of the human genome. These sequences can be classified, according to the size of their core unit pattern, into microsatellites (1–9 nucleotides), minisatellites (10–100 nucleotides) and megasatellites (>100 nucleotides). The Tandem Repeats Database (TRDB) contains over 675,000 such sequences, and among these, 53% are minisatellites enriched at chromosome ends. In particular, CEB1 (D2S90 locus) and CEB25 (D10S180 locus) are two human subtelomeric minisatellites that contain several contiguous blocks of guanines and are located in the promoter region of the NEU4 gene and in intron 9 of the LRRC27 gene, respectively. These sequences are characterized by their hypervariabiality, meaning that their overall size varies (simple or complex gains or losses of repeat units) between individuals ranging from 0.5 to 20 kb (Amrane et al. 2012). Both CEB1 and CEB25 have been shown to form stable G-quadruplexes both in vitro (Amrane et al. 2012; Adrian et al. 2014) and in vivo (Piazza et al. 2012).

6.5. Perspective: G-quadruplexes as anticancer therapeutic targets

Interestingly, putative quadruplex sequences are more frequently found within oncogenes or regulatory genes than in housekeeping, or tumor suppressor genes (Eddy and Maizels 2006; Huppert and Balasubramanian 2007). Moreover, the majority of resolved (by NMR or crystallography) promoter quadruplexes reported in the literature correspond to cancer-associated genes and are mostly parallel intramolecular folds (to date, 29 PDB entries; see Table 6.3). Well-studied examples of genes containing such quadruplex sites include c-MYC, RET, VEGF-A, KRAS, BCL-2 and c-KIT (Brooks et al. 2010). Notably, alterations in the expression of these oncogenes are recognized as hallmarks of cancer, a series of vital cellular and microenvironmental processes aberrantly regulated during oncogenic transformation and malignancy (Hanahan and Weinberg 2000). Promoters in each of these oncogenes are able to form G-quadruplexes with vast diversity in their sequence motifs (Brooks et al. 2010), making them interesting targets for drug development. The majority of these putative quadruplexes tend to be immediately upstream (within 1 kb) to the TSS in these genes, and usually contain short loop sequences, which explains the frequent parallel fold observed in the resolved structures (see Table 6.3). Note that the first evidence that G4s in promoters have an effect on gene expression came from studies of the c-MYC oncogene, for which it was shown that mutations of the G4 sequence, or treatment with G4-binding molecules that alter the stability of the quadruplex, affected transcription in living cells (Siddiqui-Jain et al. 2002).

Gene	PDB accessions	Predominant G4 fold
cMYC	5VHE, 2N6C, 2L7V, 1XAV, 6ERL, 2LBY, 2MGN, 5LIG, 5W77, 2A5R, 2A5P, 2LLH, 2N6C, 6NEB	Parallel
BCL2	2F8U	Mixed
cKIT-1	2O3M, 2KQG, 2KQH, 4WO3, 6GH0, 3QXR	Parallel
cKIT-2	2KYP, 2KYO	Parallel
VEGF	2M27, 2M53	Parallel
KRAS	5I2V	Parallel
RET	2L88	Parallel
hTERT	2KZD, 2KZE	Mixed (3+1)

Table 6.3. *Resolved G-quadruplex structures in gene promoters*

As discussed in the previous section, the abundance of G-quadruplexes at telomeres, the significant relationship between G4 sequences and transcription regulation of oncogenes and genomic instability have stimulated efforts toward the design of G4 sequence-specific molecules for cancer therapeutics (Neidle 2016). Several small-molecule ligands with a higher specificity for G4s over double-stranded DNA have been designed and evaluated for their therapeutic potential (Balasubramanian et al. 2011). A recent example is the tetra-substituted naphthalene diimide derivative MM41, and a similar computer-designed CM03 derivative, which causes a significant decrease in tumor growth in PANC-1 and MIA PaCa-2 pancreatic cancer experimental models (Ohnmacht et al. 2015; Marchetti et al. 2018). Although this effect can be explained, in part, by an associated strong downregulation in KRAS and BCL-2 oncogenes expression, other G4-related effects may also contribute to the efficacy of such treatments (Marchetti et al. 2018). The G4-binding molecules (also termed G4 ligands) pyridostatin and RHPS4 trigger DNA double-strand break formation, the activation of DNA repair pathways and the activation of poly(ADP-ribose) polymerase 1 (PARP1), which is the first responder that detects DNA damage and then facilitates the choice of an appropriate DNA repair pathway (Salvati et al. 2010; Rodriguez et al. 2012). Furthermore, ablation of the breast cancer type 2 susceptibility gene (BRCA2) in HCT116 colon cancer cells (which makes these cells deficient in specific DNA repair pathways) results in an approximately 10-fold increase in sensitivity to the G4-binding drug pyridostatin compared with DNA repair-proficient cells (McLuckie et al. 2013). These recent findings suggest that G4 ligands could be considered for cancer therapy, especially in tumors that are genetically deficient in DNA damage repair (McLuckie et al. 2013).

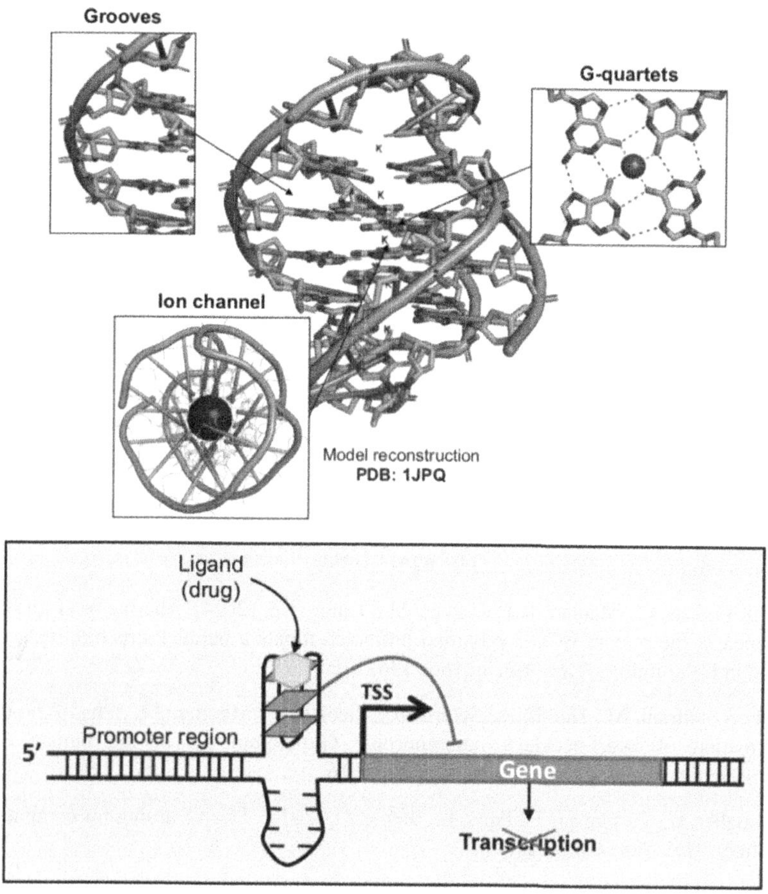

Figure 6.5. *Targeting G-quadruplexes. The model reconstruction of a G-quadruplex indicates the common features of a G4 that can be targeted by a G4-binding drug (or G4 ligand). Most of the ligands that have currently been developed target the invariant feature of the quadruplex structure: the G-quartet. The scheme below depicts a possible mechanism of action of these G4 ligands (targeting in particular an oncogene containing a G4 sequence in its promoter): if a G4 is stabilized by a ligand (i.e. made inaccessible to helicases for its unwinding), the persistence of the structure can interfere with the RNA polymerase activity (or the recruitment of transcription factors needed to effectively start transcription), thus altering gene expression at transcriptional level*

As for the clinical aspect, the G4 ligand Quarfloxin (also known as CX-3543; Drygin et al. 2009) completed phase II clinical trials for the treatment of several tumors (neuroendocrine, carcinoid, lymphoma; for additional information, see ClinicalTrials.gov identifier: NCT00780663). However, this drug did not reach Phase III due to high albumin (the main protein of human blood plasma) binding, which alters biodistribution and bioavailability of the drug. Currently, the CX-5461 compound is the only G4 ligand still undergoing a clinical trial. This drug was initially described as an inhibitor of ribosomal DNA transcription, resulting in nucleolar disruption specifically in cancer cells (Bywater et al. 2012). Treatment with this small molecule relies on selectively killing B-lymphoma cells, while maintaining the wild-type – "normal" – B-cell population. Interestingly, more recent studies have shown that CX-5461 can bind and stabilize G4 DNA structures (Xu et al. 2017) and is thus, by definition, a G4 ligand. CX-5461 is now in advanced phase I clinical trials for patients with hematological malignancies, and in particular, patients with DNA repair deficient tumors (ClinicalTrials.gov identifier: NCT02719977).

6.6. References

Adrian, M., Ang, D.J., Lech, C.J., Heddi, B., Nicolas, A., Phan, A.T. (2014). Structure and conformational dynamics of a stacked dimeric G-quadruplex formed by the human CEB1 minisatellite. *J. Am. Chem. Soc.*, 136, 6297–6305.

Agrawal, P., Lin, C., Mathad, R.I., Carver, M., Yang, D.Z. (2014). The major G-quadruplex formed in the human BCL-2 proximal promoter adopts a parallel structure with a 13-nt loop in K+ solution. *J. Am. Chem. Soc.*, 136, 1750–1753.

Amrane, S., Adrian, M., Heddi, B., Serero, A., Nicolas, A., Mergny, J.L., Phan A.T. (2012). Formation of pearl-necklace monomorphic G-quadruplexes in the human CEB25 minisatellite. *J. Am. Chem. Soc.*, 134, 5807–5816.

Angermueller, C., Pärnamaa, T., Parts, L., Stegle, O. (2016). Deep learning for computational biology. *Mol. Syst. Biol.*, 12, 878.

Beaudoin, J.D., Jodoin, R., Perreault, J.P. (2013). In-line probing of RNA G-quadruplexes. *Methods*, 64, 79–87.

Beaudoin, J.D., Jodoin, R., Perreault, J.P. (2014). New scoring system to identify RNA G-quadruplex folding. *Nucleic Acids Res.*, 42, 1209–1223.

Bedrat, A., Lacroix, L., Mergny, J.L. (2016). Re-evaluation of G-quadruplex propensity with G4Hunter. *Nucleic Acids Res.*, 44, 1746–1759.

Besnard, E., Babled, A., Lapasset, L., Milhavet, O., Parrinello, H., Dantec, C., Marin, J.M., Lemaitre, J.M. (2012). Unraveling cell type-specific and reprogrammable human replication origin signatures associated with G-quadruplex consensus motifs. *Nat. Struct. Mol. Biol.*, 19, 837–844.

Bhattacharyya, D., Mirihana, A.G., Basu, S. (2016). Metal cations in G-quadruplex folding and stability. *Front. Chem.*, 4, 38.

Biffi, G., Tannahill, D., McCafferty, J., Balasubramanian, S. (2013). Quantitative visualization of DNA G-quadruplex structures in human cells. *Nat. Chem.*, 5, 182–186.

Biffi, G., Di Antonio, M., Tannahill, D., Balasubramanian, S. (2014). Visualization and selective chemical targeting of RNA G-quadruplex structures in the cytoplasm of human cells. *Nat. Chem.*, 6, 75–80.

Blackburn, E.H. and Gall, J.G. (1978). A tandemly repeated sequence at the termini of the extrachromosomal ribosomal RNA genes in *Tetrahymena*. *J. Mol. Biol.*, 120, 33–53.

Brooks, T.A., Kendrick, S., Hurley, L. (2010). Making sense of G-quadruplex and i-motif functions in oncogene promoters. *FEBS J.*, 277, 3459–3469.

Bugaut, A. and Balasubramanian, S. (2008). A sequence-independent study of the influence of short loop lengths on the stability and topology of intramolecular DNA G-quadruplexes. *Biochemistry*, 47, 689–697.

Bugaut, A. and Balasubramanian, S. (2012). 5'-UTR RNA G-quadruplexes: Translation regulation and targeting. *Nucleic Acids Res.*, 40, 4727–4741.

Burge, S., Parkinson, G.N., Hazel, P., Todd, A.K., Neidle, S. (2006). Quadruplex DNA: Sequence, topology and structure. *Nucleic Acids Res.*, 34, 5402–5415.

Burger, A.M., Dai, F., Schultes, C.M., Reszka, A.P., Moore, M.J., Double, J.A., Neidle, S. (2005). The G-quadruplex-interactive molecule BRACO-19 inhibits tumor growth, consistent with telomere targeting and interference with telomerase function. *Cancer Res.*, 65, 1489–1496.

Bywater, M.J., Poortinga, G., Sanij, E., Hein, N., Peck, A., Cullinane, C., Wall, M., Cluse, L., Drygin, D., Anderes, K. et al. (2012). Inhibition of RNA polymerase I as a therapeutic strategy to promote cancer-specific activation of p53. *Cancer Cell*, 22, 51–65.

Campbell, N.H. and Parkinson, G.N. (2007). Crystallographic studies of quadruplex nucleic acids. *Methods*, 43, 252–263.

Capra, J.A., Paeschke, K., Singh, M., Zakian, V.A. (2010). G-quadruplex DNA sequences are evolutionarily conserved and associated with distinct genomic features in *Saccharomyces cerevisiae*. *PLoS Comput. Biol.*, 6, e1000861.

Cayrou, C., Ballester, B., Peiffer, I., Fenouil, R., Coulombe, P., Andrau, J.-C., van Helden, J., Méchali, M. (2015). The chromatin environment shapes DNA replication origin organization and defines origin classes. *Genome Res.*, 25, 1873–1885.

Chambers, V.S., Marsico, G., Boutell, J.M., Di Antonio, M., Smith, G.P., Balasubramanian, S. (2015). High-throughput sequencing of DNA G-quadruplex structures in the human genome. *Nat. Biotech.*, 33, 877–881.

Crnugelj, M., Sket, P., Plavec, J. (2003). Small change in a G-rich sequence, a dramatic change in topology: New dimeric G-quadruplex folding motif with unique loop orientations. *J. Am. Chem. Soc.*, 125, 7866–7871.

Drygin, D., Siddiqui-Jain, A., O'Brien, S., Schwaebe, M., Lin, A., Bliesath, J., Ho, C.B., Proffitt, C., Trent, K., Whitten, J.P. et al. (2009). Anticancer activity of CX-3543: A direct inhibitor of rRNA biogenesis. *Cancer Res.*, 69, 7653–7661.

Eddy, J. and Maizels, N. (2006). Gene function correlates with potential for G4 DNA formation in the human genome. *Nucleic Acids Res.*, 34, 3887–3896.

Eddy, J. and Maizels, N. (2008). Conserved elements with potential to form polymorphic G-quadruplex structures in the first intron of human genes. *Nucleic Acids Res.*, 36, 1321–1333.

Eddy, J. and Maizels, N. (2009). Selection for the G4 DNA motif at the 5' end of human genes. *Mol. Carcinog.*, 48, 319–325.

Eddy, J., Vallur, A.C., Varma, S., Liu, H., Reinhold, W.C., Pommier, Y., Maizels, N. (2011). G4 motifs correlate with promoter-proximal transcriptional pausing in human genes. *Nucleic Acids Res.*, 39, 4975–4983.

Garant, J.M., Luce, M.J., Scott, M.S. (2015). G4RNA: An RNA G-quadruplex database. *Database*, bav059.

Garant, J.M., Perreault, J.P., Scott, M.S. (2017). Motif independent identification of potential RNA G-quadruplexes by G4RNA screener. *Bioinformatics*, 33, 3532–3537.

Gellert, M., Lipsett, M.N., Davies, D.R. (1962). Helix formation by guanylic acid. *Proc. Natl. Acad. Sci.*, 48, 2013–2018.

Granotier, C., Pennarun, G., Riou, L., Hoffschir, F., Gauthier, L.R., De Cian, A., Gomez, D., Mandine, E., Riou, J.F., Mergny, J.L. et al. (2005). Preferential binding of a G-quadruplex ligand to human chromosome ends. *Nucleic Acids Res.*, 33, 4182–4190.

Gray, L.T., Vallur, A.C., Eddy, J., Maizels, N. (2014). G-quadruplexes are genome-wide targets of transcriptional helicases XPB and XPD. *Nat. Chem. Biol.*, 10, 313–318.

Guédin, A., Gros, J., Alberti, P., Mergny, J.L. (2010). How long is too long? Effects of loop size on G-quadruplex stability. *Nucleic Acids Res.*, 38, 7858–7868.

Haider, S., Parkinson, G.N., Neidle, S. (2002). Crystal structure of the potassium form of an *Oxytricha nova* G-quadruplex. *J. Mol. Biol.*, 320, 189–200.

Han, H., Hurley, L.H., Salazar, M. (1999). A DNA polymerase stop assay for G-quadruplex-interactive compounds. *Nucleic Acids Res.*, 27, 537–542.

Hanahan, D. and Weinberg, R.A. (2000). The hallmarks of cancer. *Cell*, 100, 57–70.

Hänsel-Hertsch, R., Beraldi, D., Lensing, S.V., Marsico, G., Zyner, K., Parry, A., Di Antonio, M., Pike, J., Kimura, H., Narita, M. et al. (2016). G-quadruplex structures mark human regulatory chromatin. *Nat. Genet.*, 48, 1267–1272.

Hazel, P., Huppert, J.H., Balasubramanian, S., Neidle, S. (2004). Loop length dependent folding of G-quadruplexes. *J. Am. Chem. Soc.*, 126, 16405–16415.

Heddi, B., Martín-Pintado, N., Serimbetov, Z., Kari, T.M., Phan, A.T. (2016). G-quadruplexes with (4n – 1) guanines in the G-tetrad core: Formation of a G-triad·water complex and implication for small-molecule binding. *Nucleic Acids Res.*, 44, 910–916.

Henderson, A., Wu, Y., Huang, Y.C., Chavez, E.A., Platt, J., Johnson, F.B., Brosh Jr., R.M., Sen, D., Lansdorp, P.M. (2014). Detection of G-quadruplex DNA in mammalian cells. *Nucleic Acids Res.*, 42, 860–869.

Huppert, J.L. and Balasubramanian, S. (2005). Prevalence of quadruplexes in the human genome. *Nucleic Acids Res.*, 33, 2908–2916.

Huppert, J.L. and Balasubramanian, S. (2007). G-quadruplexes in promoters throughout the human genome. *Nucleic Acids Res.*, 35, 406–413.

Kazemier, H.G., Paeschke, K., Lansdorp, P.M. (2017). Guanine quadruplex monoclonal antibody 1H6 cross-reacts with restrained thymidine-rich single stranded DNA. *Nucleic Acids Res.*, 45, 5913–5919.

Keller, H., Kiosze, K., Sachsenweger, J., Haumann, S., Ohlenschläger, O., Nuutinen, T., Syväoja, J.E., Görlach, M., Grosse, F., Pospiech, H. (2014). The intrinsically disordered amino-terminal region of human RecQL4: Multiple DNA-binding domains confer annealing, strand exchange and G4 DNA binding. *Nucleic Acids Res.*, 42, 12614–12627.

Kikin, O., D'Antonio, L., Bagga, P.S. (2006). QGRS Mapper: A web-based server for predicting G-quadruplexes in nucleotide sequences. *Nucleic Acids Res.*, 34, W676–W682.

Klimentova, E., Polacek, J., Simecek, P., Alexiou, P. (2020). PENGUINN: Precise exploration of nuclear G-Quadruplexes using interpretable neural networks. *Front. Genet.*, 11, 568546.

Koch Lerner, L. and Sale, J.E. (2019). Replication of G quadruplex DNA. *Genes (Basel)*, 10, 95.

König, S.L.B., Evans, A.C., Huppert, J.L. (2010). Seven essential questions on G-quadruplexes. *BioMol. Concepts*, 1, 197–213.

Kreig, A., Calvert, J., Sanoica, J., Cullum, E., Tipanna, R., Myong, S. (2015). G-quadruplex formation in double strand DNA probed by NMM and CV fluorescence. *Nucleic Acids Res.*, 43, 7961–7970.

Kuryavyi, V.V. and Patel, D.J. (2010). Solution structure of a unique G-quadruplex scaffold adopted by a guanosine-rich human intronic sequence. *Structure*, 18, 73–82.

Kwok, C.K. and Balasubramanian, S. (2015). Targeted detection of G-quadruplexes in cellular RNAs. *Angew. Chem. Int. Ed.*, 54, 6751–6754.

Kwok, C.K. and Merrick, C.J. (2017). G-quadruplexes: Prediction, characterization, and biological application. *Trends Biotechnol.*, 35, 997–1013.

Kwok, C.K., Sahakyan, A.B., Balasubramanian, S. (2016). Structural analysis using SHALiPE to reveal RNA G-quadruplex formation in human precursor micro-RNA. *Angew. Chem. Int. Ed.*, 55, 8958–8961.

Kwok, C.K., Marsico, G., Balasubramanian S. (2018). Detecting RNA G-quadruplexes (rG4s) in the transcriptome. *Cold Spring Harb. Perspect. Biol.*, 10, a032284.

Maizels, N. and Gray, L.T. (2013). The G4 genome. *PLoS Genet.*, 9, e1003468.

Malgowska, M., Czajczynska, K., Gudanis, D., Tworak, A., Gdaniec, Z. (2016). Overview of the RNA G-quadruplex structures. *Acta Biochim. Pol.*, 63, 609–621.

Marchetti, C., Zyner, K.G., Ohnmacht, S.A., Robson, M., Haider, S.M., Morton, J.P., Marsico, G., Vo, T., Laughlin-Toth, S., Ahmed, A.A. et al. (2018). Targeting multiple effector pathways in pancreatic ductal adenocarcinoma with a G-quadruplex-binding small molecule. *J. Med. Chem.*, 61, 2500–2517.

Marsico, G., Chambers, V.S., Sahakyan, A.B., McCauley, P., Boutell, J.M., Di Antonio, M., Balasubramanian, S. (2019). Whole genome experimental maps of DNA G-quadruplexes in multiple species. *Nucleic Acids Res.*, 47, 3862–3874.

McLuckie, K.I., Di Antonio, M., Zecchini, H., Xian, J., Caldas, C., Krippendorff, B.F., Tannahill, D., Lowe, C., Balasubramanian, S. (2013). G-quadruplex DNA as a molecular target for induced synthetic lethality in cancer cells. *J. Am. Chem. Soc.*, 135, 9640–9642.

Mergny, J.L. and Maurizot, J. (2001). Fluorescence resonance energy transfer as a probe for G-quartet formation by a telomeric repeat. *ChemBioChem*, 2, 124–132.

Mergny, J.L., Phan, A.T., Lacroix, L. (1998). Following G-quartet formation by UV-spectroscopy. *FEBS Lett.*, 435, 74–78.

Mesner, L.D., Valsakumar, V., Cieslik, M., Pickin, R., Hamlin, J.L., Bekiranov, S. (2013). Bubble-seq analysis of the human genome reveals distinct chromatin-mediated mechanisms for regulating early- and late-firing origins. *Genome Res.*, 23, 1774–1788.

Morris, M.J. and Basu, S. (2009). An unusually stable G-quadruplex within the 5'-UTR of the MT3 matrix metalloproteinase mRNA represses translation in eukaryotic cells. *Biochemistry*, 48, 5313–5319.

Mukundan, V.T. and Phan, A.T. (2013). Bulges in G-quadruplexes: Broadening the definition of G-quadruplex-forming sequences. *J. Am. Chem. Soc.*, 135, 5017–5028.

Neidle, S. (2016). Quadruplex nucleic acids as novel therapeutic targets. *J. Med. Chem.*, 59, 5987–6011.

Nguyen, G.H., Tang, W., Robles, A.I., Beyer R.P., Gray L.T., Welsh, J.A., Schetter, A.J., Kumamoto, K., Wang, X.W., Hickson, I.D. et al. (2014). Regulation of gene expression by the BLM helicase correlates with the presence of G-quadruplex DNA motifs. *Proc. Natl. Acad. Sci. USA*, 111, 9905–9910.

Ohnmacht, S.A., Marchetti, C., Gunaratnam, M., Besser, R.J., Haider, S.M., Di Vita, G., Lowe, H.L., Mellinas-Gomez, M., Diocou, S., Robson, M. et al. (2015). A G-quadruplex-binding compound showing anti-tumour activity in an in vivo model for pancreatic cancer. *Sci. Rep.*, 5, 11385.

Paeschke, K., Juranek, S., Simonsson, T., Hempel, A., Rhodes, D., Lipps, H.J. (2008). Telomerase recruitment by the telomere end binding protein-beta facilitates G-quadruplex DNA unfolding in ciliates. *Nat. Struct. Mol. Biol.*, 15, 598.

Paeschke, K., McDonald, K.R., Zakian, V.A. (2010). Telomeres: Structures in need of unwinding. *FEBS Lett.*, 584, 3760–3772.

Palumbo, S.L., Ebbinghaus, S.W., Hurley, L.H. (2009). Formation of a unique end-to-end stacked pair of G-quadruplexes in the hTERT core promoter with implications for inhibition of telomerase by G-quadruplex-interactive ligands. *J. Am. Chem. Soc.*, 131, 10878–10891.

Paramasivan, S., Rujan, I., Bolton, P.H. (2007). Circular dichroism of quadruplex DNAs: Applications to structure, cation effects and ligand binding. *Methods*, 43, 324–331.

Parkinson, G.N., Lee, M.P., Neidle, S. (2002). Crystal structure of parallel quadruplexes from human telomeric DNA. *Nature*, 417, 876–880.

Petryk, N., Kahli, M., d'Aubenton-Carafa, Y., Jaszczyszyn, Y., Shen, Y., Silvain, M., Thermes, C., Chen, C.L., Hyrien, O. (2016). Replication landscape of the human genome. *Nat. Commun.*, 7, 10208.

Piazza, A., Serero, A., Boulé, J.B., Legoix-Né, P., Lopes, J., Nicolas, A. (2012). Stimulation of gross chromosomal rearrangements by the human CEB1 and CEB25 minisatellites in *Saccharomyces cerevisiae* depends on G-quadruplexes or Cdc13. *PLoS Genet.*, 8, e1003033.

Picard, F., Cadoret, J.C., Audit, B., Arneodo, A., Alberti, A., Battail, C., Duret, L., Prioleau, M.N. (2014). The spatiotemporal program of DNA replication is associated with specific combinations of chromatin marks in human cells. *PLoS Genet.*, 10, e1004282.

Puig Lombardi, E. and Londoño-Vallejo, A. (2020). A guide to computational methods for G-quadruplex prediction. *Nucleic Acids Res.*, 48, 1–15.

Puig Lombardi, E., Holmes, A., Verga, D., Teulade-Fichou, M.P., Nicolas, A., Londoño-Vallejo, A. (2019). Thermodynamically stable and genetically unstable G-quadruplexes are depleted in genomes across species. *Nucleic Acids Res.*, 47, 6098–6113.

Rachwal, P.A. and Fox, K.R. (2007). Quadruplex melting. *Methods*, 43, 291–301.

Rachwal, P.A., Findlow, I.S., Werner, J.M., Brown, T., Fox, K.R. (2007). Intramolecular DNA quadruplexes with different arrangements of short and long loops. *Nucleic Acids Res.*, 35, 4214–4222.

Rocher, V., Genais, M., Nassereddine, E., Mourad, R. (2020). DeepG4: A deep learning approach to predict active G-quadruplexes from DNA. *bioRxiv* (preprint). doi: 10.1101/2020.07.22.215699.

Rodriguez, R., Miller, K.M., Forment, J.V., Bradshaw C.R., Nikan, M., Britton, S., Oelschlaegel, T., Xhemalce, B., Balasubramanian, S., Jackson, S.P. (2012). Small-molecule-induced DNA damage identifies alternative DNA structures in human genes. *Nat. Chem. Biol.*, 8, 301–310.

Saccà, B., Lacroix, L., Mergny, J.L. (2005). The effect of chemical modifications on the thermal stability of different G-quadruplex-forming oligonucleotides. *Nucleic Acids Res.*, 33, 1182–1192.

Sahakyan, A., Chambers, V.S., Marsico, G., Santner, T., Di Antonio M., Balasubramanian S. (2017). Machine learning model for sequence-driven DNA G-quadruplex formation. *Sci. Rep.*, 7, 14535.

Salvati, E., Scarsella, M., Porru, M., Rizzo, A., Iachettini, S., Tentori, L., Graziani, G., D'Incalci, M., Stevens, M.F., Orlandi, A. et al. (2010). PARP1 is activated at telomeres upon G4 stabilization: Possible target for telomere-based therapy. *Oncogene*, 29, 6280–6293.

Schaffitzel, C., Berger, I., Postberg, J., Hanes, J., Lipps, H.J., Plückthun, A. (2001). In vitro generated antibodies specific for telomeric guanine-quadruplex DNA react with *Stylonychia lemnae* macronuclei. *Proc. Natl. Acad. Sci.*, 98, 8572–8577.

Sen, D. and Gilbert, W. (1988). Formation of parallel four-stranded complexes by guanine rich motifs in DNA and its implications for meiosis. *Nature*, 334, 364–366.

Siddiqui-Jain, A., Grand, C.L., Bearss, D.J., Hurley, L.H. (2002). Direct evidence for a G-quadruplex in a promoter region and its targeting with a small molecule to repress c-MYC transcription. *Proc. Natl. Acad. Sci.*, 99, 11593–11598.

Todd, A.K., Johnston, M., Neidle, S. (2005). Highly prevalent putative quadruplex sequence motifs in human DNA. *Nucleic Acids Res.*, 33, 2901–2907.

Valton, A.L., Hassan-Zadeh, V., Lema, I., Boggetto, N., Alberti, P., Saintomé, C., Riou J.-F., Prioleau, M.N. (2014). G4 motifs affect origin positioning and efficiency in two vertebrate replicators. *EMBO J.*, 33, 732–746.

Víglaský, V., Bauer, L., Tlucková, K. (2010). Structural features of intra- and intermolecular G-quadruplexes derived from telomeric repeats. *Biochemistry*, 49, 2XX-20.

Wang, Y. and Patel, D.J. (1993). Solution structure of the human telomeric repeat d[AG3(T2AG3)3] G-tetraplex. *Structure*, 1, 263–282.

Warner, K.D., Chen, M.C., Song, W., Strack, R.L., Thorn, A., Jaffrey, S.R., Ferré-D'Amaré, A.R. (2014). Structural basis for activity of highly efficient RNA mimics of green fluorescent protein. *Nat. Struct. Mol. Biol.*, 21, 658–663.

Webba da Silva, M. (2007). NMR methods for studying quadruplex nucleic acids. *Methods*, 43, 264–277.

Weldon, C., Behm-Ansmant, I., Hurley, L.H., Burley, G.A., Branlant, C., Eperon, I.C., Dominguez, C. (2017). Identification of G-quadruplexes in long functional RNAs using 7-deazaguanine RNA. *Nat. Chem. Biol.*, 13, 18–20.

Williamson, J.R., Raghuraman, M.K., Cech, T.R. (1989). Monovalent cation-induced structure of telomeric DNA: The G-quartet model. *Cell*, 59, 871–880.

Wolfe, A.L., Singh, K., Zhong, Y., Drewe, P., Rajasekhar, V.K., Sanghvi, V.R., Mavrakis, K.J., Jiang, M., Roderick, JE., Van der Meulen, J. et al. (2014). RNA G-quadruplexes cause eIF4A-dependent oncogene translation in cancer. *Nature*, 513, 65–70.

Wu, F., Niu, K., Cui, Y., Li, C., Lyu, M., Ren, Y., Chen, Y., Deng, H., Huang, L., Zheng, S. et al. (2021). Genome-wide analysis of DNA G-quadruplex motifs across 37 species provides insights into G4 evolution. *Commun. Biol.*, 4, 98.

Xu, H., Di Antonio, M., McKinney, S., Mathew, V., Ho, B., O'Neil, N.J., Dos Santos, N., Silvester, J., Wei, V., Garcia, J. et al. (2017). CX-5461 is a DNA G-quadruplex stabilizer with selective lethality in BRCA1/2 deficient tumours. *Nat. Commun.*, 8, 14432.

Yu, K., Chedin, F., Hsieh, C.L., Wilson, T.E., Lieber, M.R. (2003). R-loops at immunoglobulin class switch regions in the chromosomes of stimulated B cells. *Nat. Immunol.*, 4, 442–451.

7

Satellite DNA, Microsatellites and Minisatellites

Wilhelm VAYSSE–ZINKHÖFER and Guy-Franck RICHARD
Instabilités naturelles & synthétiques des génomes,
Institut Pasteur, CNRS UMR3525, Paris, France

7.1. Satellite DNAs, origin and definition

In the late 1960s, methods for purifying total genomic DNA using a cesium chloride gradient made it possible to isolate large quantities of DNA from various organisms to study their physicochemical properties, such as density. From this, it was possible to deduce the average percentage of GC base pairs of the genome of the organism being studied. In these experiments, one or two minor DNA species could often be detected in the gradient, above or below the major band containing the main nuclear genomic DNA. For this reason, these minor species were called "satellite DNA", and their GC percentage could be lower or higher than that of the main band. These satellite DNAs have been found in many eukaryotic species in *Drosophila* (Lohe et al. 1993), plants (Sharma and Raina 2005) and most mammals (Arrighi et al. 1970). This observation, together with satellite DNA's property of renaturing quickly (but imperfectly) during denaturation–renaturation experiments (see the Introduction to this book), led to the conclusion that the composition of this DNA had to be repetitive yet simple. Remarkably, armed simply with these

---

For a color version of all figures in this chapter, see http://www.iste.co.uk/richard/function.zip.

Function and Evolution of Repeated DNA Sequences,
coordinated by Guy-Franck RICHARD.

conclusions and physicochemistry experiments, it was determined that the sequence of the guinea-pig α satellite must be CCCT and variants derived from this sequence, reproduced in millions of copies (Southern 1970). It would take several more years before DNA sequencing techniques could give us the exact sequence. The discovery of a type of DNA present in such a large quantity in animals and plants (up to 30% in some species) raised the question of its function, and the suggestion that it may offer a selective advantage during meiosis by facilitating proper segregation of homologous chromosomes was put forward (Walker 1971).

Satellite DNA is found mainly in heterochromatin regions, such as centromeres, plant telomeres, and the *Drosophila* Y chromosome, but it may also be encountered as DNA interspersed in other genomic places. Its size can reach several million nucleotides, and it is made up of DNA sequences repeated in tandem whose motif varies from five base pairs for human satellite 3 to several tens of base pairs (Tyler-Smith and Willard 1993). The motif is not repeated in a strictly identical manner and shows variations around a consensus. The consensus sequence of the human satellite 3 motif is GGAAT. It is located at the centromeres, as is the α satellite, whose base motif is 171 nucleotides long.

7.1.1. *Minisatellites*

The first minisatellite was discovered in 1980 within a 16-kilobase piece of human DNA, whose hybridization showed frequent size polymorphism. Eight alleles were detected, and in more than 75% of cases, the individuals tested were heterozygous, indicating the presence of two different-sized alleles in their genome (Wyman and White 1980). Subsequently, many hypervariable DNA regions were discovered in the human genome and called minisatellites in reference to the repeated regions of several hundred kilobases covered by satellite DNAs (Jeffreys et al. 1985). For this reason, minisatellites are also referred to as variable numbers of tandem repeats (VNTRs). More recently, a further distinction was made between minisatellites whose repeated motif was several tens of nucleotides long and those whose motif was longer than 100 base pairs. The latter were given the name "megasatellites" (Thierry et al. 2008).

7.1.2. *Microsatellites*

Microsatellites were discovered soon after by noticing the size polymorphism of the GGAT tetranucleotide repeated 165 times in tandem upstream of the human gene encoding myoglobin (Weller et al. 1984). However, the term microsatellite was only used for the first time a few years later, when a TG dinucleotide repeat located in the actin-coding gene of the cardiac muscle was shown to be highly polymorphic. PCR amplification of this sequence in genomic DNA from 37 samples of diverse origin showed the presence of 12 alleles of different lengths and a very high level of heterozygosity (Litt and Luty 1989). Subsequent studies showed that the human genome contained tens of thousands of microsatellites of variable size. They also later came to be called short sequence repeats (SSRs).

7.2. From semantics to biology

It is reasonable to ask whether the distinction between satellite DNA, minisatellites and microsatellites is not just a matter of semantics because they all involve similar genetic objects: DNA tandem repeats, showing a high level of polymorphism between individuals and characterized only by the size of the repeated motif. No consensus exists as to the limit between micro- and minisatellites. It is commonly accepted that the former are tandems whose motif is less than or equal to nine nucleotides, but some authors do not consider mononucleotide repeats (AAAAA, GGGGG, etc.) as microsatellites. Minisatellites can therefore be defined by deduction as tandem repeats whose motif is greater than or equal to 10 base pairs. However, beyond this question of semantics, can these repeats be classified on the basis of a genetic function or a specific distribution in the genomes in which they are observed?

7.2.1. *Distribution of satellite DNAs in genomes*

The era of genomics has offered us access to hundreds, and soon thousands, of genomes of different prokaryotic and eukaryotic species. There are numerous software tools dedicated to analyzing their composition, despite some obvious caveats, particularly in regard to examining repeated sequences, which are, by definition, of low complexity. An old study suggests that bacterial genomes contain ten times fewer microsatellites than eukaryotic genomes per megabase[1] of genomic DNA. Nevertheless, the genome of *Methanococcus jannaschii*, an archaea, shows an intermediate number of microsatellites between eubacteria and eukaryotes (Richard et al. 1999b). According to the endosymbiotic theory, the ancestor of mitochondria was an alphaproteobacterium, and the most likely ancestor of the eukaryotic cell nucleus was an archaea. There now exists very strong molecular evidence favoring this scenario (for a review, see Richard 2018). Although the precise order of events between the acquisition of the nuclear membrane and that of respiration is not yet

1 Megabase: 1 million base pairs.

fully clear, it seems very likely that the nucleus of the first eukaryotic cells was an Archaea of the "Asgard" clade, whose many representatives are found in hot springs and marine sediments (Zaremba-Niedzwiedzka et al. 2017). Microsatellites were perhaps more numerous in this archaeal ancestor and were propagated in the eukaryotic genome to reach, over the course of evolution, their present-day number. It can also be imagined that the constraints on genome size and composition are more relaxed in archaea and eukaryotes than in eubacteria, allowing greater latitude to accommodate repeated sequences. In all cases, eukaryotes are a far better choice than prokaryotes if we wish to study microsatellite distribution in fully sequenced genomes.

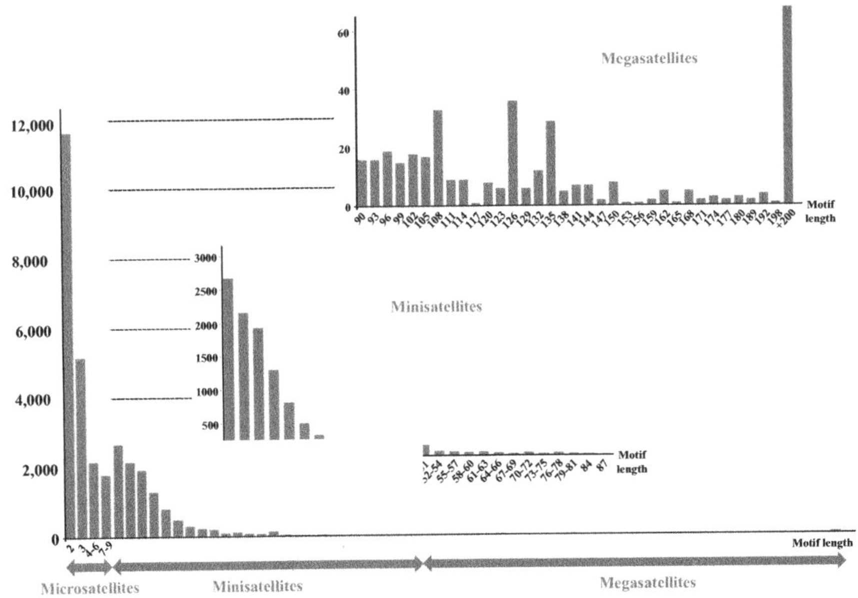

Figure 7.1. *Length distribution of the different tandem repeats. The abscissa axis shows the size of the basic motif in nucleotides, and the ordinate shows the number of each repeat in the yeast genomes studied (see text). Microsatellites are shown in blue, minisatellites in purple and megasatellites in red. The two insets are magnifications of the main graph. Note the scale change of the two inserts*

An analysis of 14 partially sequenced yeast genomes from the *Saccharomycotina* clade showed that the number of microsatellites of each species was highly variable, both in quantity and in quality. The genomes of *Saccharomyces servazzi* and *Kluyveromyces marxianus* contain two to three times more microsatellites per DNA megabase than brewer's yeast *Saccharomyces cerevisiae*, which itself contains many more than *Pichia angusta*, a methylotrophic yeast able to use methanol or methane as a carbon source. In the latter, ACG and AGG trinucleotide repeats are more frequent than the average, whereas they are underrepresented in most other species. Overall, dinucleotide repeats are by far the most frequent, followed by trinucleotides and then tetranucleotides (see Figure 7.1).

It should be noted that microsatellites whose base motif is a multiple of three (tri-, hexa- and nona-nucleotide repeats) are significantly more frequent than others (except dinucleotides, which remain the most frequent). As multiple-of-3 microsatellites never alter the reading of the genetic message, they are weakly counterselected in coding regions, unlike other microsatellites.

In contrast, tandem repeats with a base motif 10–12 nucleotides long are more common than hepta-, octa- and nona-nucleotide repeats combined (see Figure 7.1). Minisatellites therefore have a slightly different distribution than microsatellites, and as we will see below, their function in genomes is also dissimilar because they are found in genes of distinct functions. The number of minisatellites slowly decreases as the base motif increases in size to approximately 90–100 nucleotides. A study of completely sequenced genomes of 21 fungal species (including several yeasts) showed that beyond this limit (which will vary slightly from species to species), some minisatellites are found more commonly than others of similar size. This is the case for minisatellites with a motif of 108, 126 or 135 base pairs long, which are twice as numerous as the minisatellites close in size (see Figure 7.1). This observation was made for the first time in the opportunistic pathogenic yeast *Candida glabrata* and was subsequently extended to many other fungi (Tekaia et al. 2013). These large minisatellites were called "megasatellites" so as to differentiate them from classical minisatellites (Thierry et al. 2008). As we will see below, the functions of their encoding genes are more diversified than those of minisatellites, and their nucleotide motif length is constrained by the structure of their encoded protein motifs.

7.2.2. Polymorphic genetic markers

The discovery of the length variability of tandem-repeat DNA sequences inspired geneticists to use them as genetic markers. In the 1980s, genotyping methods were limited to the analysis of size polymorphisms between restriction fragments of genomic DNA extracted from different individuals[2]. In practice, this

---

2 RFLP: Restriction fragment length polymorphism.

consisted of extracting DNA from several individuals, digesting it with a restriction endonuclease, and migrating in an agarose gel to separate the different fragments according to their molecular weight. The gel content was then transferred to a nitrocellulose (or later, nylon) membrane, and DNA was covalently bound to the membrane and hybridized with a probe specific to the locus studied. The entire experimental protocol took approximately 1 week, and the result was rarely informative because the chance of a polymorphism between two individuals found by accident at a restriction site is very low[3]. When the first minisatellites were discovered, it became immediately apparent that their size polymorphism could be used as a genetic marker by the RFLP technique, particularly by those which appeared to be "hypervariable" in the human population (Jeffreys et al. 1985). Soon, improved techniques for genotyping these minisatellites were developed, particularly in Alec Jeffreys' laboratory in England. By using partial (instead of total) restrictions of genomic DNA, it became possible to accurately count the number of motifs of a given minisatellite in each genome analyzed (Jeffreys et al. 1990). The size polymorphism of minisatellites was used very early on in paternity tests (Helminen et al. 1988) as well as in forensic research (Gill et al. 1985). The invention of DNA polymerase chain reaction (PCR[4]) allowed this genotyping method to be applied to very small quantities of DNA, such as the amount retrieved from a postage stamp stuck on an envelope (Hopkins et al. 1994). Nevertheless, the use of minisatellites as efficient genetic markers was limited by an intrinsic defect: their low frequency in genomes. That of brewer's yeast, a model organism in genetics, contains one minisatellite every 100,000 nucleotides, and it is estimated that the human genome contains approximately one per megabase (Richard et al. 2008). This low density explains why minisatellites were soon abandoned as genotyping markers were in favor of microsatellites.

As seen above, microsatellites are 5–10 times more common in genomes than minisatellites (see Figure 7.1). Their short length (several tens of nucleotides for the most part) and high levels of polymorphism thus make them ideal markers that can be amplified by PCR if the sequence flanking the microsatellite is known. It is within this perspective that they very quickly came to replace minisatellites as of the early 1990s (see Figure 7.2).

3 Of the order of 3%, counting a nucleotide polymorphism every 200 base pairs on average and using restriction enzymes recognizing a 6 base-pair site.

4 Polymerase chain reaction, an enzymatic reaction making it possible to amplify DNA present in very small quanties, in a tube, in a few hours.

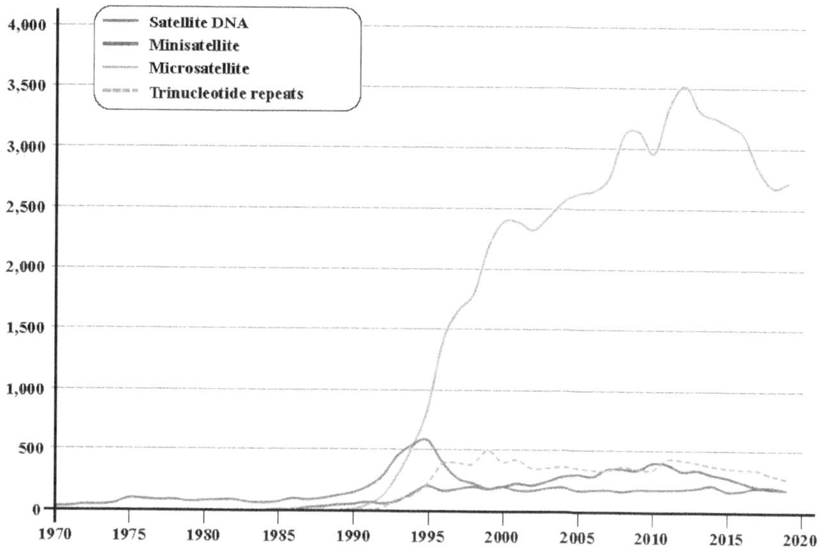

Figure 7.2. *Number of citations in the PubMed database for different tandem repeats. We searched for the following keywords: "Microsatellite", "Minisatellite", "Satellite DNA" and "Trinucleotide repeat"*

They were, and still are, used in forensic medicine (Hagelberg et al. 1991), as well as for paternity tests. In France, the use of microsatellite genotyping in the context of criminal investigations is very strictly regulated by law[5]. DNA extracted from saliva, blood, semen or a hair with its bulb is amplified by PCR on 13 different polymorphic loci. The chances of two different individuals possessing the same genotype at these 13 loci are considered to be null. This can be possible with monozygotic twins, but in the context of a criminal investigation, this kind of information is normally known to investigators. Genetic fingerprints are kept in a national database, the FNAEG[6], where they can be consulted by police authorities within the framework of judicial investigations. They are kept for 40 years, but a person appearing on the register may request their communication and deletion. As of 2018, the FNAEG contained the fingerprints of approximately 3.5 million individuals, that is, 5% of the French population.

5 Available at: http://www.assemblee-nationale.fr/12/rap-info/i0504.asp#P132_9790.

6 *Fichier National Automatisé des Empreintes Génétiques*: Automated national database of genetic fingerprints, created in France in 1998.

Paternity tests also use microsatellite markers, but their application is subject to local legislation. In the United States and the United Kingdom, these tests are perfectly legal and anyone can request one. In France, they are authorized only in the context of legal proceedings to establish or contest paternity. It is illegal to carry out such a test outside this framework, and any violation is punishable by 1 year in prison and a €15,000 fine[7]. However, French legislation on this point will likely evolve rapidly, given the popularity of genetic tests offered for under €100 by many companies. Ancestry, 23andMe and MyHeritage are the three most successful companies offering this service, but others also exist since this highly competitive field is undergoing continuous expansion. Between January 2015 and January 2018, the number of genotyped individuals increased from approximately 1 to 3 million at 23andMe and 6 million at Ancestry. This increase, by continuing at the same rate, will see an estimated 60 million Americans genotyped by 2025 (Khan and Mittelman 2018). The impact in terms of personalized medicine and public health will be considerable and reach far beyond the scope of the topics discussed in this book.

Among the most unorthodox uses of microsatellite genotyping is the identification of the remains of Josef Mengele, the “Angel of Death” from the Auschwitz concentration camp. After fleeing to South America at the end of World War II, he lived in Argentina, Paraguay, and Brazil, where he died in 1979. His remains were found and exhumed in 1985. Microsatellites from 10 different loci were amplified by PCR and compared with those of his still-living son, showing that the remains found in Brazil had a 99.94% chance of being those of the infamous doctor (Jeffreys et al. 1992). Another historical debate was also resolved through the use of microsatellites: DNA analysis of the third president of the United States, Thomas Jefferson, showed that he was the biological father of the last child of Sally Hemmings, one of his slaves. This discovery offered a new perspective on the personality of the man considered one of the greatest American presidents, known for his liberal ideas and for the abolition of the slave trade (Foster et al. 1998).

Microsatellite length polymorphism, associated with a greater frequency than that of minisatellites in complex genomes, has enabled their use on a massive scale as genetic markers. The human genome contains approximately 168,000 CA dinucleotide repeats, that is, one every 35 kb of DNA, on average. It is the most common microsatellite in the human genome (International Human Genome Sequencing Consortium 2001). For this reason, it was the marker most used in mapping genes of medical interest, as well as for drawing up the first complete physical map of the human genome, based on more than 5,000 CA dinucleotide

7 Articles 226-25 to 226-30 of the French Penal Code.

repeats, by the team of Jean Weissenbach, within the framework of the human genome sequencing project (Dib et al. 1996). Subsequently, with sequencing gaining momentum, microsatellite genotyping was gradually replaced by single nucleotide polymorphisms (SNPs), which are polymorphisms of a nucleotide resulting in allelic variations between individuals. These are much more common (1–10/kb) and therefore enable better resolution. However, it should be noted that an SNP can only contain a maximum of four different alleles (A, T, C or G), and most of the time, only two or three different bases are associated with a given SNP. In comparison, a microsatellite can theoretically exhibit a large number of sizes depending on the length of the repeat, with long microsatellites showing more polymorphism. In practice, not all possible sizes for a given microsatellite are found in the human population.

7.2.3. *Trinucleotide repeat expansions*

In the very early 1990s, when microsatellites served only as genetic markers and were still considered as “junk DNA”, a succession of discoveries greatly increased the interest placed in their study. Fragile X syndrome, the most common cause of hereditary intellectual disability in humans, was correlated with the massive expansion of a CCG trinucleotide repeat in the 5’ non-coding region of the *FMR1* gene (Verkerk et al. 1991). This gene, of unknown function, shows a significant length polymorphism within the population, and the pathology is always associated with triplet repeat expansion and hypermethylation of the region. Chromosomal fragility is reflected by constrictions or breaks on one or two sister chromatids, visible on metaphase-plate chromosomes isolated from cells grown in the presence of folic acid. Soon after, an avalanche of diseases related to trinucleotide repeat expansion was discovered.

Remarkably, it was observed very early on that CGG and CAG (or CTG, its complement) triplet repeats were overrepresented in the human genome (Han et al. 1994), suggesting the possibility of other diseases involving expansions of these repeats. However, what some authors had called the “rule of three” would become obsolete when expansions of other microsatellites came to be discovered (Mandel 1997). Tetra-, penta- and hexanucleotide repeat expansions were soon identified and were associated with other neurological diseases. At the present time, approximately 30 human diseases are associated with microsatellite expansions, some developmental, others neurological or neurodegenerative (see Figure 7.3).

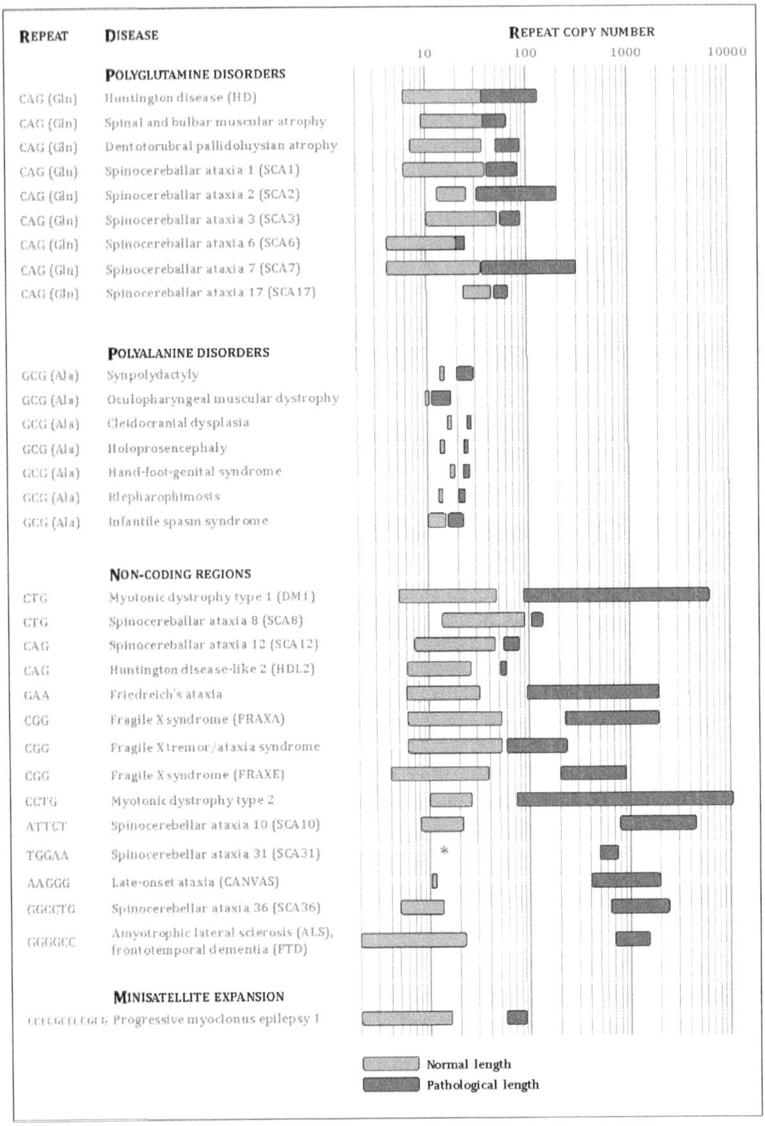

Figure 7.3. *Microsatellite expansion disorders. On the left: diseases are classified according to their different repeat types. In coding regions, polyglutamine and polyalanine diseases are found; in non-coding regions, other diseases involve tri-, tetra-, penta- or hexanucleotide repeats or a minisatellite. On the right: blue rectangles represent the size distribution of each repeat in the disease-free population, and red rectangles represent the population affected by the disease. The size scale is logarithmic*

The size of these expansions can be moderate, as in Huntington's chorea, or massive, as in myotonic dystrophy type 1, reaching several hundred, or even thousands of triplets in one single generation. The effect of these expansions on the metabolism of the genes containing them depends on their genetic location, as will be discussed below. Note that exonic microsatellite expansions, whether in the case of polyglutamine or polyalanine diseases, are generally much more moderate than expansions outside coding sequences. The reason for this difference is not known, but it is possible that excessive expansions cause inactivation of the corresponding gene and are consequently counterselected. The molecular mechanisms at the origin of these expansions include DNA slippage during replication, repair and homologous recombination, but many questions on this subject remain unresolved. What is known about these mechanisms is drawn mainly from studies conducted in model organisms and will be discussed in the next section.

7.2.4. *Microsatellites regulate gene expression*

Microsatellites are indeed very common in eukaryotic genomes and are found both in non-coding and coding regions. Mono- or dinucleotide repeats contained in exons are short and mostly correspond to genes of unknown function or whose existence is not confirmed by any experiment (Young et al. 2000). Among the genes with an identified function, only repeats whose base motif is a multiple of three (trinucleotide or hexanucleotide repeats) are generally encountered, since changing the length of any other motif would inactivate the open reading frame. It is precisely this property that is used by many bacteria to escape their hosts' immune system by a mechanism called "phase variation". Although microsatellites are much less common in prokaryotic genomes, many commensal or pathogenic bacteria contain microsatellites whose base motif is one to seven nucleotides long within genes encoding surface antigens. These microsatellite length changes affect gene expression differently depending on their location. A repeat length change within a promoter will modify gene expression, whereas in the coding region, it will lead to gene inactivation by changing the reading frame. The effect will therefore be transcriptional or translational (van der Woude and Bäumler 2004). In all cases, the expression of the surface antigen will be modified, an active gene will be inactivated, and a previously inactive gene may become activated. By switching genes encoding surface antigens, bacteria thus escape the immune system in a permanent battle for survival.

The regulation of transcription by microsatellites has also been studied in the brewer's yeast *Saccharomyces cerevisiae*. The promoters of yeast genes are rich in AT base pairs and commonly contain (AT)n or (AAT)n di- or trinucleotide repeats.

Like all microsatellites, they are polymorphic, exhibiting varying lengths between different yeast isolates. It has been shown that the expression level of these genes was correlated with the repeat size, with maximum expression centered on a certain value, contractions or expansions of the microsatellite decreasing gene expression. This observation was confirmed for reporter genes whose promoter contained repeats of different lengths (Vinces et al. 2009).

At the present time, it seems that this property to regulate gene expression is also observed in mammals. The prairie vole (*Microtus ochrogaster*) is one of the few mammals to be monogamous, forming bonded pairs and showing deep family attachment, with a high level of social interaction. In comparison, a closely related species, the mountain vole (*Microtus montanus*), shows strong social indifference, with the males not caring for their offspring and not remaining attached to the same female. These very different social behaviors are due to the presence of a CAGA microsatellite located in the 5' regulatory region of the V1aR gene, encoding the vasopressin receptor. The longest allele is associated with a higher level of transcription and more sociable behavior (Hammock and Young 2005). The exact role of vasopressin, an antidiuretic hormone, in the regulation of this sociability is still poorly understood.

The expression of many human genes also appears to be regulated by microsatellites. This regulation operates through various mechanisms, including the fixation of transcription factors to the microsatellite, variable spacing between promoter elements, selection of the transcription start site, RNA half-life, modification of histones or packaging into nucleosomes (for review, see Bagshaw 2017).

Outside promoter regions, many eukaryotic genes contain microsatellites, mainly trinucleotide repeats. These triplet repeats encode some amino acids more frequently than others. In *Saccharomycotina*, budding yeasts, it is mainly charged amino acids, glutamine, glutamic acid, asparagine and aspartic acid that are encoded by microsatellites. The products of these genes are preferentially nuclear proteins. In particular, transcription factors are overrepresented within proteins encoded by trinucleotide repeat-containing genes (Richard and Dujon 1997; Alba et al. 1999; Malpertuy et al. 2003). In the human genome, trinucleotide repeats identified in open reading frames encode polyglutamine (20% of cases), polyalanine (15%), polyglutamate (14%) and polyleucine (12%) repeat tracts for the four most common amino acids. Transcription factors are also overrepresented in proteins encoded by these genes, possibly suggesting a common property in the living world of trinucleotide repeat genes (Kozlowski et al. 2010).

Among transcription factors containing microsatellites, developmental genes of the multigene families Alx, Hox, Runx, Six, etc., contain imperfect trinucleotide repeats, mostly encoding polyglycine or polyalanine tracts. An imperfect repeat is interrupted by one or more triplets different from the repeat consensus. The length of an imperfect repeat is more stable than that of a perfect repeat (see next section). In the congenic lines of dogs bred by humans through inbreeding, these repeats have lost their interruptions and have thus become perfect. As a result, they show a significant degree of polymorphism between individuals of the same race. The Pyrenean mountain dog exhibits bilateral polydactyly of the first hind toe. In this dog, the Alx4 gene allele shows a deletion of 51 nucleotides (17 amino acids) within a microsatellite encoding a proline and glutamine repeat. This mutation inactivates gene function because in the mice where the Alx4 gene has been inactivated demonstrate the same polydactyly. Similarly, the remarkably rapid evolution of the skulls of Saint-Bernards, Bull terriers and Newfoundlands is linked to the change in size of a microsatellite encoding an alanine and glutamine repeat in the Runx-2 gene (Fondon and Garner 2004). It should be noted that several of these genes are involved in serious developmental diseases in humans. Trinucleotide repeat expansions in the Hox-D13, Runx-2 and Zic-2 genes are involved in synpolydactyly (Akarsu et al. 1996; Goodman et al. 1997), cleidocranial dysplasia (Mundlos et al. 1997) and holoprosencephaly (Brown et al. 2001), respectively. All these serious developmental diseases are part of the trinucleotide expansion disorders, whose molecular mechanisms will be discussed in the next section.

Lastly, a comprehensive analysis of 12 genomes of *Drosophila* species shows that the most common triplet in coding repeats is CAG, followed by CAA, two codons specifying glutamine. The genes encoding these polyglutamine segments are more commonly associated with DNA binding and transcription regulatory functions, confirming what was observed in yeasts and in humans (Huntley and Clark 2007).

Thus, it appears in light of current genomic studies, even if the precise function of all microsatellites is not yet known, and even if it is possible that not all of them have a function, that these repeats play a central role in regulating genetic expression throughout the living world, whether they are located in gene-promoter or -regulatory regions or in regions where they will be translated into proteins.

7.2.5. *Minisatellites are important in cell adhesion*

Minisatellites are less common in genomes than their smaller microsatellite brethren, which explains why they were quickly replaced by the latter as genetic markers. In humans, they are mainly found in non-coding regions for the same reason as microsatellites. Any change in the size of a nucleotide number that is not a multiple of three would result in gene inactivation. Therefore, only minisatellites with a basic motif that is a multiple of three are found in genes, such as the dodecamer repeat of the EPM1 gene, whose expansion leads to progressive myoclonic epilepsy (see Figure 7.3).

The first detailed studies of the distribution and function of minisatellites were conducted on the genome of the yeast *Saccharomyces cerevisiae*, mainly because it was the first eukaryotic organism whose good quality, sequence and assembly allowed for an extensive study of these repeated elements. This genome contains 84 minisatellites (compared to more than 1,800 microsatellites; Malpertuy et al. 2003), 55 of which are found in genes, 11 in non-coding regions and 18 in repeated elements in subtelomeres[8]. Half of the minisatellite-containing genes encode proteins involved in cell wall formation (Richard and Dujon 2006). It is worth pointing out here that yeasts are surrounded by a thick wall composed of glycoproteins protecting them from attacks from the external environment. These proteins have a C-terminus transmembrane region embedded into the plasma membrane and an N-terminus region of variable size that is generally heavily glycosylated. The type of sugars and their chain length depend on the yeast species (Latgé and Calderone 2005). These sugars form bonds among themselves, stiffening the wall but also allowing interactions with other yeast cells. Among the genes containing minisatellites, several belong to the *FLO* gene family, which is involved in yeast flocculation, that is, the ability to aggregate in clumps of several thousand, or even several million cells that sediment in liquid culture. This property is widely used in the brewing industry to clarify fermentation tanks. *FLO1*, the main gene involved in fermentation, contains a minisatellite with a 135-nucleotide long base motif. It encodes many serine and threonine residues, which are the glycosylation sites of the Flo1 protein. This motif is repeated 10 times in tandem in the reference strain, but this number varies from one strain to another. If strains containing a variable number of FLO motifs are constructed, their ability to flocculate in liquid medium is positively correlated with the number of motifs. The length of the minisatellite is therefore directly responsible for the adhesion property of yeasts among themselves but also to substrates such as polystyrene (Verstrepen et al. 2005). Similar experiments have been conducted on the *FLO11* gene, which contains two minisatellites, whose base motifs are 30 and 36 nucleotides. The length of the second minisatellite is highly variable from one strain to another, particularly in yeasts used in the fermentation of sherry, a specialty from southern Andalusia. The maturation of these wines occurs in the presence of air in oak barrels. A veil of

8 Yeast subtelomeres are defined as the region immediately upstream of telomeres that do not contain an essential gene. They extend over approximately 30 kb in *S. cerevisiae*.

yeast, called "*flor*", grows on the surface of the wine, protecting it from oxidation. The ability of yeasts to form this veil depends directly on the size of the second *FLO11* minisatellite. Increasing its length will increase the hydrophobicity of the yeast cell wall, allowing them to float to the surface of the wine and form a biofilm that will both give them access to the oxygen present in the barrel and protect the wine from overly rapid oxidation (Fidalgo et al. 2006).

Unlike microsatellites, which are mainly involved in the regulation of gene expression, minisatellites, at least those of yeasts, appear to have specialized cell adhesion functions. These functions are also represented in genes that contain megasatellites.

7.2.6. *Function of megasatellites*

Systematic research of minisatellites in sequenced genomes has shown that their number varies widely from one yeast species to another. *S. cerevisiae* contains only a few of them in comparison to *Candida glabrata*, an opportunistic pathogen. This yeast is the second most common nosocomial disease[9] after *Candida albicans*. It is responsible for superficial infections affecting the mucosal membranes and skin, mainly the oral cavity, vaginal mucosa, and esophagus. In the most severe cases, it is responsible for systemic candidiasis, which can affect all organs. Its genome was fully sequenced in 2004 (Dujon et al. 2004), and its analysis showed that it contained 3.6 times more minisatellites than *S. cerevisiae* for genomes of equivalent sizes. Sixty percent of the 238 minisatellites are found in genes, and in these cases, the size of the repeated motif is always a multiple of three, such that any change in the number of motifs does not alter the reading frame. Surprisingly, some minisatellites are made up of two motifs of different sizes, interspersed among themselves like two different decks of cards shuffled together. Most noteworthy, however, is the presence of 44 minisatellites with a very large base motif (126–429 nucleotides), which can be repeated a large number of times, the longest covering more than 20 kb of DNA. For this reason, they were called "megasatellites" (Thierry et al. 2008). Because *C. glabrata* genes are less well characterized than those of *S. cerevisiae*, the precise function of most of them is unknown. However, several megasatellite-containing genes seem to encode membrane proteins involved in yeast adhesion to its host cells or contain known motifs of transmembrane domains, suggesting their location at the cell surface. Subsequent studies in other yeasts and fungi suggest that most other megasatellites, detected in their genomes, are

9 Nosocomial diseases are diseases contracted in hospital. They are mainly caused by bacterial or fungal infections.

contained in genes encoding membrane proteins, probably glycosylated and involved in intracellular interactions (Tekaia et al. 2013).

In view of these observations, it could be thought that all mini- and megasatellites have cell adhesion functions. However, observations made in the genomes of other organisms suggest otherwise. Vertebrate genomes, for example, contain a large number of repeats that can be classified in the megasatellite category. The genes containing them often belong to multigene families, such as zinc finger genes, whose tandem motifs are 90 nucleotides long. Each finger contains a combination of histidine and cysteine residues that coordinate a zinc atom and recognize a sequence of three nucleotides. Zinc finger proteins are DNA-binding proteins involved most of the time in transcription regulation. As another example, the ankyrin motif is encoded by a repeat with a basic motif of 99 nucleotides (33 amino acids). This megasatellite is found in the three kingdoms of life, but it is particularly common in eukaryotes. The folding of this protein allows interactions between proteins exhibiting very different functions. The WD40 protein domain is encoded by a motif of 120 base pairs, repeated in tandem four to 16 times, forming a kind of toroidal blade around a central axis. Genes containing this megasatellite encode proteins that serve as a rigid structure for the assembly of other complex proteins. These three (non-exhaustive) examples demonstrate to what extent megasatellite-encoding genes in vertebrates exhibit much more diverse functions than those found in fungi and yeasts. This is probably due in part to their unicellular or micellar nature, which requires more labile intercellular interactions than that of metazoans, which in return have diversified the functions encoded by their megasatellites.

7.2.7. Centromeric satellite DNA, complexity of structure–function studies

In mammalian chromosomes, satellite DNA is found mainly in centromeric regions and on the Y chromosome. The partial sequencing of these regions in the 1990s demonstrated that they were mainly composed of repeated DNA whose basic motif size varies between five nucleotides for satellites 2 and 3 present on almost all chromosomes and up to 171 nucleotides for the α satellite present on all human chromosomes (Tyler-Smith and Willard 1993). The latter is the fixation site of the CENP-B protein, via a motif of 17 base pairs, the "CENP-B box". Centromeric satellite DNA therefore plays a role in centromere function and thus in the segregation of mitotic chromosomes. This complex mechanism is explained in detail in Chapter 4.

In the 1990s, much research was focused on satellite DNA (see Figure 7.2) but was quickly confronted with a significant problem. Centromeric satellite DNA extends over huge distances, covering more than 1 million base pairs in humans. In the mouse genome, the centromeres consist of a major satellite that is a tetranucleotide repeat and a minor satellite whose base motif is 120 nucleotides long, both regions altogether covering more than six megabases of DNA. The problem was barely less complex in *Drosophila* or in plants. Centromeres of the vinegar fly are made of satellite DNA containing two pentanucleotide repeats of different sequences scattered between transposable elements, altogether covering 400 kb of DNA. *Arabidopsis thaliana* centromeres consist of a 180-nucleotide tandem repeat covering more than 100 kb of DNA (Muller et al. 2019). Thus, it soon became clear that with classical sequencing techniques, which generated 150–200 base reads at best, it was impossible to fully sequence and assemble the centromeres of mammals, plants or *Drosophila*. The next generation of sequencers[10] did not solve this problem either because the reads are just as short, generally approximately 200 base pairs, and it is impossible to assemble overly long repeated sequences. With third-generation sequencers[11], the read size reaches several tens of kilobases. Using a mix of short, long and very long reads, the first fully complete human genome was recently achieved and sequenced from telomeres to telomeres, including previously unsequenced centromeres (Nurk et al. 2022). Many other complete centromere sequences will hopefully become available in the coming years, revealing, at last, structures and properties that have thus far remained inaccessible to us.

7.3. The evolutionary mechanisms of tandem repeats

Several mathematical models have been proposed to explain the evolution of microsatellites over time. These can be divided into two major families: incremental models and saltatory models.

The stepwise mutation model (SMM) describes the evolution of microsatellites as a succession of additions and subtractions, with equal probability, of a single motif of the repeat. A repeat of length n motifs will therefore evolve into a repeat of length $n - 1$ or $n + 1$ motifs. The corollary of this model is that a given repeat retains the memory of its prior state(s), and the length of an allele differs very little from that of its progenitor allele. However, according to the SMM model, the addition or subtraction of a repeat unit is equally probable and would theoretically allow the

10 Mainly the technology developed by *Illumina*.

11 The two main technologies are *PacBio* and *Oxford Nanopore Technology*.

presence of infinitely long or zero-length repeated sequences. Moreover, this model does not converge on a stable state of repeat lengths over time.

The infinite alleles model (IAM) is a saltatory model which postulates that a microsatellite of size *n* can give birth to a microsatellite of size *m*, very different from *n*. This model suggests that a random process of mutations can produce, with equal probability, any allele size at a given microsatellite locus. A variant of this model suggests that only a limited number, *k*, of alleles can be obtained from a microsatellite of size *n* (Jarne and Lagoda 1996).

A study of more than 100 tetranucleotide repeats from the human genome showed that their expansion rate was constant regardless of the microsatellite length, while the contraction rate increased with it (Xu et al. 2000). There therefore exists a size around which a given microsatellite is at the equilibrium between expansion and contraction. Nevertheless, it is not clear whether this size is identical for all microsatellites or whether it is subject to the same constraints in all regions of the genome or between genomes of different organisms.

In any case, experiments conducted mainly in model organisms have fortunately helped us to identify the molecular mechanisms by which microsatellite expansions and contractions occur.

7.3.1. *Historical model of slippage during replication*

Historically, the first molecular model proposed was slippage of the two DNA strands during replication (slipped-strand mispairing (SSM)). This involves local denaturation and incorrect re-annealing of the two complementary strands of DNA following slippage along the repeated sequence, downstream or upstream of its original pairing site. If the newly synthesized strand re-anneals downstream, a DNA loop will form on the template strand, leading to a contraction of the repeat by deletion of one or more motifs. If the newly synthesized strand re-anneals upstream, it forms a loop, leading to expansion through the addition of one or more motifs (see Figure 7.4). Because the repeated sequence is its own instability substrate, the more repeated motifs it contains, the greater its probability of mutation. In addition, microsatellites whose sequence can form secondary structures that may slow or hinder the progression of the replication fork will be more likely to trigger this slippage. Experiments conducted in the yeast *Saccharomyces cerevisiae* show that this mechanism is a source of instability that can add one to two motifs to a dinucleotide repeat (Strand et al. 1993). This observation suggests that length changes might be generally incremental rather than saltatory.

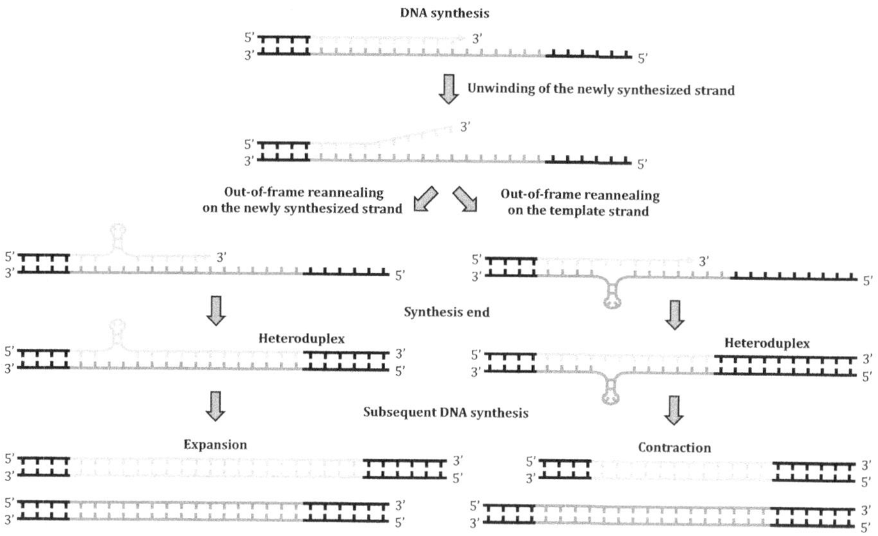

Figure 7.4. *Replication slippage model. During DNA synthesis in a tandemly repeated region (in blue and red), the newly synthesized strand can temporarily unwind from the matrix strand and then re-anneal in a shifted manner. If this slippage occurs on the newly synthesized strand, the formation of a loop followed by DNA synthesis will result in a longer strand because the same DNA region will be replicated twice. The molecule obtained is a DNA heteroduplex, the two strands of which carry two different pieces of genetic information. Replication of this heteroduplex will result in two molecules of different lengths, one normal and the other carrying a repeat expansion. If slippage occurs on the template strand, this mechanism will lead to repeat contraction. It should be noted that a DNA sequence capable of forming a secondary structure, such as a hairpin, will tend to favor this replication slippage mechanism*

An interruption along a repeated sequence influences DNA strand slippage and/or affects the stability of the possibly formed secondary structures. There naturally exist 1–3 CAT/ATG interruptions within the (CAG)n trinucleotide repeats responsible for spinocerebellar ataxia type 1 and 1–3 AGG/CCT interruptions within the (CGG)n repeat responsible for fragile X syndrome. These interruptions prevent the formation of an excessively long repeated DNA segment that could trigger expansions. It has been experimentally shown that an interruption placed in a microsatellite makes it more stable than the same sequence perfectly repeated (Petes et al. 1997; Rolfsmeier and Lahue 2000).

7.3.2. *Slippage during DNA repair*

By extension, it can be considered that any mechanism involving de novo DNA synthesis in a repeated sequence could generate instability. In particular, it was shown that oxidative damage repair increased the instability of CAG trinucleotide repeats in mice.

The base excision repair (BER) mechanism is the primary mechanism for repairing oxidative damage. Free radicals are the by-products of cellular metabolism, including superoxides, hydrogen peroxide and hydroxyl radicals. These generate the most DNA damage. Guanines are the bases most sensitive to oxidation and notably form 8-oxo-guanines, the other three bases being modified into 8-oxo-adenines, uracil or dihydro 5-6 thymine. Modifying a base alters its pairing property with the complementary base. This modification can cause a mutation of the modified base after genome replication, which may lead to the development of diseases such as some cancers. In patients with neurodegenerative disorders such as Alzheimer's or Huntington's disease, the BER is altered, reflected by an increase in the number of 8-oxo-guanines.

Two mechanisms exist for BER. The first mechanism, single nucleotide BER (SN-BER), enables elimination of a nucleotide through the activity of the glycosylase OGG1, which excises 8-oxoguanines. After displacement of the incised strand, β polymerase will resynthesize the missing DNA. Long patch BER (LP-BER) enables the repair of oxidative damage by introducing a combination of polymerases and replicative factors. A short fragment of 2–15 nucleotides is excised, unlike the single base removed in the SN-BER. The stability of CAG trinucleotide repeats is increased in mutant mice for the OGG1 glycosylase. In these animals, expansions of CAG microsatellites, which occur spontaneously with age, disappear in the absence of OGG1, suggesting that the mechanism of these somatic expansions involves BER (Kovtun et al. 2007).

7.3.3. *Repeat expansions and contractions during homologous recombination*

Homologous recombination is one of the mechanisms by which a cell will be able to repair a double-strand break in its DNA. Long considered a faithful mechanism because it calls on a homologous sequence for repair, it has been shown to be a possible source of instability and mutations, especially in tandem repeats. Several mechanisms exist for homologous recombination repair. All of them use, in yeast, the mediator *RAD52*, whose functional homolog in humans is *BRCA2* (Sung et al. 2003). Homologous recombination mechanisms dependent on the RecA/Rad51 recombinase, such as gene conversion, can be differentiated from mechanisms that are Rad51-independent but Rad52 dependent, such as single-strand annealing (SSA). Lastly, some mechanisms are independent of Rad51 and Rad52, such as microhomology end-joining (MMEJ), which, as its name suggests, makes use of DNA microhomologies. Although these different mechanisms each have their molecular specificities, an initial step of resection of the double-strand break, producing 3'-OH single-stranded ends, initiates them (Krogh and Symington 2004).

7.3.3.1. *Two gene conversion repair mechanisms*

Gene conversion is the most common form of homologous recombination initiated after double-strand break. It corresponds to the exchange of non-reciprocal genetic material between a homologous template sequence and the region carrying the double-strand break. In the recombination model proposed by Szostak, Orr-Weaver, Rothstein and Stahl almost 40 years ago, gene conversion can be associated with an exchange of flanking sequences called "crossover" (Szostak et al. 1983). This exchange is derived from the resolution of a four-stranded DNA junction called "Holliday junctions" (see Figure 7.5). They were named after Robin Holliday, who in the 1960s was the first to suggest such a structure as a recombination intermediate leading to crossover. Physical proof of the existence of meiotic double Holliday junctions was demonstrated decades later by Tony Schwacha and Nancy Kleckner (Schwacha and Kleckner 1995).

An alternative model was proposed in 1994 by Nassif and colleagues, who studied the repair events of double-strand breaks in *Drosophila* following the excision of a P element (Nassif et al. 1994). In this alternative model (called synthesis-dependent strand annealing (SDSA)), once DNA synthesis is initiated, the two newly synthesized strands can dissociate themselves from their homologous template and re-anneal with each other. There is no Holliday junction resolution, and SDSA is therefore a conservative gene conversion mechanism that enables a double-strand break to be repaired without generating crossover. The existence of SDSA was formally demonstrated in the yeast *S. cerevisiae* by Pâques and Haber using an elegant double-strand break repair mechanism. A truncated *LEU2* gene in its central part could only be repaired by recopying two complementary pieces of information present on two different chromosomes, thus excluding the possibility of generating a crossover and the obligation to unwind at least one of the newly synthesized strands during repair to hybridize it with its complementary sequence (Pâques et al. 1998).

7.3.3.2. *Slippage during gene conversion*

Changes in minisatellite length were long believed to be due to unequal crossover occurring between two alleles during meiotic recombination. The study of gene conversion associated with the instability of several minisatellites, in yeast and humans, showed unambiguously that it was independent of crossover (Richard and Pâques 2000).

In *S. cerevisiae*, gene conversion is associated with a non-reciprocal exchange of genetic material at the *CUP1* tandemly repeated locus, leading to expansions or contractions of large tandem repeats (Welch et al. 1990). The study of the human hypervariable MS32 minisatellite by Alec Jeffreys' laboratory showed that 74% of mutations observed in sperm were expansions. These are preferentially polarized on one side of the repeats and are due to inter- and intra-allelic exchanges of genetic information (Jeffreys et al. 1994). In the majority of cases, the absence of exchange of flanking sequences indicates that these rearrangements are derived from gene conversion independent of crossover. Analysis of the rare crossover events shows that a meiotic recombination hotspot is situated upstream of the MS32 minisatellite and promotes instability. The observed frequency of gene conversion decreases as one moves away from the breakpoint, explaining the observed mutational polarity (Jeffreys et al. 1998). These results explain the instability of the MS32 minisatellite in humans, as well as the fact that this instability is greater in meiosis than in mitosis when the same minisatellite was integrated into the genome of *S. cerevisiae* (Appelgren et al. 1997). When the human CEB1 minisatellite is inserted near the *ARG4* locus, known to be a meiotic recombination hotspot in the yeast *S. cerevisiae*, its instability increases during meiosis. The inactivation of the *SPO11* gene, which is responsible for the formation of double-strand breaks in meiosis, eliminates this instability. The insertion of CEB1 in a region that is not subject to double-strand breaks also stabilizes the tandem repeat, demonstrating that minisatellite stability depends on the generation of a double-strand break in its vicinity (Debrauwère et al. 1999). All these experiments show that a minisatellite is strongly destabilized by gene conversion initiated by a double-strand break in its proximity and that this instability is very rarely associated with crossover. It is therefore quite natural that the SDSA recombination model was proposed to explain these observations (see Figure 7.5). This mechanism was also proposed to explain rearrangements of tandem-repeat 5S ribosomal DNA in *Drosophila* (Pâques et al. 1996) and yeast (Pâques et al. 1998).

Subsequent experiments on CTG microsatellites showed that microsatellite expansions and contractions occur very frequently during gene conversion initiated by a double-strand break. These length changes are under the control of the MRX complex, a protein complex conserved from bacteria to humans responsible for the early stages of double-strand break repair (Richard et al. 2000). It was subsequently shown that crossover did not occur at a detectable frequency in cells that had undergone these rearrangements (Richard et al. 2003).

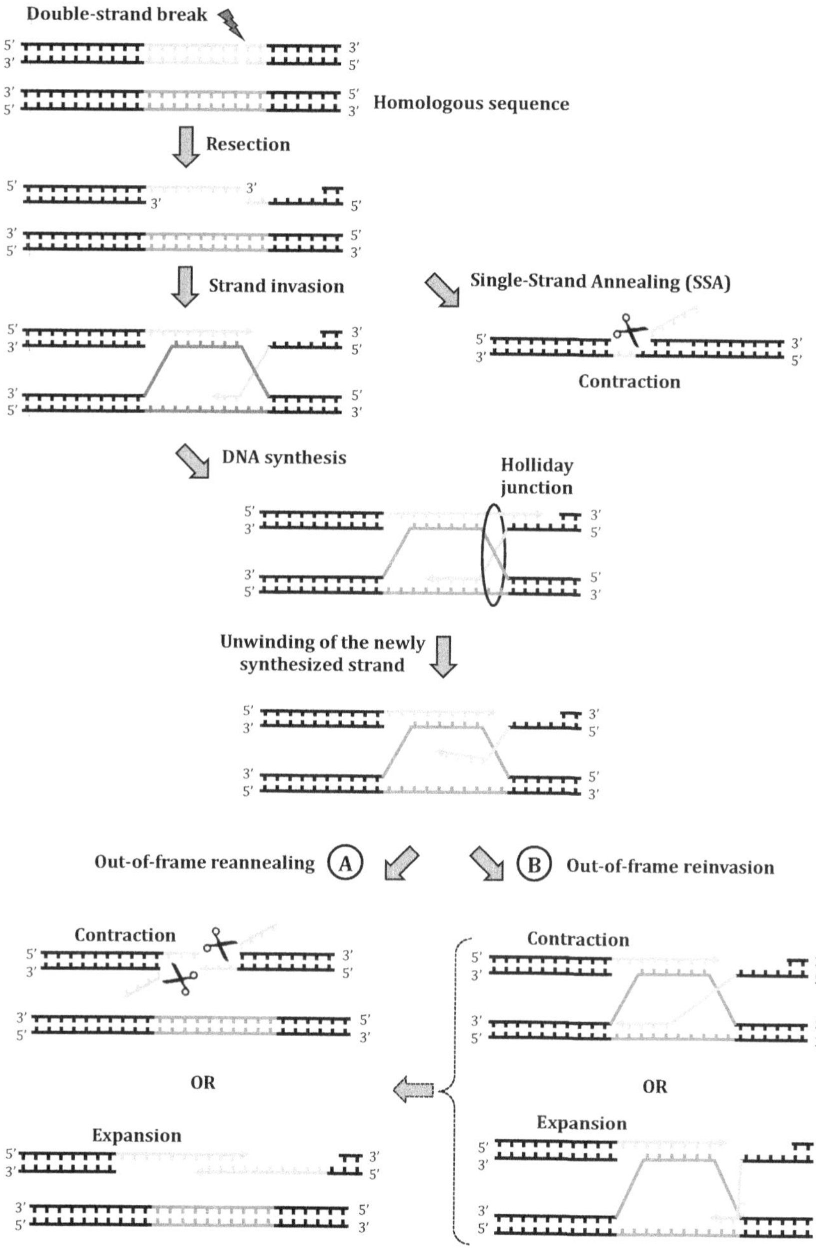

Figure 7.5. *Model of slippage during homologous recombination*

NOTES ON FIGURE 7.5.– *A DNA molecule carrying a tandem repeat (in red) receives a double-strand break. The break is resected by specific enzymes that release 3'-OH ends that will be able to recombine with a homologous DNA molecule containing the same repeated sequence (in blue). The two single-stranded ends can be paired by single-strand annealing (SSA), leading to repeat contraction. Alternatively, they can invade the homologous molecule to repair the break by gene conversion. During the DNA synthesis associated with this repair, a Holliday junction is formed. The newly synthesized strand can temporarily unwind from its template, similar to what occurs during replication-associated slippage (see Figure 7.4). It can re-anneal in a shifted manner with its complementary strand, leading to a contraction or expansion of the repeat (A). Alternatively, it can reinvade homologous DNA in a shifted manner, leading to an expansion or contraction of the repeat (B). It is possible to pass from B to A at any time when the newly synthesized strand unwinds from the template strand. The scissors represent DNA endonucleases that digest the 3' end flaps to complete the repair.*

In conclusion, microsatellites as well as minisatellites may undergo expansions and contractions during gene conversion associated with the repair of a double-strand break in their vicinity, and this mechanism is very rarely associated with crossovers.

7.3.3.3. *Trinucleotide repeat contraction by single-strand annealing*

SSA enables the repair of a double-strand break through hybridization of two homologous regions flanking the break. The region between the two homologous sequences is deleted (Pâques and Haber 1999), thus rendering repair by SSA mutagenic. Similar to gene conversion, it is dependent on the resection of both ends of the break, releasing 3' single-stranded ends that can anneal with each other. When a double-strand break is initiated within a CTG trinucleotide repeat, both ends of the break anneal to form a shorter repeat (Richard et al. 1999a; Richard et al. 2014). The genetic study of this phenomenon shows that, like SSA, it is dependent on Rad52 and on the MRX complex but not on Rad51 (Mosbach et al. 2018) (see Figure 7.5).

In conclusion, homologous recombination very strongly destabilizes all tandem repeats, micro- and minisatellites. If the break is made near but outside of the repeat, it will be repaired by gene conversion, leading to contractions and expansions of the repeat tract. If the break occurs inside the repeated region, it will be repaired by SSA, leading to repeat contraction.

7.4. Microsatellites in human diseases

7.4.1. *Triplet repeat expansion disorders*

As discussed above, several neurological and developmental diseases are due to expansions of a particular family of microsatellites called trinucleotide repeats (see Figure 7.3). This group of human disorders, mainly affecting the nervous system, are linked to the massive, uncontrolled expansion of certain repeated sequences. The first diseases discovered exclusively concerned triplet repeats, CAG (or its complement, CTG), CCG and GAA. Expansions of other tetra-, penta-, and hexanucleotide repeats were subsequently found which were associated with myotonic dystrophy type 2, spinocerebellar ataxias, and amyotrophic lateral sclerosis (ALS), respectively. Nevertheless, they are generally referred to as trinucleotide repeat expansion disorders, even if other microsatellites are affected (for review, see Brown and Brown 2004; Cummings and Zoghbi 2000). It has been shown that some of these microsatellites are able to form secondary structures in vitro and probably in vivo (Poggi and Richard 2021). It was proposed very early on that these secondary structures were at the origin of the mechanism triggering expansions (McMurray 1999).

We previously reviewed several mechanisms affecting microsatellite stability. These mechanisms of slippage during replication, DNA repair, or homologous recombination generate frequent expansions and contractions of tandem repeats. However, expansions due to these mechanisms are moderate, generally not exceeding twice the initial size of the repeat. In the case of certain diseases, such as myotonic dystrophy type 1, expansions can reach several thousand triplets in one single generation. The proposed molecular mechanisms do not explain the size of such expansions unless they are assumed to occur iteratively in the same cell.

Among these disorders, Huntington's chorea, a dramatic dominant neurodegenerative disorder, is related to the expansion of a CAG trinucleotide repeat into the first exon of the gene encoding huntingtin (The Huntington's Disease Collaborative Research Group 1993). A CTG triplet repeat expansion in the 3' untranslated region of the *DMPK* gene is responsible for the development of myotonic dystrophy type 1 (or Steinert's disease) (Brook et al. 1992). Friedreich ataxia, a recessive neurodegenerative disease due to abnormal iron metabolism, is caused by the expansion of a GAA triplet repeat in the first intron of the FRDA gene (Campuzano et al. 1996), and an expansion of CGG triplets into the 5' UTR of the *FRM1* gene is responsible for fragile X syndrome (see Figure 7.3). Expansions of triplets located in coding regions lead to polyglutamine or polyalanine expansions, altering the function of the proteins that contain them. Expansions in transcribed but non-coding regions will result in modifications of RNA properties through the gain or loss of function or will negatively interact with premessenger RNA splicing or protein synthesis. Huntington's disease and spinocerebellar ataxia types 1, 2, 3, 6, 7, and 17 express a toxic protein after the exonic expansion of a CAG triplet encoding a glutamine. Spinocerebellar ataxia types 8 and 12 as well as myotonic dystrophy type 1 express an RNA toxic to the cell (gain of function) due to expansions in the 5' or 3' UTR of the genes concerned. Lastly, fragile X syndrome and Friedreich's ataxia are due to expansions within the 5' UTR or the first intron of the affected genes, leading to a decrease in protein expression.

These diseases present the so-called "anticipation" phenomenon, the increasingly early appearance and aggravation of symptoms during transmission to offspring. A positive correlation is observed between repeat size, the appearance of symptoms and the severity of the disease, while triplet length increases at each generation. Repeats generally become unstable at approximately 35-50 triplets, and the observed changes in size are small until reaching a threshold (usually approximately 50 triplets), beyond which expansions sometimes greater than 10,000 triplets can be observed during the parental transmission.

In addition to these neurological or neurodegenerative diseases, another category of trinucleotide expansion diseases affects embryonic development. These are polyalanine diseases triggered by the expansion of a GCN triplet into the coding region of developmental genes, such as Hox-D13, Runx-2 and Zic-2 (see section 7.2.4 above). It should be noted that expansions in these diseases are more moderate in size than in polyglutamine diseases. This may be due to the nature of the repeats that are imperfect and suggests that the mechanism leading to these expansions is probably different (Brown and Brown 2004).

7.4.2. *Colorectal cancers and the mismatch repair system*

The DNA mismatch repair (MMR) system is a mechanism involved in a set of metabolic functions and is found in all living organisms. First highlighted in its role against the accumulation of mutations during replication, it promotes genomic stability. The main substrates of MMR are base substitutions and small insertions/deletions but also base oxidative damage. MMR loss leads to a rise in the number of spontaneous mutations, and we observe, notably in families carrying mutations in the *MSH2, MLH1, MSH6, MSH3, PML2,* PMS1 and/or *PMS2* genes, the appearance of hereditary non-polyposis colorectal cancers (HNPCC, or Lynch syndrome) (Fishel 2001). Microsatellite instability is observed in more than 90% of HNPCC cases. A set of genes has been identified as being mutated in these cancers. A change in the size of mono- or dinucleotide repeats within their coding sequence inactivates these genes, deregulating the cell cycle or apoptosis. These metabolic alterations confer an advantage to the tumor cell, which can divide beyond control.

The MMR system has been extensively studied in the bacterium *Escherichia coli* and the yeast *Saccharomyces cerevisiae*. It is based on the recognition by the MutS complex(es) of the errors left on DNA after replication. In the case of microsatellites, these errors are due to slippages during replication, as seen previously (see Figure 7.4). The MutL complex(es) then attach(es) to MutS, triggering DNA incision at the mutation site and its repair. Inactivation of the MMR system increases the mutation rate in microsatellites by several hundred-fold (Strand et al. 1993). In some trinucleotide repeats, the MMR system is disturbed by the presence of the repeat and specifically causes expansions (reviewed in Richard 2021).

Lastly, it should be noted that the stabilizing effect of interruptions within a microsatellite is independent of the mismatch repair machinery. Inactivation of *MSH2*, *MSH3* or *PMS1* does lead to an increase in small insertions or deletions of one or two base pairs in a dinucleotide repeat, but it does not impact stabilization of the same microsatellite interrupted by a motif different from the consensus (Petes et al. 1997).

7.4.3. *Fragile sites*

Fragile sites are drug-sensitive chromosomal loci that induce replication fork slowdown and lead to the appearance of single- and double-strand breaks during the S phase. They are classified as common or rare according to their proportion in the population and have been found and documented in numerous mammals. They are involved in chromosomal rearrangements such as translocations, sister chromatid exchanges, oncogenic virus insertions, deletions or expansions found in certain cancers such as adenocarcinoma, myeloma, leukemia and lymphoma (Richards 2001).

Common fragile sites (CFSs) are found in all individuals and are notably sensitive to treatment with aphidicolin, inhibiting the activity of the α, δ and ε polymerases. They are found to be late replicating during the cell cycle. In particular, this phenotype is observed in the case of the common fragile site FRA3B induction with aphidicolin, where the locus is not yet fully replicated upon entry into the G2 phase of the cell cycle. Moreover, this fragility is specific to a type of tissue (Letessier et al. 2011).

Rare fragile sites (RFSs) are caused by the expansion of microsatellites or minisatellites, whose secondary structures hinder replication. The folate-sensitive group is the major group of RFSs, the expression of which is due to a CGG trinucleotide repeat expansion in the *FRM1* (FRAXA site) or *FRM2* (FRAXE site) gene. Other RFSs are not sensitive to folate but to other drugs. This is the case for FRA10B, which is induced by bromodeoxyuridine (BrdU), and FRA16B, which is induced by distamycin A, an antibiotic peptide that binds to AT-rich sequences in the small DNA groove. FRA10B and FRA16B are induced by the expansion of AT-rich minisatellites. FRA16B is a 33-nucleotide motif that is amplified, and FRA10B is an imperfect motif of 35-43 nucleotides that undergoes expansion. In both cases, the hypervariable minisatellites are located in a region rich in minisatellites. Three groups of RFSs can be distinguished: a folate-sensitive group (FRA11B, FRA16A, FRAXA, FRAXE, FRAXF), a group sensitive to distamycin A (FRA16B, FRA17A, FRA8E, FRA11I) and a third group sensitive to BrdU (FRA10B, FRA12C) (Sutherland et al. 1998). The fragility of these sequences appears proportional to their size.

The insertion of CAG/CTG or CCG/CGG repeats into an artificial yeast chromosome behaves as a fragile site in the presence of hydroxyurea (HU), a drug that decreases the cell's dNTP poll, leading to replication fork slowdown. The longer the trinucleotide repeat is, the higher the break frequency (Freudenreich et al. 1998). Notably, CAG/CTG repeats, even those that are large, do not behave as fragile sites in humans. This difference from what is observed in yeast therefore suggests subtle differences in trinucleotide repeat metabolism between these two eukaryotes.

7.5. De novo formation and evolution of tandem repeats

7.5.1. *Birth and death of microsatellites*

While microsatellite expansion and contraction mechanisms have been widely studied as well as the relevance of the models explored with dedicated experiments, the same is not true of the origin of these repeated sequences. The chances of a repeated sequence of n motifs being randomly found in a genome can be calculated by knowing the frequency of each individual motif. Since the first complete genome sequences, it has been known that this frequency does not depend solely on the GC content. The *S. cerevisiae* genome contains approximately 38% GC base pairs, and open reading frames are slightly richer (40%). When the distribution of all dinucleotides in the genome was calculated, it was noted that some were significantly overrepresented compared to what would be expected. For example, the dinucleotide TG is 10 times more common than would be expected given the GC content. Conversely, the dinucleotide TA is 20 times less common than expected (Dujon et al. 1994). This observation is also true for trinucleotides. Consequently, any statistical study on the distribution of di- or trinucleotide repeats must take into account these composition biases and not just the GC richness, as is unfortunately too often the case in numerous studies. When the trinucleotide composition of the genome is taken into account to calculate the theoretical number of repeats of each triplet, it is observed that repeats containing 4- to 6-triplets are underrepresented, with the exception of CAG repeats. In contrast, long repeats (8 triplets or more) are generally overrepresented, although they are low in number in the yeast genome (Richard and Dujon 1996). For mono-, di- and tetranucleotide repeats, there also appears to exist a size limit beyond which these microsatellites are overrepresented (Rose and Falush 1998). These observations suggest a model in which short repeats are rare, but the length of some of them increases gradually, possibly under the effect of selection.

However, how is a microsatellite born? The alignment of gene sequences containing trinucleotide repeats between several yeast species shows that some of these microsatellites are present only in certain species and absent in others, while the rest of the gene sequence aligns perfectly (Malpertuy et al. 2003). This suggests that there exist(s) one or more mechanisms actively producing a microsatellite from an unrepeated sequence. There are three models to explain the creation and evolution of microsatellites.

– The slippage model:

Slippages during replication or homologous recombination (see above) can cause expansions of a microsatellite, but the frequency of the slippage depends on its length, with longer repeats changing in size more frequently through this mechanism (Rose and Falush 1998). In addition, this model postulates that at least two tandem repeat motifs are required to obtain an expansion (or contraction) of the repeat (see Figures 7.4 and 7.5). Therefore, it cannot explain the appearance of a microsatellite in an unrepeated DNA sequence.

– The substitution model:

A second mechanism proposes that intermittent substitutions of one or more bases could create a repeat of two (or more) motifs. This duplicate (or protomicrosatellite) will then be able to undergo expansion by slippage or by another mechanism (see Figure 7.6).

– The insertion–deletion model:

Lastly, a third model suggests that microindels[12] could create duplicates that would later evolve into microsatellites (Dieringer and Schlötterer 2003). These microindels would occur with a constant frequency, regardless of microsatellite length, in contrast to classical slippages (see Figure 7.6). A study of several thousand mutations observed in the human genome shows that the majority of the identified duplicates come from nucleotide insertions due to microindels: 29% of dinucleotides, 37% of trinucleotides and 88% of tetranucleotides. In comparison, only 16% of dinucleotide duplicates, 6% of trinucleotide duplicates and 3% of tetranucleotide duplicates are derived from base substitutions (Zhu et al. 2000). It should be noted that these microindels can occur by microduplication, using an identical flanking nucleotide, or by microinsertion, without an identical nucleotide (see Figure 7.6), the first mechanism probably being the most used.

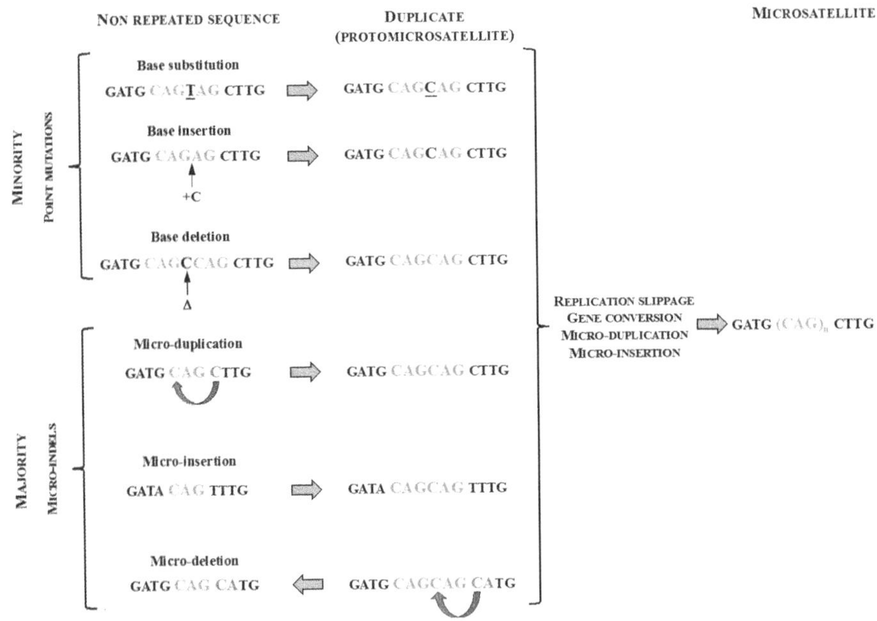

Figure 7.6. *Mechanisms of microsatellite formation by mutation. An unrepeated DNA sequence first becomes a duplicate, either by point mutation or by microinsertion, microdeletion or microduplication. The duplicate thus created can now become a microsatellite by either of the slippage mechanisms described above*

12 Indel: contraction of "insertion–deletion", commonly used in genetics to characterize insertions or deletions of one or several nucleotides.

A study comparing the microsatellites of the human genome and that of chimpanzees comes to the same conclusions. The insertions of a number of nucleotides equal to the size of the basic motif are not random and are the consequence of a local microduplication of the motif, producing a duplicate. These microduplications can be initiated from a single repeated nucleotide flanking the motif. Similarly, microdeletions were identified but required two nucleotides flanking the repeat (see Figure 7.6). Lastly, this study also showed that microsatellite contractions were twice as frequent as their expansion into the human genome, raising the question of the existence and maintenance of long microsatellites (Leclercq et al. 2010).

It was also proposed that protomicrosatellites could be created from a double-strand DNA break. This break, which could occur during DNA replication, for example, would generate single-stranded ends that can be repaired in two different ways. A first mechanism involves mispairing of the two ends of the break containing a microhomology, followed by ligation using microhomology-mediated end joining (MMEJ), a variant of non-homologous end joining, a non-homologous repair mechanism involving religation of the two ends of a double-strand break (see Figure 7.7).

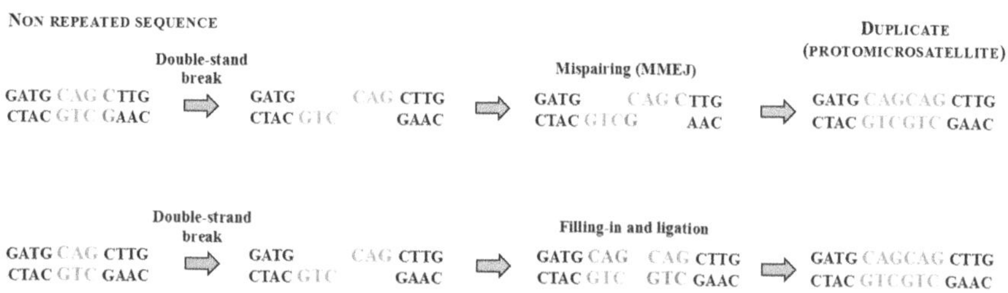

Figure 7.7. *Mechanisms of microsatellite formation by end-joining. A double-strand break in an unrepeated DNA sequence can be repaired by joining the ends between two microhomologies or by local DNA synthesis and religation, leading to a duplicate, which can now be amplified in a microsatellite by one or more of the slippage mechanisms described above*

The second mechanism postulates the filling of single-stranded ends by a DNA polymerase, followed by re-ligation of both ends (see Figure 7.7). Note that these mechanisms remain theoretical for now because it is impossible to differentiate them from other mechanisms involving microindels or point mutations (Leclercq et al. 2010).

Lastly, from a duplicate, a microsatellite of three (or more) repeated motifs can be obtained by microindel or by replication slippage or gene conversion, as seen above.

7.5.2. *Formation of minisatellites*

While many authors have addressed the issue of the origin of microsatellites, there are unfortunately fewer studies on minisatellites. Although the problem remains fundamentally the same – going from a sequence containing no repeats to a duplicate, then to a tandem repeat of three or more repeated motifs – its complexity is different. While we have a clear understanding of how a mono-, di- or trinucleotide repeat can be created by point mutation of one, two or three base pairs, arriving at the same result with basic motifs 10, 20 or 100 nucleotides long seems unlikely. The probability of finding, by chance, a duplicate of 10 nucleotides or more in a genome is close to zero. By carefully studying the sequence of several minisatellites discovered in yeast and humans, Haber and Louis made the remarkable observation that each repeated motif was flanked by a short sequence of five to nine nucleotides that could be used to create the initial duplicate by replication slippage (Haber and Louis 1998). From there, it becomes easy to amplify this duplicate again by slippage or by homologous recombination between duplicated motifs to produce a minisatellite (see Figure 7.8).

A similar observation was made in the mitochondrial DNA of different populations of guppies (*Poecilia reticulata*). In these fish, there is a duplication of 11 nucleotides flanks, a mitochondrial minisatellite that is not present in all guppies. Observing the genome of fish that do not possess tandem repetition, it was seen that the two 11-nucleotide sequences differ at two positions. This observation suggests that, initially, two point mutations made these two sequences identical. Then, DNA slippage during replication or homologous recombination between these two duplications created the minisatellite (Taylor and Breden 2000).

A comprehensive study of all the minisatellites of *Saccharomyces cerevisiae* showed that 49 out of 55 of them contain two short flanking sequences, three to seven nucleotides long (most often a duplicated pentamer) (Richard and Dujon 2006). A similar study of the genome of the opportunistic pathogenic yeast *Candida glabrata* arrived at a similar conclusion, since half of the minisatellites detected are flanked by two short sequences, most often pentamers (Thierry et al. 2008). These studies strongly suggest that the mechanism initially proposed by Louis and Haber could be used mostly to create duplicates that will give birth to minisatellites, although in *Candida glabrata,* it is possible that another mechanism may also be at work.

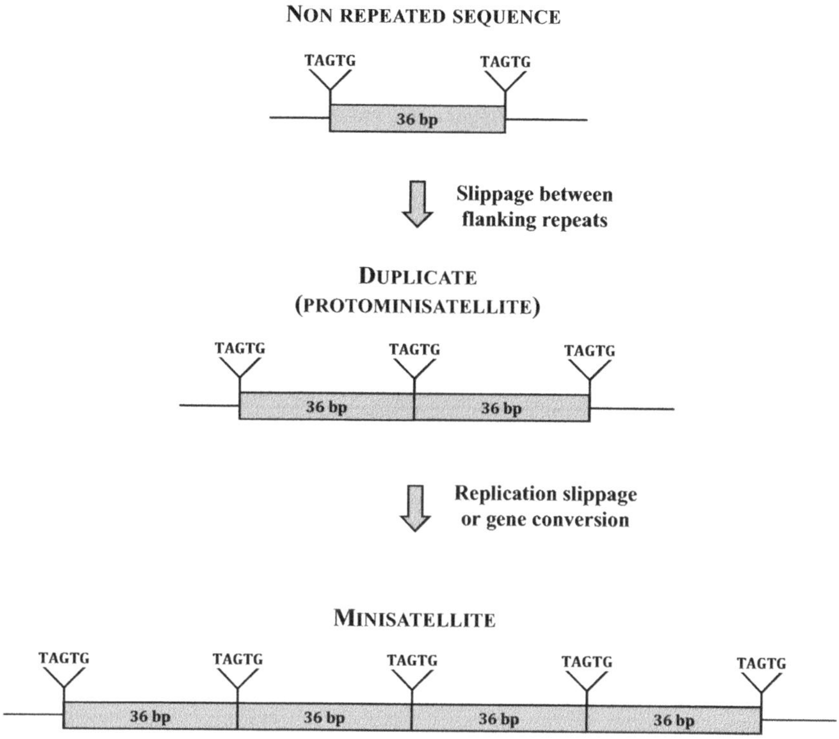

Figure 7.8. *Mechanism of formation of a minisatellite. A slippage between two tandemly repeated short regions will generate a duplicate, which can become a minisatellite by one or more of the slippage mechanisms described above*

Most of the megasatellites of this yeast are grouped into two major families, called SHITT and SFFIT, according to the sequence of these five amino acids, which are strictly conserved in each motif. A phylogenetic analysis of all of the motifs comprising these two families showed that *Candida glabrata* megasatellites evolved by three distinct mechanisms: (i) the duplication of a gene already containing a megasatellite, creating a paralogous gene; (ii) homologous recombination resulting in the transfer of a motif from one megasatellite to another by gene conversion; (iii) another mechanism, resembling a "motif jump" from one megasatellite to another presenting only little identity, such as a transposable element (see Figure 7.9). Analyses of other genomes should determine whether this observation is specific to *Candida glabrata* or whether similar events can be traced in other branches of the tree of life.

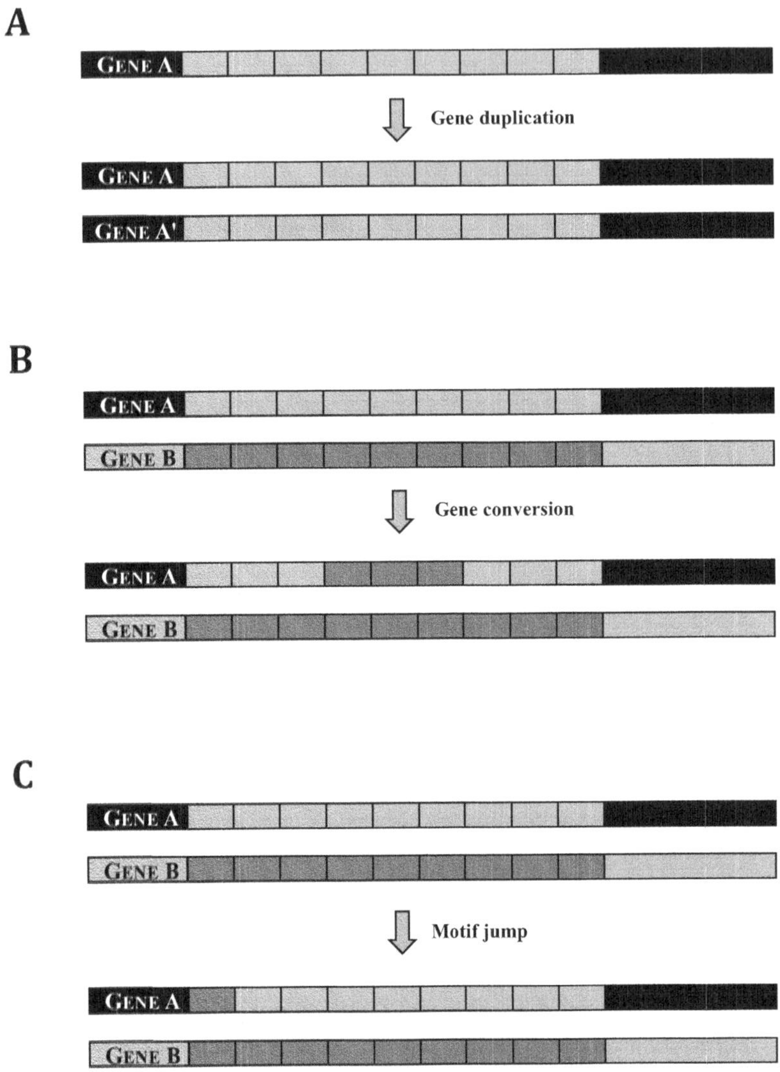

Figure 7.9. *Mechanisms of evolution of mini- and megasatellites. (A) The duplication of a gene leads to the duplication of the tandem repeat it contains. (B) Gene conversion between two tandem repeats of similar sequences leads to the replacement of one or more motifs by one or more similar motifs. The number of motifs can also vary during recombination events (indeed, this is often the case). (C) One or more motifs can move from one tandem repeat to another through a mechanism that is still unknown but perhaps involves retrotransposition of an RNA intermediate*

7.6. Perspectives

The current avalanche of prokaryotic and eukaryotic genome sequences produces an amount of data that is unprecedented in the history of biology. Several hundred eukaryotic genomes have already been sequenced, and project sequencing of all eukaryotes has been proposed (Pennisi et al. 2017). This is much to the delight of geneticists; however, they are then faced with a major challenge: how can we store and analyze all these data? The sequence is currently produced more rapidly than the increase in computing power and data storage. Moreover, researchers interested in tandemly repeated sequences face three additional problems: incomplete genomes, inadequate software tools, and the lack of consensus on the definition of these repeated elements. The first problem arises from the fact that long tandem repeats in particular are complicated to sequence, even with third-generation techniques that generate long reads, but with high error rates. When good-quality sequences are obtained, it is not always possible to assemble them because of their repeated nature. The experiment involving whole-genome sequencing of *Arabidopsis thaliana* using third-generation technology showed that the entire genome could be assembled in a single experiment (already impressive in itself!), with the exception of centromeres, which are almost entirely made up of long tandem repeats, impossible to assemble (Michael et al. 2018). In 2022, for the first time, a human genome was completely finished using three different sequencing technologies: short, long and very long reads. The "telomere-to-telomere" genome was the first to cover previously unsequenced regions, such as centromeres. However, the complexity of such repeated regions is such that 46 ambiguities still remain in the final assembly (Nurk et al. 2022). There is little doubt that other complete genome sequences will be achieved in the coming years, allowing us for the first time to truly study and compare long tandem repeats, whose sequence has thus far remained elusive.

7.6.1. *Inadequacy of software tools*

The second problem lies in the inadequacy of software tools in identifying tandem repeats. More than a dozen algorithms have been published to search for tandem repeats, particularly microsatellites, in genomes. They are based on three different types of approaches. Programs using the statistical approach scan the genome to detect regions that can be microsatellites according to predefined statistical rules and then submit these regions to validation tests. The region is validated as a microsatellite if the validation score exceeds a threshold set by the user. The best known and most representative program of this category is the tandem repeats finder (TRF), developed by Benson and Waterman (1999). Programs using the combinatorial approach scan the genome to detect sequences according to predefined construction rules. Two sequences form part of the same repeat if they diverge from each other by a maximum of user-predefined differences. MREPS (Kolpakov et al. 2003) and Sputnik (Abajian 1994) belong to this category. Lastly, the alignment approach uses a known motif (or more commonly, a database of known motifs) to scan the genome. Sequences with an alignment score above a user-defined threshold are validated as microsatellites. RepeatMasker (Smit et al. 1996) and STAR (Delgrange and Rivals 2004) are good representatives of this category of software. We may then wonder whether one of these approaches is more efficient than another for detecting tandem repeats and whether one of these programs is more powerful.

A study of this type, comparing five different programs based on the three types of approaches, showed that the results obtained were very different from one program to another (Leclercq et al. 2007). Microsatellites were searched for in the whole genomes of *Saccharomyces cerevisiae*, *Neurospora crassa* and *Drosophila melanogaster*, as well as in the human X chromosome sequence, using MREPS, TRF, Sputnik, RepeatMasker and STAR. The number of detections depends largely on the parameters and detection threshold set by the user and increases as the threshold decreases. However, detected microsatellites are increasingly shorter as the detection threshold gradually lowers. It should be noted that the detection parameters strongly influence the type of microsatellites detected. A perfect repeat will have a higher score than a repeat of the same size containing one or more nucleotides different from the consensus. The score of these imperfect repeats will be reduced, and if this score is below the threshold set by the user, the sequence will not be conserved by the program. An example of this limitation is shown in Figure 7.10.

A perfect microsatellite $(CAG)_7$ will be correctly detected at a given detection threshold (case 1), as will two shorter microsatellites interrupted by one or two nucleotides different from the consensus (cases 2 and 3). However, a third nucleotide not aligned with the consensus will lower the score below the selected threshold, and the microsatellite will not be detected by the program (case 4). By reducing the detection threshold very slightly, two shorter microsatellites will be detected (case 5). In addition, by altering the respective weights of the nucleotides aligned (match weight) or not aligned (mismatch weight) with the consensus, the score can be significantly modified, and therefore, the number of microsatellites above or below the set threshold. Note that RepeatMasker and STAR, using motif databases, can only detect repeats that align perfectly on these motifs. This type of algorithm can therefore only detect perfect repeats[13].

13 This was true in the article published by Leclercq et al. (2007). However, since 2013, RepeatMasker uses TRF to detect microsatellites and other imperfect repeats.

Comparing the results of the five programs showed that Sputnik found three times as many microsatellites as TRF and 5–20 times as many as the other three, but these were one-third shorter. The major difference in efficacy is correlated with the detection of small microsatellites. Indeed, most microsatellites are short, and the threshold values of most programs prevent their efficient detection. Sputnik is by far the most efficient in detecting these short microsatellites.

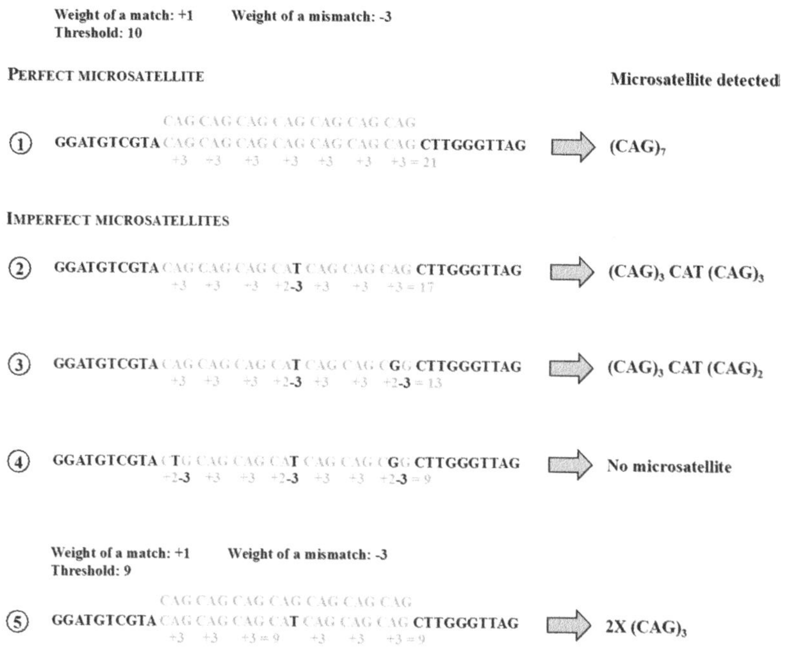

Figure 7.10. *The limits of microsatellite detection algorithms. Depending on the number of nucleotides different from the consensus, a microsatellite may be detected as perfect (1), imperfect in certain cases (2 and 3) or undetected (4). Depending on the detection threshold set, the same microsatellite will be detected as imperfect (2) or as two shorter microsatellites (5)*

STAR and RepeatMasker find approximately ten times fewer microsatellites than TRF because they are unable to detect imperfect microsatellites. Lastly, MREPS is the most efficient at finding imperfect microsatellites, no doubt because it is capable of detecting microsatellites composed of more than one motif, for example, $(CAG)_5(CAA)_4$. The other programs would not be able to find them because they take the two motifs into account separately: the final scores will therefore be too low to be retained (Leclercq et al. 2007).

To add to the complexity of the problem, other algorithms have been developed, such as ACMES (Reneker et al. 2004) and MREPATT (Roset et al. 2003), dedicated to detecting perfect repeats, and TandemSWAN (Boeva et al. 2006), specialized in searching for imperfect repeats, not to mention the many "in-house programs" whose algorithms are not always well described (and often not published). However, because the efficacy of all of these programs has not been evaluated in comparison with other better-known and more widely used programs, it is difficult to draw conclusions as to their respective qualities and shortcomings.

All of these observations highlight the importance of using the same tools, particularly in comparative genomics, to be able to compare results with each other. From one program to another, the number of microsatellites can vary by a factor of 20 for the same genome. One software program will detect a large number of imperfect microsatellites, while another will find none. Beyond this problem of choice of approach, largely dependent on the initial question, the problem of tools that are not adapted to the search for tandem repeats is posed. Not one program is available at present, not even among the most efficient, is capable of detecting all microsatellites in a complex genome. This question is not only related to an algorithmic problem but also depends on the limits imposed on these genetic objects.

7.6.2. *The importance of definitions in biology*

As was seen at the beginning of this chapter, no consensus exists on the definition of a microsatellite. Some authors do not consider mononucleotide repeats to be microsatellites, while others place a limit on the size of the basic motif, which can be set between five and ten nucleotides. Lastly, the minimum number of tandem repeat motifs considered to be a microsatellite varies from one author to another, with some setting this limit to three, four or five repeated motifs. These differences in definition make comparisons between publications a complex matter.

The comparison between different works focusing on the *Saccharomyces cerevisiae* genome shows, unsurprisingly, that the definition used for each study has a considerable influence on the results. Using the same program (TRF), when searching for microsatellites (perfect or imperfect) of at least four repeated motifs, 1,769 of them are found, while only 924 are detected if the minimum length is set to five motifs (Richard and Dujon 1997). In another study using TRF on the same genome, by placing the lower size limit at five motifs, but only searching for perfect repeats, only 621 microsatellites were found (Malpertuy et al. 2003). Lastly, by considering perfect repeats of at least nine repeated motifs, TRF detected only 141 of them in the yeast genome (Astolfi et al. 2003).

It would therefore be good to try to harmonize, to some extent, what is considered a microsatellite. There is no reason not to consider mononucleotide repeats as microsatellites, as they follow the same biological rules as other microsatellites with longer motifs. Lastly, the minimum number of repeated motifs could be set to three, with duplicates needing to be considered as protomicrosatellites which are able to easily disappear or amplify to become *bona fide* microsatellites. However, what about imperfect microsatellites? Depending on the penalty given to nucleotides different from the consensus and the validation threshold, a sequence can be validated as containing or not containing one or two microsatellite(s) (see Figure 7.10). It therefore seems complicated to define a microsatellite in an unambiguous way, and the difficulty of detecting such objects by existing algorithms partly reflects the difficulty in clearly defining them. This question arises just as keenly for other tandem repeats, mini- and megasatellites, whose repeated motifs may be identical to a lesser or greater extent and sometimes interrupted by other repeated motifs different from the main minisatellite consensus (Thierry et al. 2008).

One solution could be to try, on the one hand, to define a set of sequences that are considered microsatellites or minisatellites (perfect or imperfect) and to use learning algorithms based on these repeat databases to detect all similar sequences in genomes. By using the same algorithm on the entire living world, we should finally be able to define a set of micro- and minisatellites representative of each monophyletic group to answer questions as to their formation and evolution, as well as their role in the plasticity of eukaryotic genomes.

7.7. Acknowledgments

We would like to thank Sébastien Leclercq (INRA, Poitiers) for his help in modeling mechanisms involved in the appearance of microsatellites.

7.8. References

Abajian, C. (1994). Sputnik [Online]. Available at: http://espressosoftware.com/pages/sputnik.jsp.

Akarsu, A.N., Stoilov, I., Yilmaz, E., Sayli, B.S., Sarfarazi, M. (1996). Genomic structure of HOXD13 gene: A nine polyalanine duplication causes synpolydactyly in two unrelated families. *Hum. Mol. Genet.*, 5(7), 945–952.

Alba, M.M., Santibañez-Koref, M.F., Hancock, J.M. (1999). Amino acid reiterations in yeast are overrepresented in particular classes of proteins and show evidence of a slippage-like mutational process. *J. Mol. Evol.*, 49, 789–797.

Appelgren, H., Cederberg, H., Rannug, U. (1997). Mutations at the human minisatellite MS32 integrated in yeast occur with high frequency in meiosis and involve complex recombination events. *Molecular & General Genetics: MGG*, 256(1), 7–17.

Arrighi, F.E., Mandel, M., Bergendahl, J., Hsu, T.C. (1970). Buoyant densities of DNA of mammals. *Biochemical Genetics*, 4(3), 367–376.

Astolfi, P., Bellizzi, D., Sgaramella, V. (2003). Frequency and coverage of trinucleotide repeats in eukaryotes. *Gene*, 317, 117–125.

Bagshaw, A.T.M. (2017). Functional mechanisms of microsatellite DNA in eukaryotic genomes. *Genome Biology and Evolution*, 9(9), 2428–2443.

Benson, G. (1999). Tandem repeats finder: A program to analyze DNA sequences. *Nucleic Acids Res.*, 27(2), 573–580.

Boeva, V., Regnier, M., Papatsenko, D., Makeev, V. (2006). Short fuzzy tandem repeats in genomic sequences, identification, and possible role in regulation of gene expression. *Bioinformatics*, 22(6), 676–684.

Brook, J.D., McCurrach, M.E., Harley, H.G., Buckler, A.J., Church, D., Aburatani, H., Hunter, K., Stanton, V.P., Thirion, J.P., Hudson, T. et al. (1992). Molecular basis of myotonic dystrophy: Expansion of a trinucleotide (CTG) repeat at the 3' end of a transcript encoding a protein kinase family member. *Cell*, 68, 799–808.

Brown, L.Y. and Brown, S.A. (2004). Alanine tracts: The expanding story of human illness and trinucleotide repeats. *Trends in Genetics*, 20, 51–58.

Brown, L.Y., Odent, S., David, V., Blayau, M., Dubourg, C., Apacik, C., Delgado, M.A., Hall, B.D., Reynolds, J.F., Sommer, A. et al. (2001). Holoprosencephalie due to mutations in ZIC2: Alanine tract expansion mutations may be caused by parental somatic recombination. *Human Molecular Genetics*, 10, 791–796.

Campuzano, V., Montermini, L., Molto, M.-D., Pianese, L., Cossée, M., Cavalcanti, F., Monros, E., Rodius, F., Duclos, F., Monticelli, A. et al. (1996). Friedreich's ataxia: Autosomal recessive disease caused by an intronic GAA triplet repeat expansion. *Science*, 271, 1423–1427.

Cummings, C.J. and Zoghbi, H.Y. (2000). Fourteen and counting: Unraveling trinucleotide repeat diseases. *Hum. Mol. Genet.*, 9, 909–916.

Debrauwère, H., Buard, J., Tessier, J., Aubert, D., Vergnaud, G., Nicolas, A. (1999). Meiotic instability of human minisatellite CEB1 in yeast requires double-strand breaks. *Nat. Genet.*, 23, 367–371.

Delgrange, O. and Rivals, E. (2004). STAR: An algorithm to search for tandem approximate repeats. *Bioinformatics*, 20(16), 2812–2820.

Dib, C., Faure, S., Fizames, C., Samson, D., Drouot, N., Vignal, A., Millasseau, P., Marc, S., Hazan, J., Seboun, E. et al. (1996). A comprehensive genetic map of the human genome based on 5,264 sequences. *Nature*, 380, 152–154.

Dieringer, D. and Schlötterer, C. (2003). Two distinct modes of microsatellite mutation processes: Evidence from the complete genomic sequences of nine species. *Genome Research*, 13(10), 2242–2251.

Dujon, B., Alexandraki, D., André, B., Ansorge, W., Baladron, V., Ballesta, J.P.G. (1994). Complete DNA sequence of yeast chromosome XI. *Nature*, 369, 371–378.

Dujon, B., Sherman, D., Fischer, G., Durrens, P., Casaregola, S., Lafontaine, I., de Montigny, J., Marck, C., Neuvéglise, N., Talla, E. et al. (2004). Genome evolution in yeasts. *Nature*, 430(6995), 35–44.

Fidalgo, M., Barrales, R.R., Ibeas, J.I., Jimenez, J. (2006). Adaptive evolution by mutations in the FLO11 gene. *Proc. Natl. Acad. Sci. USA*, 103, 11228–11233.

Fishel, R. (2001). The selection for mismatch repair defects in hereditary nonpolyposis colorectal cancer. *Cancer Research*, 61(20), 7369–7374.

Fondon, J.W. and Garner, H.R. (2004). Molecular origins of rapid and continuous morphological evolution. *Proceedings of the National Academy of Sciences*, 101(52), 18058–18063.

Foster, E.A., Jobling, M.A., Taylor, P.G., Donnelly, P., de Knijff, P., Mieremet, R., Zerjal, T., Tyler-Smith, C. (1998). Jefferson fathered slave's last child. *Nature*, 396, 27–28.

Freudenreich, C.H., Kantrow, S.M., Zakian, V.A. (1998). Expansion and length-dependent fragility of CTG repeats in yeast. *Science*, 279, 853–856.

Gill, P., Jeffreys, A.J., Werrett, D.J. (1985). Forensic application of DNA "fingerprints". *Nature*, 318, 577–579.

Goodman, F.R., Mundlos, S., Muragaki, Y., Donnai, D., Giovannucci-Uzielli, M.L., Lapi, E., Majewski, F., McGaughran, J., McKeown, C., Reardon, W. et al. (1997). Synpolydactyly phenotypes correlate with size of expansions in HOXD13 polyalanine tract. *Proc. Natl. Acad. Sci. USA*, 94(14), 7458–7463.

Haber, J.E. and Louis, E.J. (1998). Minisatellite origins in yeast and humans. *Genomics*, 48, 132–135.

Hagelberg, E., Gray, I.C., Jeffreys, A.J. (1991). Identification of the skeletal remains of a murder victim by DNA analysis. *Nature*, 352, 427–429.

Hammock, E.A.D. and Young, L.J. (2005). Microsatellite instability generates diversity in brain and sociobehavioral traits. *Science*, 308(5728), 1630–1634.

Han, J., Hsu, C., Zhu, Z., Longshore, J.W., Finley, W.H. (1994). Over-representation of the disease associated (CAG) and (CGG) repeats in the human genome. *Nuc. Acids Res.*, 22, 1735–1740.

Helminen, P., Ehnholm, C., Lokki, M.L., Jeffreys, A., Peltonen, L. (1988). Application of DNA "fingerprints" to paternity determinations. *Lancet*, 1(8585), 574–576.

Hopkins, B., Williams, N.J., Webb, M.B., Debenham, P.G., Jeffreys, A.J. (1994). The use of minisatellite variant repeat-polymerase chain reaction (MVR-PCR) to determine the source of saliva on a used postage stamp. *Journal of Forensic Science*, 39, 526–531.

Huntley, M.A. and Clark, A.G. (2007). Evolutionary analysis of amino acid repeats across the genomes of 12 Drosophila species. *Mol. Biol. Evol.*, 24(12), 2598–609.

International Human Genome Sequencing Consortium (2001). Initial sequencing and analysis of the human genome. *Nature*, 409, 860–921.

Jarne, P. and Lagoda, J.L. (1996). Microsatellites, from molecules to populations and back. *Trends in Ecology and Evolution*, 11, 424–429.

Jeffreys, A.J., Wilson, V., Thein, S.L. (1985). Hypervariable "minisatellite" regions in human DNA. *Nature*, 314, 67–73.

Jeffreys, A.J., Neumann, R., Wilson, V. (1990). Repeat unit sequence variation in minisatellites: A novel source of DNA polymorphism for studying variation and mutation by single molecule analysis. *Cell*, 60(3), 473–485.

Jeffreys, A.J., Allen, M.J., Hagelberg, E., Sonnberg, A. (1992). Identification of the skeletal remains of Josef Mengele by DNA analysis. *Forensic Sci. Int.*, 56, 65–76.

Jeffreys, A.J., Tamaki, K., MacLeod, A., Monckton, D.G., Neil, D.L., Armour, J.A. (1994). Complex gene conversion events in germline mutation at human minisatellites. *Nature Genetics*, 6(2), 136–145.

Jeffreys, A.J., Murray, J., Neumann, R. (1998). High-resolution mapping of crossovers in human sperm defines a minisatellite-associated recombination hot spot. *Mol. Cell*, 2, 267–273.

Khan, R. and Mittelman, D. (2018). Consumer genomics will change your life, whether you get tested or not. *Genome Biology*, 19(1), 120.

Kolpakov, R., Bana, G., Kucherov, G. (2003). mreps: Efficient and flexible detection of tandem repeats in DNA. *Nucl. Acids. Res.*, 31, 3672–3678.

Kovtun, I.V., Liu, Y., Bjoras, M., Klungland, A., Wilson, S.H., McMurray, C.T. (2007). OGG1 initiates age-dependent CAG trinucleotide expansion in somatic cells. *Nature*, 447(7143), 447–452.

Kozlowski, P., de Mezer, M., Krzyzosiak, W.J. (2010). Trinucleotide repeats in human genome and exome. *Nuc. Acids Res.*, 38(12), 4027–4039.

Krogh, B.O. and Symington, L.S. (2004). Recombination proteins in yeast. *Annual Review of Genetics*, 38(1), 233–271.

Latgé, J.-P. and Calderone, R. (2005). The fungal cell wall. In *The Mycota XIII*, Esser, K., Fischer, R. (eds). Springer, Berlin.

Leclercq, S., Rivals, E., Jarne, P. (2007). Detecting microsatellites within genomes: Significant variation among algorithms. *BMC Bioinformatics*, 8(1), 125.

Leclercq, S., Rivals, E., Jarne, P. (2010). DNA Slippage occurs at microsatellite loci without minimal threshold length in humans: A comparative genomic approach. *Genome Biology and Evolution*, 2, 325–335.

Letessier, A., Millot, G.A., Koundrioukoff, S., Lachagès, A.-M., Vogt, N., Hansen, R.S., Malfoy, B., Brison, O., Debatisse, M. (2011). Cell-type-specific replication initiation programs set fragility of the FRA3B fragile site. *Nature*, 470(7332), 120–123.

Litt, M. and Luty, J.A. (1989). A hypervariable microsatellite revealed by in vitro amplification of a dinucleotide repeat within the cardiac muscle actin gene. *Am. J. Hum. Genet.*, 44, 397–401.

Lohe, A.R., Hilliker, A.J., Roberts, P.A. (1993). Mapping simple repeated DNA sequences in heterochromatin of *Drosophila melanogaster. Genetics*, 134, 1149–1174.

Malpertuy, A., Dujon, B., Richard, G.-F. (2003). Analysis of microsatellites in 13 hemiascomycetous yeast species: Mechanisms involved in genome dynamics. *Journal of Molecular Evolution*, 56, 730–741.

Mandel, J.-L. (1997). Breaking the rule of three. *Nature*, 386, 767–769.

McMurray, C.T. (1999). DNA secondary structure: A common and causative factor for expansion in human disease. *Proc. Natl. Acad. Sci. USA*, 96, 1823–1825.

Michael, T.P., Jupe, F., Bemm, F., Motley, S.T., Sandoval, J.P., Lanz, C., Loudet, O., Weigel, D., Ecker, J.R. (2018). High contiguity Arabidopsis thaliana genome assembly with a single nanopore flow cell. *Nature Communications*, 9(1), 541.

Mosbach, V., Poggi, L., Viterbo, D., Charpentier, M., Richard, G.-F. (2018). TALEN-induced double-strand break repair of CTG trinucleotide repeats. *Cell Reports*, 22, 2146–2159.

Muller, H., Gil, J., Drinnenberg, I.A. (2019). The impact of centromeres on spatial genome architecture. *Trends in Genetics: TIG*, 35(8), 565–578.

Mundlos, S., Otto, F., Mundlos, C., Mulliken, J.B., Aylsworth, A.S., Albright, S., Lindhout, D., Cole, W.G., Henn, W., Knoll, J.H. et al. (1997). Mutations involving the transcription factor CBFA1 cause cleidocranial dysplasia. *Cell*, 89(5), 773–779.

Nassif, N., Penney, J., Pal, S., Engels, W.R., Gloor, G.B. (1994). Efficient copying of nonhomologous sequences from ectopic sites via P-element-induced gap repair. *Molecular and Cellular Biology*, 14(3), 1613–1625.

Nurk, S., Koren, S., Rhie, A., Rautiainen, M., Bzikadze, A., Mikheenko, A., Vollger, M.R., Altemose, N., Uralsky, L., Gershman, A. et al. (2022). The complete sequence of a human genome. *Science*, 376(6588), 44–53.

Pâques, F. and Haber, J.E. (1999). Multiple pathways of recombination induced by double-strand breaks in *Saccharomyces cerevisiae. Microbiology and Molecular Biology Reviews*, 63(2), 349–404.

Pâques, F., Bucheton, B., Wegnez, M. (1996). Rearrangements involving repeated sequences within a P element preferentially occur between units close to the transposon extremities. *Genetics*, 142, 459–470.

Pâques, F., Leung, W.-Y., Haber, J.E. (1998). Expansions and contractions in a tandem repeat induced by double-strand break repair. *Molecular and Cellular Biology*, 18(4), 2045–2054.

Pennisi, E. (2017). Biologists propose to sequence the DNA of all life on Earth. *Science* [Online]. Available at: http://www.sciencemag.org/news/2017/02/biologists-propose-sequence-dna-all-life-earth [Accessed 27 August 2018].

Petes, T.D., Greenwell, P.W., Dominska, M. (1997). Stabilization of microsatellite sequences by variant repeats in the yeast *Saccharomyces cerevisiae*. *Genetics*, 146(2), 491–498.

Poggi, L. and Richard, G.-F. (2021). Alternative DNA structures in vivo: Molecular evidence and remaining questions. *Microbiology and Molecular Biology Reviews*, 85(1) [Online]. Available at: https://mmbr.asm.org/content/85/1/e00110-20 [Accessed 1 March 2021].

Reneker, J., Shyu, C.R., Zeng, P., Polacco, J.C., Gassmann, W. (2004). ACMES: Fast multiple-genome searches for short repeat sequences with concurrent cross-species information retrieval. *Nuc. Acids Res.*, 32(Web Server issue), W649–W653.

Richard, G.-F. (2018). *Des comètes à l'homme synthétique.* EDP Sciences, Les Ulis.

Richard, G.-F. (2021). The startling role of mismatch repair in trinucleotide repeat expansions. *Cells*, 10(5), 1019.

Richard, G.-F. and Dujon, B. (1996). Distribution and variability of trinucleotide repeats in the genome of the yeast *Saccharomyces cerevisiae*. *Gene*, 174(1), 165–174.

Richard, G.-F. and Dujon, B. (1997). Trinucleotide repeats in yeast. *Res. Microbiol.*, 148, 731–744.

Richard, G.-F. and Dujon, B. (2006). Molecular evolution of minisatellites in hemiascomycetous yeasts. *Mol. Biol. Evol.*, 23(1), 189–202.

Richard, G.-F. and Pâques, F. (2000). Mini- and microsatellite expansions: The recombination connection. *EMBO Reports*, 1(2), 122–126.

Richard, G.-F., Dujon, B., Haber, J.E. (1999a). Double-strand break repair can lead to high frequencies of deletions within short CAG/CTG trinucleotide repeats. *Mol. Gen. Genet.*, 261, 871–882.

Richard, G.-F., Hennequin, C., Thierry, A., Dujon, B. (1999b). Trinucleotide repeats and other microsatellites in yeasts. *Res. Microbiol.*, 150, 589–602.

Richard, G.-F., Goellner, G.M., McMurray, C.T., Haber, J.E. (2000). Recombination-induced CAG trinucleotide repeat expansions in yeast involve the MRE11/RAD50/XRS2 complex. *EMBO J.*, 19, 2381–2390.

Richard, G.-F., Cyncynatus, C., Dujon, B. (2003). Contractions and expansions of CAG/CTG trinucleotide repeats occur during ectopic gene conversion in yeast, by a MUS81-independent mechanism. *Journal of Molecular Biology*, 326, 769–782.

Richard, G.-F., Kerrest, A., Dujon, B. (2008). Comparative genomics and molecular dynamics of DNA repeats in eukaryotes. *Microbiol. Mol. Biol. Rev.*, 72(4), 686–727.

Richard, G.-F., Viterbo, D., Khanna, V., Mosbach, V., Castelain, L., Dujon, B. (2014). Highly specific contractions of a single CAG/CTG trinucleotide repeat by TALEN in yeast. *PLOS ONE*, 9(4), e95611.

Richards, R.I. (2001). Fragile and unstable chromosomes in cancer: Causes and consequences. *Trends in Genetics*, 17(6), 339–345.

Rolfsmeier, M.L. and Lahue, R.S. (2000). Stabilizing effects of interruptions on trinucleotide repeat expansions in *Saccharomyces cerevisiae. Mol. Cell. Biol.*, 20, 173–180.

Rose, O. and Falush, D. (1998). A threshold size for microsatellite expansion. *Molecular Biology and Evolution*, 15(5), 613–615.

Roset, R., Subirana, J.A., Messeguer, X. (2003). MREPATT: Detection and analysis of exact consecutive repeats in genomic sequences. *Bioinformatics*, 19(18), 2475–2476.

Schwacha, A. and Kleckner, N. (1995). Identification of double Holliday junctions as intermediates in meiotic recombination. *Cell*, 83, 783–791.

Sharma, S. and Raina, S.N. (2005). Organization and evolution of highly repeated satellite DNA sequences in plant chromosomes. *Cytogenetic and Genome Research*, 109, 15–26.

Smit, A., Hubley, R., Green, P. (1996). RepeatMasker [Online]. Available at: http://repeatmasker.org/.

Southern, E.M. (1970). Base sequence and evolution of guinea-pig α-satellite DNA. *Nature*, 227(5260), 794–798.

Strand, M., Prolla, T.A., Liskay, R.M., Petes, T.D. (1993). Destabilization of tracts of simple repetitive DNA in yeast by mutations affecting DNA mismatch repair. *Nature*, 365(6443), 274–276.

Sung, P., Krejci, L., Van Komen, S., Sehorn, M.G. (2003). Rad51 recombinase and recombination mediators. *J. Biol. Chem.*, 278(44), 42729–42732.

Sutherland, G.R., Baker, E., Richards, R.I. (1998). Fragile sites still breaking. *Trends in Genetics*, 14(12), 501–506.

Szostak, J.W., Orr-Weaver, T.L., Rothstein, R.J., Stahl, F.W. (1983). The double-strand-break repair model for recombination. *Cell*, 33(1), 25–35.

Taylor, J.S. and Breden, F. (2000). Slipped-strand mispairing at noncontiguous repeats in Poecilia reticulata: A model for minisatellite birth. *Genetics*, 155, 1313–1320.

Tekaia, F., Dujon, B., Richard, G.-F. (2013). Detection and characterization of megasatellites in orthologous and nonorthologous genes of 21 fungal genomes. *Eukaryot Cell*, 12(6), 794–803.

The Huntington's Disease Collaborative Research Group (1993). A novel gene containing a trinucleotide repeat that is expanded and unstable on Huntington's disease chromosomes. *Cell*, 72, 971–983.

Thierry, A., Bouchier, C., Dujon, B., Richard, G.-F. (2008). Megasatellites: A peculiar class of giant minisatellites in genes involved in cell adhesion and pathogenicity in Candida glabrata. *Nucl. Acids Res.*, 36(18), 5970–5982.

Tyler-Smith, C. and Willard, H.F. (1993). Mammalian chromosome structure. *Current Opinion in Genetics & Development*, 3(3), 390–397.

Verkerk, A.J.M.H., Pieretti, M., Sutcliffe, J.S., Fu, Y.-H., Kuhl, D.P.A., Pizzuti, A., Reined, O., Richards, S., Victoria, M.F., Zhang, F. et al. (1991). Identification of a gene (FMR-1) containing a CGG repeat coincident with a breakpoint cluster region exhibiting length variation in fragile X syndrome. *Cell*, 65, 905–914.

Verstrepen, K.J., Jansen, A., Lewitter, F., Fink, G.R. (2005). Intragenic tandem repeats generate functional variability. *Nature Genetics*, 37, 986–990.

Vinces, M.D., Legendre, M., Caldara, M., Hagihara, M., Verstrepen, K.J. (2009). Unstable tandem repeats in promoters confer transcriptional evolvability. *Science*, 324(5931), 1213–1216.

Walker, P.M.B. (1971). Origin of satellite DNA. *Nature*, 229(5283), 306–308.

Welch, J.W., Maloney, D.H., Fogel, S. (1990). Unequal crossing-over and gene conversion at the amplified CUP1 locus of yeast. *Mol. Gen. Genet.*, 222(2–3), 304–310.

Weller, P., Jeffreys, A.J., Wilson, V., Blanchetot, A. (1984). Organization of the human myoglobin gene. *EMBO J.*, 3, 439–446.

van der Woude, M.W. and Bäumler, A.J. (2004). Phase and antigenic variation in bacteria. *Clinical Microbiology Reviews*, 17(3), 581–611.

Wyman, A.R. and White, R. (1980). A highly polymorphic locus in human DNA. *Proc. Natl. Acad. Sci. USA*, 77, 6754–6758.

Xu, X., Peng, M., Fang, Z., Xu, X. (2000). The direction of microsatellite mutations is dependent upon allele length. *Nat. Genet.*, 24, 396–399.

Young, E.T., Sloan, J.S., Van Riper, K. (2000). Trinucleotide repeats are clustered in regulatory genes in *Saccharomyces cerevisiae*. *Genetics*, 154, 1053–1068.

Zaremba-Niedzwiedzka, K., Caceres, E.F., Saw, J.H., Bäckström, D., Juzokaite, L., Vancaester, E., Seitz, K.W., Anantharaman, K., Starnawski, P., Kjeldsen, K.U. et al. (2017). Asgard archaea illuminate the origin of eukaryotic cellular complexity. *Nature*, 541 [Online]. Available at: http://www.nature.com/nature/journal/vaop/ncurrent/full/nature21031.html [Accessed 13 January 2017].

Zhu, Y., Strassmann, J.E., Queller, D.C. (2000). Insertions, substitutions, and the origin of microsatellites. *Genetics Research*, 76(3), 227–236.

8

CRISPR-Cas: An Adaptive Immune System

Marie TOUCHON
Génomique évolutive des microbes,
Institut Pasteur, CNRS UMR3525, Paris, France

In genetics, *clustered regularly interspaced short palindromic repeats*, more commonly referred to as *CRISPRs*, are a family of repeated sequences of nucleotides widely distributed in archaeal and bacterial genomes. This family is characterized by series of short, direct repeats regularly interspaced by non-repetitive sequences derived from exogenous DNA. These CRISPRs in combination with specific genes, the *cas* (CRISPR-associated) genes, together form a *CRISPR-Cas system*, which reveals itself to be a defense system against invasive elements in prokaryotes. Who would have imagined that repeated sequences discovered in microorganisms would reveal, 20 years later, the existence of a sophisticated and unsuspected *adaptive immune system* in these unicellular organisms? And who would have bet that studying these repeats would result in revolutionary genetic engineering tools 30 years later?

8.1. A brief history of the discovery of CRISPR-Cas systems

This repeated structure was first observed back in 1987 by Ishino and his colleagues during sequencing of the *iap* gene of the bacterium *Escherichia coli*. On that occasion, they came across a surprising non-coding region composed of

14 identical *repeats* of 29 nucleotides (nt), partly palindromic, separated by variable sequences of 32 and 33 nt, which would later come to be named as *spacers* (see Figure 8.1). At that time, it was hard to predict the biological function of these repeated sequences owing to the limited number of DNA sequences available.

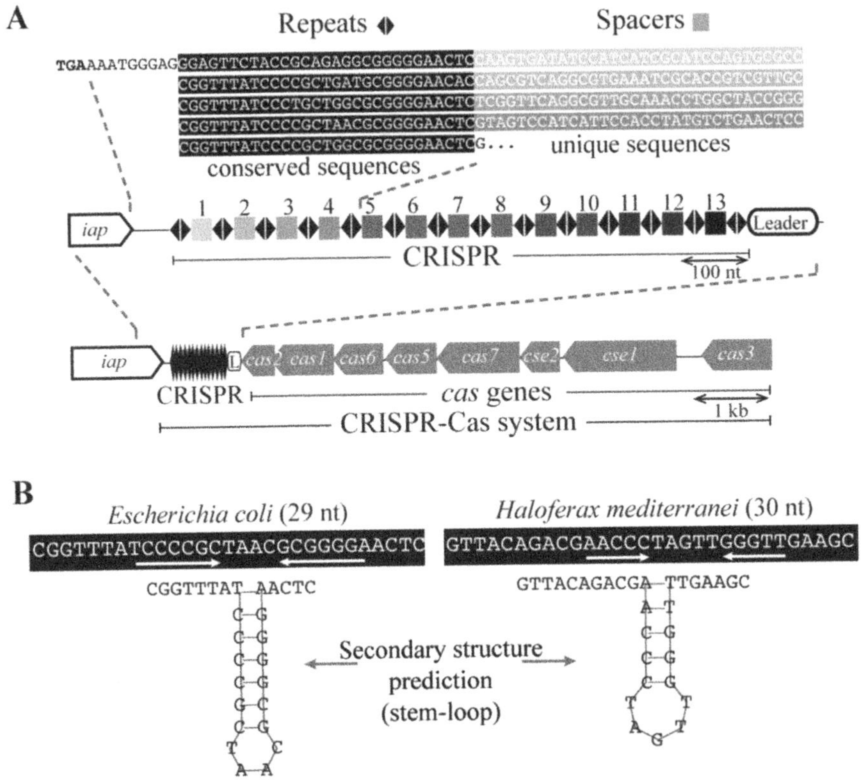

Figure 8.1. *(A) Representation of the CRISPR genomic sequence with its succession of repeats (in black) and spacers (in gray) (top), its organization (middle) and the associated cas genes (bottom) identified near the iap gene in* E. coli. *(B) The* E. coli *and* H. mediterranei *repeats are partially palindromic and form a secondary stem-loop structure*

In 1995, similar structures were first observed in archaeal genomes, and in *Haloferax mediterranei* (Mojica et al. 1995) in particular (see Figure 8.1). This was the beginning of the genomic era, which would experience considerable technical advances in the 2000s. As sequencing techniques progress, these repeats are detected in an ever-increasing number of bacterial and archaeal genomes. The frequency and conservation of these very particular repeated sequences in two of the three domains of life are attracting increasing interest. Various acronyms are used to describe these repeats, such as TREP (Tandem REPeats), SRSR (short regularly spaced repeats), LCTR (large clusters of tandem repeats) and SPIDR (spacer interspaced direct repeats). Faced with the multitude of acronyms used to qualify this same family of repeats, and to avoid any confusion, the community agreed in 2002 to use the acronym *CRISPR* (clustered regularly interspaced short palindromic repeats) (Jansen et al. 2002). At this time, the biological function of these repeats was still unknown.

In 2002, analysis of the genomic environment of CRISPRs in different genomes revealed that these repeats are very often located close to a particular group of gene encoding proteins presenting nuclease, helicase or polymerase domains, and more generally, for proteins involved in DNA and RNA manipulation. As these genes were exclusively associated with CRISPRs, they came to be referred to as *cas (CRISPR-associated) genes* (Jansen et al. 2002). The association between CRISPRs and *cas* genes was now established, and these two components together formed what would be called the *CRISPR-Cas system* (see Figure 8.1). With the *cas* genes being diverse and numerous, a classification of CRISPR-Cas systems was proposed (Jansen et al. 2002; Haft et al. 2005).

It was not until 2005 that three independent groups noted that spacers exhibited strong sequence similarities with the *mobile genetic elements* infecting prokaryotes, such as phages (bacterial and archaeal viruses) or plasmids (extrachromosomal DNA capable of autonomous replication) (Bolotin et al. 2005; Mojica et al. 2005; Pourcel et al. 2005). They suggested that spacers were of *foreign origin* (viral or plasmid) and that they were acquired in a polarized manner in the CRISPR array over time during the course of infections. Thus in 2005 it was proposed, for the first time, that CRISPR-Cas systems could be an *adaptive immune system*, that is, a system that memorizes the DNA of the invasive agents that it has encountered, to guard against them during new attacks. At that time, this major discovery, revealed mostly by in silico approaches, was largely underestimated by the rest of the scientific community.

The first experimental evidence confirming this hypothesis was provided in 2007 in the groundbreaking article by Barrangou et al. (2007). By exposing the lactic bacterium *Streptococcus thermophilus* to phages, the authors discovered that certain bacterial clones become resistant to a phage after the incorporation of new spacers into the CRISPR array, whose sequences perfectly match those of the phage used (see Figure 8.2). Thus, for the first time, they observed the *immunity acquired* in action and showed that CRISPR-Cas systems confer resistance to phages, whose specificity is determined by the similarity between the spacer sequences and those of the phages. They also confirmed that spacers are acquired in a polarized manner over the course of infections during the lifetime of the clone. A year later, it was shown experimentally that these systems also interfere with the conjugation and transformation of plasmids in *Staphylococcus epidermidis*, reinforcing the initial hypothesis that these systems not only limit the spread of phages, but that of all invasive elements (Marraffini and Sontheimer 2008). These CRISPR-Cas systems prove to be, on the one hand, *adaptive* and *heritable*, with the acquired immunity being transmitted vertically to the next generations (Koonin and Wolf 2009), but also *transferable*, with acquired immunity being transmittable horizontally between unrelated strains.

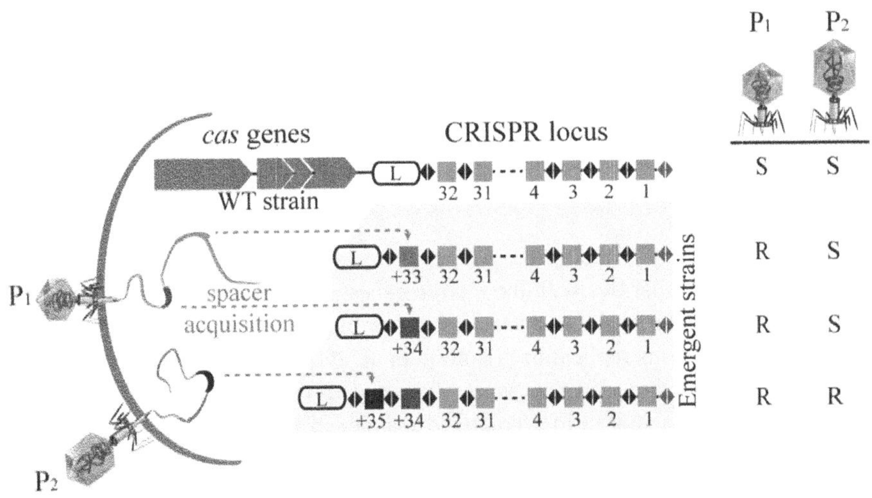

Figure 8.2. *Evolution of the CRISPR array following the infection of a sensitive wild strain (WT) by two phages (P1,P2). Polarized acquisition of new spacers (+) in the array leads to the emergence of resistant strains. The sensitivity (S) and resistance (R) to the two phages of the strains are reported on the right*

In the years since this key breakthrough, research has evolved into a highly dynamic and competitive field of microbiology. The molecular mechanisms have been partly revealed, and the structural and functional characteristics of the different proteins encoded by the *cas* genes partly described (Nussenzweig and Marraffini 2020), paving the way for the design of tools considerably revolutionizing epidemiology, biotechnology and molecular engineering (Pickar-Oliver and Gersbach 2019). The scope of these new technologies, inspired by these natural CRISPR-Cas systems, allowing targeted genome modifications, is immense: from the creation of antimicrobials or antivirals to gene therapy to treat genetic diseases, as well as the improvement of animal or plant species to develop new varieties. Interest in CRISPR-Cas has also extended to the public domain, where the societal and ethical implications and legislation surrounding CRISPR applications are a matter of lively debate.

This chapter, for its part, focuses on the fundamental aspects associated with research dedicated to CRISPR-Cas systems and discusses the diversity of CRISPR-Cas systems, their evolutionary dynamics, their role as an adaptive immune system, their impact on the co-evolution of microbes and their parasites, and their significance in nature.

8.2. General characteristics of CRISPR-Cas systems

CRISPR-Cas is a *two-component* system comprising a group of *cas* genes located close to one or more (in 40% of the cases analyzed) CRISPR arrays. Each *CRISPR array* is characterized by a succession of short, direct repeats and constant-sized spacers, flanked at one end by a sequence rich in A+T, called the *leader* sequence (see Figure 8.1). The presence of multiple CRISPR arrays is common in genomes, and the greatest number is observed in *Methanocaldococcus jannaschii*, with 18, representing nearly 1% of the size of this archaeal genome. The CRISPRs analyzed are relatively short and contain on average about 20 repeats and spacers. However, this number varies considerably from one CRISPR to another and can reach several hundred, as in the case of the marine bacterium *Haliangium ochraceum*, which has a CRISPR with 588 repeats that spans 40 kb (Pourcel et al. 2020). The *cas gene* clusters, meanwhile, are very diverse and are organized into one or more operons comprising between three and 10 or so genes. The number of known *cas* genes is substantial, and bioinformatics analyses are constantly revealing new ones (Makarova et al. 2020). The number of *cas* gene clusters present in a genome is often positively correlated with the number of CRISPR arrays. Most genomes contain only one CRISPR-Cas system, that is, a single cluster of *cas* genes associated with CRISPRs. However, 25% have several, with a number reaching up to eight distinct systems (Bernheim et al. 2020).

The development of algorithms allowing the automatic detection of CRISPRs and *cas* genes from genomic sequences and the analysis of all fully sequenced genomes has enabled the different characteristics of repeats, spacers and associated *cas* genes to be determined (Biswas et al. 2016; Couvin et al. 2018; Mitrofanov et al. 2021).

8.2.1. *Diversity of repeats*

Sequences of repeats are generally identical along a CRISPR, with only rare substitutions observed. Thus, for each CRISPR it is possible to define a *representative repeat* that corresponds to the repeat sequence most frequently observed in the array. Note that in most CRISPRs, differences between repeats and representative repeats gradually increase along the array. The last repeat, which is often truncated, is the most degenerated, and as we will see later, happens to be the most ancient repeat of the array and the most distant repeat from the leader sequence. The sequences of representative repeats are extremely diverse from one CRISPR to another, and several thousand different sequences are currently identified (Pourcel et al. 2020).

Despite this sequence diversity, these repeats present some common characteristics. They are *short*, varying in size between 23 and 50 nt among the different arrays. Most repeats (96%) are between 24 and 37 nt in size, and can be classified into three categories based on their size (small 24–25 nt, medium 28–30 nt, or large 36–37 nt). They are generally partially palindromic, that is, they have a partial internal symmetry that can lead to the formation of a *secondary stem-loop structure* (see Figure 8.1). A large fraction of repeats begins or ends with a GTTTC or GAAAC motif. On the basis of their similarities, the repeats have been classified into at least 12 different groups (Kunin et al. 2007). Half of these groups contain repeats potentially capable of forming stable RNA secondary structures, while the other groups contain repeats predicted to be unstructured. Analysis of the predicted stable secondary structures reveals the presence of multiple *compensatory mutations* in the stem region, that is, pairings in the structure are conserved while the nature of the bases involved is different (see Figure 8.3). Thus, in some cases, rather than their sequences, it is the secondary structures of the repeats that are conserved, suggesting that they play an important biological role (Kunin et al. 2007). Each group is characterized by a different stem length and loop size. We will describe below how these repeats are essential for the acquisition of new spacers, and how these conserved stem-loop structures are involved in the maturation of small RNAs in genomes that encode type I and type III CRISPR-Cas systems (see section 8.4.2).

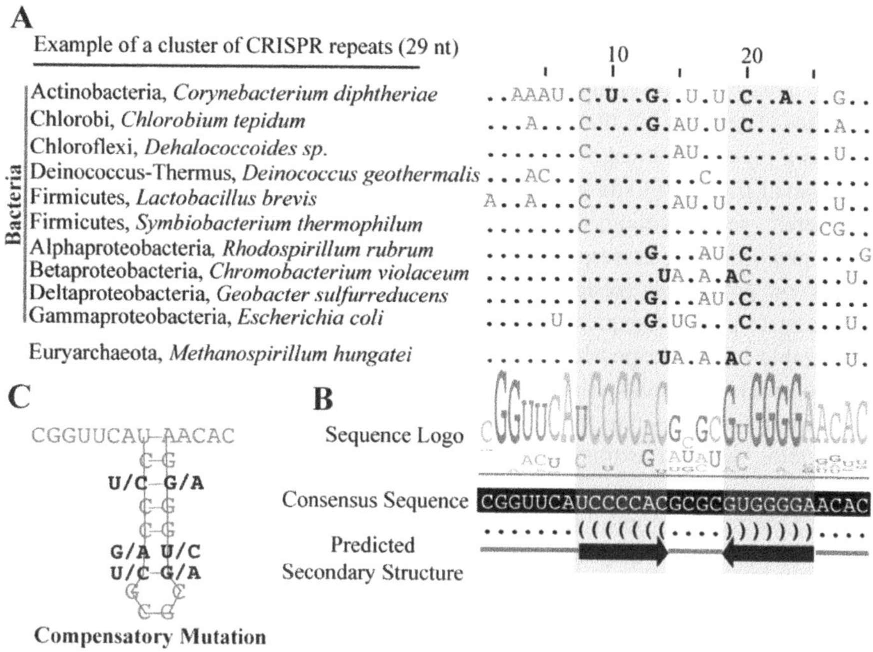

Figure 8.3. *Conservation of the secondary structure of repeats of the same group. (A) Multiple alignment of a subset of repeats of the group. Only nucleotides that are different from the consensus sequence of the group are represented, as well as the compensatory mutations (in bold) located at the stem (shaded areas). (B) Logo, consensus and predicted secondary structure of the group. (C) Representation of the predicted stem-loop structure; the compensatory mutations are indicated in bold*

8.2.2. *Diversity and origin of spacers*

Unlike repeats, spacer sequences are generally unique in the CRISPR array(s) of the same genome. Indeed, cointegrations, duplications and transfers of spacers within an array or between different arrays of the same genome are rare. The size of spacers is conserved within an array, although again it can be very different from one CRISPR to another. The variations are large, ranging from 27 to 72 nt, although in the majority of the cases studied (97%) the size is between 29 and 43 nt. The size of newly acquired spacers is not dependent on the size of the spacers already existing in the array, but is determined by a mechanism inherent in the protein machinery involved in spacer acquisition (Yosef et al. 2012).

A key step in understanding the function of CRISPRs was the identification of the origin of spacers, whose diversity is staggering. To do this, the sequences of the spacers are generally compared with all the sequences available in the public databases. For most of them, no sequence similarity is found, and their origin remains a mystery (Shmakov et al. 2017). However, in the different studies carried out, a small fraction of spacers, ranging from 1% to 19% depending on the phylum considered, shares strong similarities with known sequences. These sequences, called *protospacers*, correspond in the majority of cases (90%) to sequences of phages or prophages (phages integrated into the host chromosome), and the remainder to plasmid sequences, or to sequences of integrative and conjugative or mobilizable or transposable elements (elements integrated into the host chromosome). Thus, almost all spacers with identifiable protospacers are of *foreign origin* and target *mobile genetic elements* (MGE) known to frequently infect prokaryotes, whether extrachromosomal or integrated into the host chromosome. This foreign origin, initially inferred from their homology with sequences of invasive elements by in silico approaches, was then experimentally confirmed in vitro and in vivo. Acquisitions of spacers in response to infection by phages or plasmids have been observed in various species, such as *S. thermophilus* (Barrangou et al. 2007), *Streptococcus mutans* (van der Ploeg 2009), *S. epidermidis* (Marraffini and Sontheimer 2008), *Staphylococcus aureus* (Heler et al. 2015; Modell et al. 2017), *Pseudomonas aeruginosa* (Westra et al. 2015), *Roseburia intestinalis* (Cornuault et al. 2020) and *Sulfolobus solfataricus* (Erdmann and Garrett 2012).

Several hypotheses have been put forward to explain the difficulty in identifying the origin (the protospacer) of a large number of spacers. The first, by far the most likely, hypothesis focuses on the striking lack of phage sequences available in public databases. Because most sequencing efforts have focused on viruses that infect humans, animals and plants, only a few thousand phages are fully sequenced and available, even though they are by far the most abundant biological entities on the planet. It is worth remembering that phages are extremely diverse and numerous, and kill between 20 and 40% of marine bacteria every day. The lack of similarity between the spacers and the sequences of the available databases would thus reflect this until now barely explored diversity of phages, and more generally prokaryotic MGEs. The second envisages that spacers correspond to MGEs that no longer exist or that have mutated until they are no longer recognizable (see section 8.5.1). Thus, a lack of similarity would be inevitable, as current databases contain only functional and recent MGEs. The last hypothesis suggests that spacers correspond to MGEs that are rare and therefore less likely to be sequenced, and thus less present in public databases (Westra et al. 2015).

8.2.3. *Diversity and evolutionary classification of cas genes*

The CRISPR-Cas systems are a universal immune mechanism that, at least in principle, can adapt to defend the host against any MGE. Because of this universal adaptability, CRISPR-Cas systems, like all defense systems, are engaged in a perpetual arms race with their parasites resulting in a rapid evolution of *cas* genes and substantial diversity in the composition of *cas* gene clusters. Such diversity, resulting from pressure induced by MGEs (see section 8.5), but also from their frequent horizontal transfers (see section 8.3.2), poses considerable challenges for the evolutionary classification of CRISPR-Cas systems (Makarova et al. 2020). It also reflects the diversity of molecular mechanisms involved in this immune response (see section 8.4.2).

The *cas* gene clusters have a particular modular organization. The two main components are the *adaptation module* and the *effector module* (see Figure 8.4). In most systems, the *adaptation module* consists of at least two genes encoding the Cas1 and Cas2 proteins. These two proteins are the only two proteins conserved in the virtual majority of the CRISPR-Cas systems described, and are involved in the first phase of the immune response, namely, the adaptation phase (see section 8.4.1). Together, they form the *adaptation complex* in which Cas1 is the active enzyme subunit (integrase) involved in cleavage of the source (the protospacer) and its integration into the CRISPR array, whereas Cas2 forms the structural scaffolding of the complex. In the absence of Cas2, Cas1 is unable to bind to the CRISPR array, making these two proteins essential during adaptation. It has been shown that a single adaptation module can act in *trans* on several CRISPR arrays present in the genome as long as they contain a CRISPR repeat and a leader sequence that are compatible, whether they are close to or distant from the *cas* gene cluster, or even located on MGE. In some systems, additional Cas proteins, such as Cas4, Csn2, reverse transcriptase (RT) or Cas3 and Cas9, also contribute to the adaptation phase (Amitai and Sorek 2016).

Unlike the relatively simple and uniform architecture of the adaptation module, *effector modules* are very diverse, and their variation forms the basis of current classification (Makarova et al. 2020). We can distinguish between *two classes of systems*, depending on the nature of the effector modules. Class 1 systems have *multi-subunit* effector complexes comprising several Cas proteins, whereas in class 2 systems, the effector corresponds to a single large *multi-domain* protein, such as Cas9, Cas12 or Cas13 proteins. The classification of systems into *types* and *subtypes* uses several criteria: the presence of particular cas genes, the organization and composition of *cas* operons, and lastly the phylogeny of conserved Cas proteins.

Thus, each class currently comprises three types, namely types I, III and IV for class 1 and types II, V and VI for class 2. Most types are easy to distinguish owing to the presence of specific genes: *cas3* for type I, *cas9* for type II and *cas10* for type III, *cas12* for type V and *cas13* for type VI (see Figure 8.4). The types are then divided into subtypes according to the organization of the *cas* genes in the cluster and the presence or absence of certain other accessory or specialized genes. Currently, there are 36 different subtypes, but new subtypes are frequently proposed.

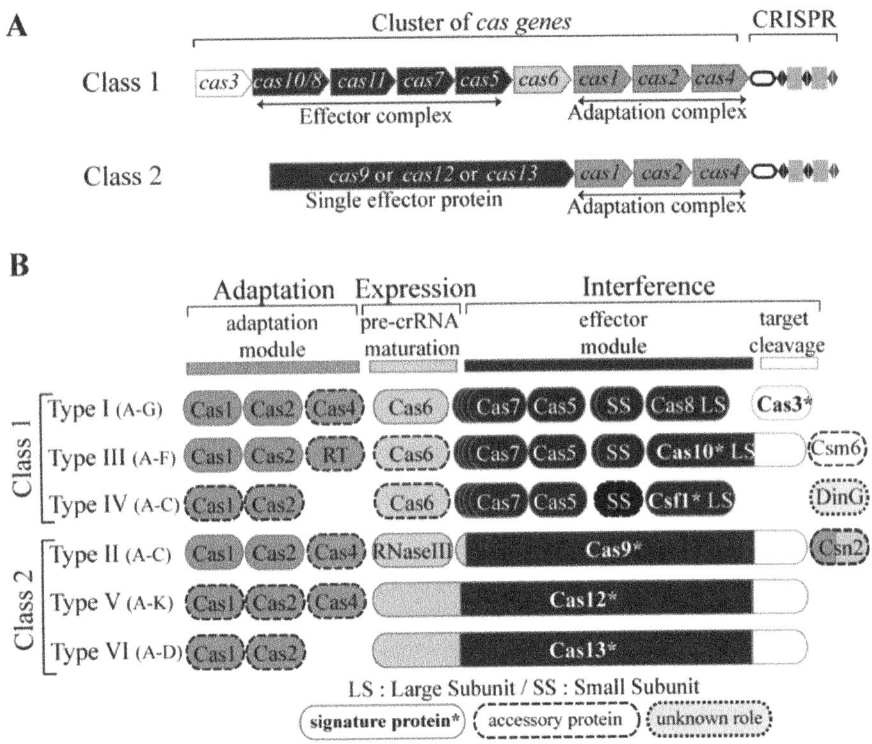

Figure 8.4. *(A) Genetic organization of class 1 and class 2 systems. (B) Modular organization of the six types defined in the current classification. Each color corresponds to a particular stage in immunity: adaptation (dark gray), expression (light gray), interference (black) and cleavage of the target (white)*

The adaptation and effector modules seem relatively independent not only in terms of functions and structures, but also from an evolutionary point of view (Koonin and Makarova 2019). Exchanges of modules between CRISPR-Cas systems of different types and subtypes are common. This partly explains why the topology of the phylogenetic tree of Cas1, the key protein of all adaptation complexes, is not consistent with the overall classification of effector modules and the phylogenies of individual effector proteins. In most cases, the adaptation module, Cas6 protein and associated CRISPR repeats are conserved in nearby genomes, while the effector module genes are often rearranged (especially in type III systems). Consequently, CRISPR repeat sequences are not taken into consideration during classification, as the relations between the repeat groups and the different types or subtypes are complex (see section 8.2.1). In many genomes, the number of effector modules exceeds that of the adaptation modules. This multiplicity of effector modules and their frequent exchange are probably a consequence of the pressure induced by MGEs (see section 8.5.2).

8.2.4. *Origin of CRISPR-Cas systems*

One of the key questions is to understand how these CRISPR-Cas systems emerged and then diversified. This complex question currently remains unanswered, but according to the prevailing theory, the different modules are derived from various MGEs (Faure et al. 2019; Koonin and Makarova 2019). In archaeal and bacterial lineages, a new superfamily of self-synthesizing transposons, called *casposons*, has recently been characterized. These casposons correspond to MGEs of 12–18 kb, flanked by terminal inverted repeats that invariably encode a DNA polymerase, suggesting that they direct their own replication during their transposition via a copy-and-paste mechanism (Krupovic et al. 2017). They have the particularity of encoding a protein homologous to Cas1, casposase, whose integrase activity has been experimentally validated. Comparison of the mechanistic characteristics of casposase-catalyzed reactions and Cas1 integrases of CRISPR-Cas systems reveals close similarities, and suggests that the repeats and the leader sequence could have evolved from the target sites of ancestral casposons. Although current casposons do not encode the Cas2 protein, some encode Cas4 and additional nucleases. It was therefore proposed that the capsosons are the ancestor of the adaptation module. Under the proposed evolutionary scenario, CRISPR-Cas systems would have emerged following the insertion of a casposon near an innate immunity locus containing a protein analogous to Cas10-like. As Cas2 proteins have structural similarities to VapD, a *Haemophilus influenzae* toxin, it has been suggested that Cas2 could be derived from a *toxin-antitoxin system*. Cas2 would thus have been coopted by the ancestral casposon, this recruitment being crucial for the transformation of casposase into Cas1, that is, for the enzyme to specifically integrate short fragments of DNA. It should be noted that casposons and toxin-antitoxin systems are not the only MGEs to have contributed to the emergence and diversification of CRISPR-Cas systems. Notably, reverse transcriptases of type-III adaptation modules would have derived from *group II introns*, and type II and V effector proteins would have emerged independently from different groups of *non-autonomous transposons*. As for the class 1 effector modules, being the most widespread, they would correspond to the ancestral form of CRISPR-Cas systems. They would have emerged in thermophilic archaea and then spread to bacteria. The origin of class 1 effector modules remains unclear. However, the recent discovery of a signaling pathway integrated with certain type III systems suggests that they may have evolved from a signal transduction system involved in stress-induced programed cell death. The subsequent evolution of the class 1 effector complexes by a series of duplication events of an ancestral gene containing at least one RNA recognition motif would have led to the relatively complex architecture of these modules. Class 2 effector nucleases may have been subsequently coopted by class 1 CRISPR-Cas systems resulting in the replacement of the typical *cas* genes of class 1 effector modules, thus explaining the emergence of two distinct CRISPR-Cas classes (Mohanraju et al. 2016; Koonin and Makarova 2019).

8.3. Evolution of CRISPR-Cas systems

8.3.1. *Scattered distribution of CRISPR-Cas systems*

CRISPR-Cas systems are not uniformly distributed between bacterial and archaeal phyla. Analysis of 13,116 complete genomes showed that CRISPR-Cas systems are present in the majority of archaea (85%), including almost all hyperthermophiles (97%), but in only 42% of bacteria (Makarova et al. 2020). To date, they have never been found in eukaryotes. Many metagenomics studies have revealed their presence in environments as diverse as in saliva, the intestine, rumen, bioreactors, groundwater, the ocean, hypersaline lakes and hot springs. However, their prevalence varies widely from one environment to another. In particular, high-temperature environments are enriched in CRISPR with thermophilic microorganisms that generally have more numerous and longer CRISPRs than mesophilic microorganisms (Anderson et al. 2011). Enrichment has also been reported in microbial metagenomes associated with marine sponges (Horn et al. 2016), whereas they appear very rare in groundwater filtrates (Burstein et al. 2016). In fact, entire lineages of uncultivable microorganisms, widespread in many environments, appear essentially devoid of CRISPR (Burstein et al. 2016). These observations suggest that ecological factors have an impact on the distribution of these systems in genomes, as detailed in section 8.7. They also raise the question of the real importance of these systems in nature, which remains an open question for the time being.

Class 1 systems are by far the most widespread in bacteria and archaea, and account for about 90% of the systems detected. Class 2 systems are less common and are found almost exclusively in bacteria. The absence of class 2 in archaea, at least in part, can be explained by the absence of RNase III, a ubiquitous bacterial enzyme essential for the functioning of type II and certain type V systems. Clear trends are observed in the distributions of certain types and subtypes. Type I systems are most common in the majority of phyla, followed by types II and III. However, notable exceptions exist: in Tenericutes only type II systems have been identified to date; in Chlamydiae, which predominantly group together obligate intracellular microorganisms, genomes are almost devoid of CRISPR-Cas systems. Type IV, V, and VI systems are extremely rare and are concentrated in distinct clades including Proteobacteria, Fusobacteria and Bacteroidetes. Although the biological foundations for this non-uniform dissemination of CRISPR-Cas systems remain to be elucidated, several avenues, including the impact of the host genetic context, are currently being explored (see section 8.6).

8.3.2. *Massive transfer of CRISPR-Cas systems*

It is not uncommon to identify similar repeats in phylogenetically distant genomes, as shown in Figure 8.3. The same is true for the different types and subtypes, which are mostly found in different bacterial and archaeal phyla. This wide dispersion of systems is likely the result of massive *horizontal transfers* between different species during evolution. It should be recalled that there are three main mechanisms of horizontal gene transfer (HGT) in prokaryotes: *conjugation* mediated by conjugative plasmids, which consists of a DNA transfer through direct contact between the donor and the recipient cell; *transduction* mediated by transducer phages (i.e. having accidentally encapsulated the DNA of the donor cell in the viral particle), which consists of a DNA transfer following infection of the recipient cell by the donor-produced transducer phages; and lastly, natural *transformation*, which corresponds to the uptake of naked DNA from the environment by competent cells.

Multiple studies support this high mobility of CRISPR-Cas systems by HGT. Of note are (1) the studies reporting the incongruence of the phylogenetic trees of *cas* genes and of species, revealing that CRISPR-Cas systems and species have distinct evolutionary histories; (2) studies showing that spacers have a different G+C content and codon bias from the rest of the chromosome, reinforcing their exogenous origin; and (3) those studies identifying these systems in MGEs, the main vectors of HGTs, such as plasmids, transposons and, more rarely, phages (Faure et al. 2019). Note that complete systems have been identified in only two phages, ICP1 and D-1873, infecting *Vibrio cholerae* and *Clostridium*, respectively, and very recently in several giant viruses. While the type I system of the ICP1 phage targets and neutralizes an anti-phage defense mechanism encoded by MGEs integrated into the host chromosome, the systems present in the giant viruses enable them to defend themselves against virophages (viruses infecting viruses). Thus, surprisingly, in a viral context, CRISPR-Cas systems have been hijacked to the benefit of the virus and appear to be involved in virus-MGE or virus-virus interactions/competitions. A comprehensive in silico study, performed on 11,805 plasmids, 2,498 phages and 26,987 prophages, has shown that CRISPR-Cas, although rare in MGEs, are much more common in plasmids (3%). The different types of systems have all been observed at least once on a plasmid, some even appearing exclusively on a plasmid (types IV), consistent with the idea that they can be transferred between species via conjugation or natural transformation (Bernheim et al. 2020).

8.3.3. *Commonly lost systems*

The number, types and even subtypes of systems observed in strains of the same species or between closely related species can be very different, suggesting that the *turnover* of these systems is significant, with frequent successive acquisitions and losses. Such losses have been observed in different bacterial species. In particular, in *Mycoplasma gallisepticum*, inactivation and rapid loss of the CRISPR-Cas system follows a host shift (from poultry to finches) (Delaney et al. 2012). In *S. epidermidis*, the loss follows the acquisition of a plasmid essential to the survival of the bacterium (i.e. that carries an antibiotic-resistance gene). Surprisingly, this system loss, even when through large deletions, seems to have little or no effect on the fitness of the resulting strains in vitro (Jiang et al. 2013). In *P. aeruginosa*, the system loss occurs after infection by temperate phages (i.e. with the ability to integrate into the host chromosome as prophages) (Rollie et al. 2020). Thus, CRISPR-Cas systems appear to be in a continuous state of flow. They could be lost when they are neutral or when they incur a cost (see section 8.6), and then be reacquired by HGT in environments or conditions where phages or other parasites are a major source of mortality. An important challenge for future work will be to understand to what extent MGEs impose selection for the loss of these systems in natural environments.

8.3.4. *Evolutionary dynamics of CRISPR arrays*

A CRISPR array can evolve very rapidly due to the acquisition of new spacers, their occasional loss, or the deletion of the whole array, which can lead strains of the same species to carry unique combinations of spacer content. This property has otherwise been exploited for typing pathogenic strains (e.g. *Mycobacterium tuberculosis* and *Yersinia pestis*), even before the discovery of the role of these arrays. Direct analysis of streptococcal CRISPRs over several months in saliva samples from healthy individuals reports that spacer content is very dynamic over time, and that it is individual-specific with a limited fraction of common spacers (Pride et al. 2011). Similar results have been observed in various human microbiomes with very rapidly evolving spacer contents, largely specific to an individual, or even to a given body site (Rho et al. 2012). Many analyses of metagenomes from a variety of natural environments report that the observed wide diversity of spacers is predominantly specific to a given geographic site, noting that it depends on the local MGE population. These observations are in agreement with experimental studies showing that the acquisition of new spacers can be rapid (less than 24 hours, under laboratory conditions) in different species, including *S. solfataricus*, *S. thermophilus* and *P. aeruginosa*. The acquisition of new spacers thus appears to be a widespread microbial response to overcome predation exerted by coexisting phages. Rapidly evolving CRISPR arrays in nature are therefore likely to play an important role in bacteria-phage interactions.

In a set of strains of the same species, the spacers are often more diversified at one end of the CRISPR array than at the other. The reason for this is simple: the acquisition of spacers is polarized, and takes place at the 3' end of the leader sequence just after the first repeat (see Figure 8.2). Spacers are therefore all the more recent, the closer they are to the leader sequence. Repeats, meanwhile, are more likely to accumulate mutations the further away they are, as they will be all the more ancient. This polarity, initially deduced from in silico analyses from complete genomes, then experimentally demonstrated in several bacterial species in vitro and in vivo, has also been highlighted in metagenomic analyses, where only the most recent spacers match coexisting phages and only more ancient spacers are conserved over long periods of time (Andersson Banfield 2008; Stern et al. 2012). The CRISPR arrays are therefore shown to be a *temporal memory of the infections* suffered by the microorganism and its ancestors during their lifetime.

The fact that CRISPR-Cas systems can acquire new spacers permanently may, however, become a real burden, as microbes cannot expand their genomes indefinitely at the risk of diminished fitness. In addition, if the MGE targeted by

CRISPR has mutated or been exterminated, there may not be much benefit in maintaining the old, obsolete spacers. It is probably for this reason that losses of spacers by homologous recombination events between repeats are frequently observed along arrays. These losses may be due to natural selection or the spontaneous bias toward deletion, typical of bacteria to promote genome compaction by eliminating neutral regions (Kuo and Ochman 2009). These spacer losses allow, on the one hand, the size of the CRISPR arrays to be limited, thus probably limiting the costs associated with the maintenance of very large arrays (see section 8.6), and, on the other hand, the efficiency of the immune response to be increased, since the proximal part of the CRISPR array is transcribed at a higher rate during the expression phase (see section 8.4.1). The CRISPR arrays would thus be a *partial* temporal memory of past infections.

The evolutionary dynamic of CRISPRs is of great interest to *epidemiologists*. The analysis of the spacer composition, by a genotyping technique called spoligotyping, allows them to distinguish different strains from the same species with great precision. By comparing the chronological organization of spacers in the CRISPR arrays, they can retrace the origin of an epidemic caused by a particular strain and track its dynamics and evolution over time. CRISPRs also represent a wealth of information for *ecologists*. They can be used to partially reconstruct the history of past infections, identify MGEs and their hosts, and thus gain a better understanding of the evolutionary dynamics of microbial populations and their parasites over time in various environments.

8.4. An adaptive immune system

8.4.1. *A three-stage immune response*

The molecular mechanisms involved in the immune response mediated by CRISPR-Cas differ significantly depending on the type of system. Nevertheless, the general operating principle is relatively stable: following a primary infection, CRISPRs integrate fragments of the invasive agent as spacers; they are then transcribed and matured into small RNA molecules that will guide Cas proteins to specifically bind and cleave complementary foreign nucleic acids. This immune response can thus be broken down into three distinct stages: adaptation, expression and interference (see Figure 8.5).

During the *adaptation phase*, the adaptation complex comprising the Cas1 and Cas2 proteins binds to an exogenous DNA molecule, such as that of a phage infecting the cell. This complex introduces two double-strand breaks in the foreign DNA, at the protospacer, generally upon recognition of a short motif (2–7 nt) known as *PAM* (protospacer-adjacent motif). In some cases, invasive DNA fragmentation may be achieved by RecBCD (when replication fork is stalled) or by restriction enzymes (from restriction-modification (RM) systems) present in the genome. The resulting fragment is then inserted at the 5' end of the CRISPR at the leader sequence and becomes a new spacer. The leader sequence and a single repeat are required during this stage. The presence of recognition sites, both in the leader sequence and in the repeat, located within 10 bp of the integration site makes it possible to direct the adaptation complex to this precise position (Yosef et al. 2012; Wei et al. 2015). The CRISPR array is then repaired by the cellular repair machinery, resulting in the duplication of the proximal repeat. Some CRISPR-Cas systems use an alternative adaptation mechanism, that is, the acquisition of spacers from the RNAs of the invasive element via the action of reverse transcriptases (RTs) that reverse-transcribe viral RNA to DNA, genetic information that can then be integrated into the CRISPR array under the action of the adaptation complex. These RTs are generally encoded by the system itself and are often fused to *cas1* (Amitai and Sorek 2016).

During *the expression phase*, the CRISPR array is transcribed into a single long pre-messenger, called pre-crRNA (precursor CRISPR RNA), generally from a promoter located in the leader sequence upstream of the CRISPR array. It is then matured by Cas proteins, whether or not part of the interference complex, to generate small crRNAs (CRISPR RNAs), which each contain the complementary sequence of part or the entirety of the repeat (named *tag crRNA*) and that of a spacer. In association with the interference complex, these crRNAs serve as guides and act as surveillance complexes within the cell awaiting invasion by a foreign nucleic acid (Charpentier et al. 2015).

During the *interference phase*, using a complementary pairing, the guide RNA bound to the interference complex recognizes the protospacer, which is then cleaved and inactivated by nucleases. These may be (depending on the type of systems) either a component of the interference complex or a distinct Cas protein (van der Oost et al. 2014). Guide RNAs typically show no preference for coding or non-coding strands during the interference phase, with most CRISPR-Cas systems targeting the DNA of the invasive element. In fact, only type III and VI systems are found to be capable of targeting RNA.

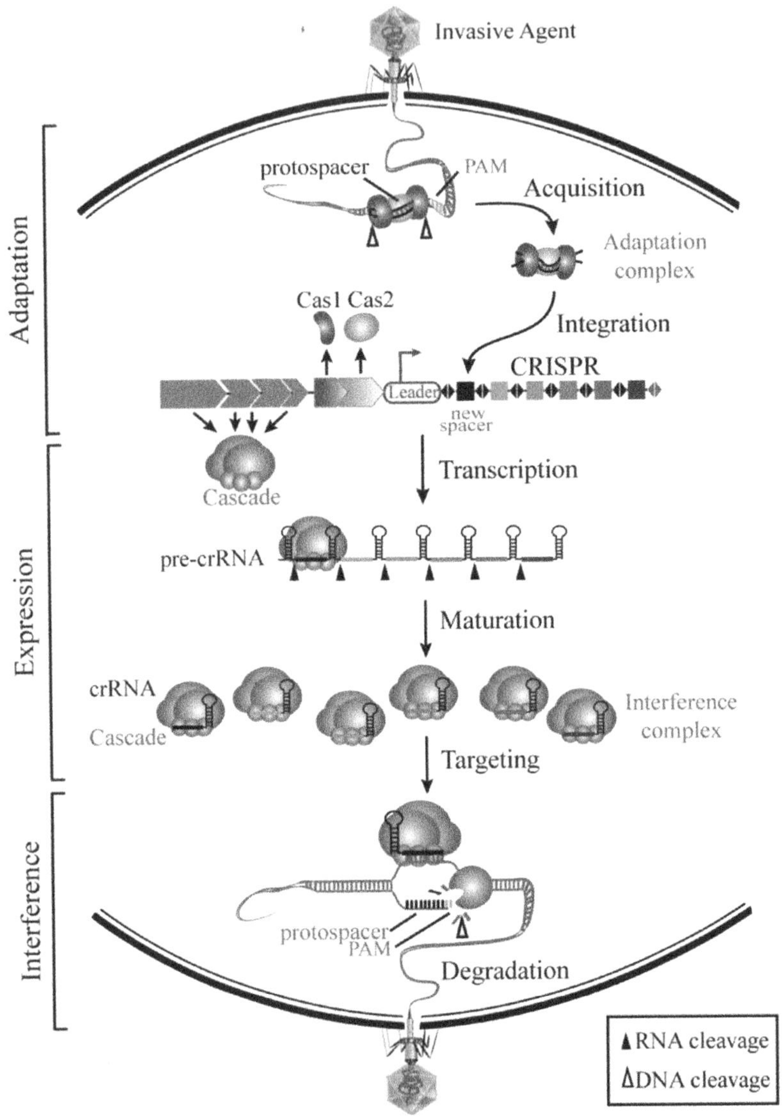

Figure 8.5. *The three stages of the immune response. Adaptation corresponds to the acquisition of a fragment of the invasive agent following a primary infection and its integration into the CRISPR array. Expression corresponds to the biogenesis of guide RNAs (crRNA): CRISPR is transcribed and then matured, the resulting crRNAs are charged by effector complexes forming interference complexes. Interference corresponds to the sequence-specific recognition of the invasive agent by the interference complex inducing its destruction under the action of nucleases*

8.4.2. *Diversity of CRISPR-Cas molecular mechanisms*

The short description above is a very simplified illustration which, as a consequence, omits many details concerning the functioning of CRISPR-Cas systems (featured in Nussenzweig and Marraffini (2020)). In this section, we will discuss the major differences observed during the maturation and interference phase of the different types of CRISPR-Cas systems.

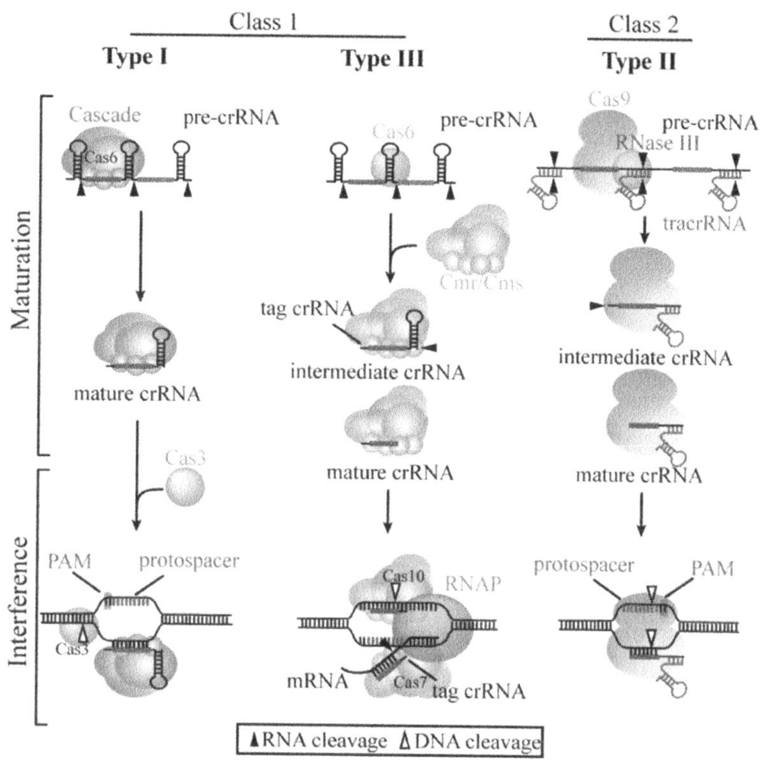

Figure 8.6. *Diversity of molecular mechanisms of type I, III and II systems during the maturation and interference phase*

The mechanisms involved in the *maturation* of pre-crRNA differ considerably in class 1 and class 2 systems (see Figure 8.6). In class 1 systems, especially types I and III, pre-crRNA maturation is catalyzed by Cas6 endoribonuclease (or, more rarely, by Cas5d). In type I systems, Cas6 is part of the interference complex, whereas it is recruited in types III. In these systems, the repeats are partially palindromic and form stem-loop structures along the pre-crRNA that are recognized by Cas6. Thus, the same enzyme may act in *trans* to mature pre-crRNAs with conserved stem-loop structures (see Figure 8.3), and thus potentially on all CRISPR arrays with repeats belonging to the same group (see section 8.2.1). In general, recognition of the hairpin by Cas6 induces a conformational rearrangement that allows cleavage of the pre-crRNA downstream of each hairpin, thereby releasing the different mature crRNAs. The stages of maturation of type III systems have not yet all been elucidated, and it appears that additional, still unknown, steps are involved in the cleavage of part of the crRNA stem-loop structure. These mature crRNAs form an interference complex in association with the effector module encoded by the CRISPR-Cas system itself, or with that of other systems present in the genome. For example, the effector modules of III-B systems that are known to use crRNAs produced by type I systems attest to integration between the different CRISPR-Cas systems. Regarding type IV systems, these are not yet functionally characterized. They are devoid of an adaptation module, and are rarely associated with CRISPR arrays. However, it is possible that they use mature crRNAs from other systems present in the genome once available, in the same way as III-B systems.

In class 2 systems, more specifically in type II and V-B systems, maturation requires the presence of additional RNAs presenting partial complementarity with the repeat sequence, known as *trans-activating crRNAs* (tracrRNAs). The trans-activating RNAs are encoded by the system; but, their positions in the array, their sequences and their structures are extremely diverse (Chylinski et al. 2013). Through sequence complementarity tracrRNAs will pair with repeats along the pre-crRNA. The resulting RNA duplex will be cleaved by a ubiquitous endogenous bacterial enzyme, the endoribonuclease RNase III, in the presence of Cas9 effector protein for type II. In this maturation stage, Cas9 nuclease has no catalytic activity but a stabilizing role in the interactions between tracrRNAs and pre-crRNA repeats. Repeats of type II systems are rarely palindromic and do not appear to have the ability to form stem-loop structures, unlike class 1 system repeats. Pairings of tracrRNAs with pre-crRNA repeats could compensate for this absence by providing an intermolecular structure inducing Cas9 recruitment and RNase III-mediated cleavage. In addition, unlike class 1 systems, mature crRNAs do not contain the entire sequence corresponding to the spacer, but usually a truncated form. In systems V and VI, cleavage does not appear to depend on RNase III, but on a nuclease activity specific to the effector protein, namely, Cas12 and Cas13, respectively, which remains to be characterized.

Major differences also exist in the *interference* phase (see Figure 8.6). In type I systems, the interference complex (called Cascade) in combination with the protospacer (the target), recruits an additional protein, the Cas3 protein, which contains a helicase domain that catalyzes the unwinding of the double-stranded target DNA, and a nuclease domain directly responsible for the *cleavage of one strand of the target DNA*. In type III systems, the Cas10 nuclease involved in cleavage of the target DNA forms an integral part of the interference complex (*Csm* or *Cmr*, depending on the subtype). These systems are unique because they require target transcription. The interference complex, in combination with the target transcripts, induces *cleavage of the target RNAs* by a ribonuclease of the Cas7 family (Csm3 or Cmr4, depending on the subtype), which will activate the *cleavage of one strand of the target DNA* (the non-transcribed strand) by Cas10. Type III systems are thus able to cleave both invasive DNA and its RNAs. In III-A systems, Cas10 also appears to be able, under conditions that remain to be determined, to synthesize a molecule that will activate a non-specific Csm6 ribonuclease capable of degrading all RNAs present in the vicinity of the targeted RNA.

In type II and V systems, the interference complex consists of the single effector protein (Cas9 or Cas12, depending on the type) associated with the crRNA/tracrRNA duplex. Upon recognition of a target DNA, the nuclease domain(s) of the effector protein ensure(s) *the double-strand cleavage* of this DNA. In type VI systems, the interference complex consists of the effector protein Cas13 associated with a crRNA. The Cas13 protein has the peculiarity of being a guide RNA-directed ribonuclease that recognizes and then cleaves target RNAs. It has recently been demonstrated that Cas13, once activated by binding to the target, *exhibits non-specific cleavage of any transcripts near the target*: in *cis* at other regions of the targeted RNA, but also in *trans* on non-target RNA located nearby. How do Cas13 and, to a lesser extent, Csm6 distinguish host-derived transcripts from those of the phage during collateral RNA cleavage? This question remains open. However, because a non-specific RNA cleavage may affect host cell functioning, it is possible that these systems can induce a *cell dormancy* or *a programed cell death* in order to limit the spread of infection in the population.

Although class 2 systems are less common than class 1 systems, they have recently received much more attention because of their applications in genome editing, gene expression regulation and detection (localization) of particular nucleic acids. Class 2 systems are particularly suitable for these applications since only one effector protein is required for the interference stage. Cas9, Cas12 and Cas13 can be easily reprogrammed to target any nucleic acid sequence, allowing the study of the precise function of a gene, modulating its expression, replacing, suppressing or locating it. These effector proteins with different molecular requirements (e.g. PAM, tracrRNA, RNaseIII) and different targets (i.e. DNA or RNA) allow the development of biotechnology tools that are increasingly specific, efficient and versatile (Pickar-Oliver and Gersbach 2019).

8.4.3. *Self- and none self-discrimination: avoiding self-targeting by CRISPR*

Prokaryotes have a diverse arsenal of tools to resist attacks by invasive agents, including *RM systems* and CRISPR-Cas systems, which are presented as innate- and adaptive-immunity mechanisms, respectively. The success of any defense system requires a mechanism to distinguish foreign nucleic acids (the non-self) from its own nucleic acids (the self). RM systems are based on the cleavage of exogenous DNA by endonucleases (restriction enzymes) upon recognition of specific sites (restriction sites), and on methylation of the prokaryote genome by methylases (modification) that protects it from the action of endonucleases. The signature of methylation is specific to each RM system, and distinguishes the self (methylated DNA) from the non-self (non-methylated DNA). For CRISPR-Cas systems, this distinction between the self and the non-self represents a particularly significant challenge since the antigenic sequences they recognize are already present as spacers in the genome. To avoid autoimmunity, the self-targeting of the chromosome, which inevitably leads to cell death (Gomaa et al. 2014), CRISPR-Cas systems have developed two different strategies. The first, used by DNA-targeting systems, is based on none self-recognition (Mojica et al. 2009; Semenova et al. 2011), while the second, used by RNA-targeting systems, is primarily based on self-recognition (Marraffini and Sontheimer 2010; Meeske and Marraffini 2018).

In all CRISPR-Cas systems studied to date, the sequences that flank protospacers in invasive DNA are fundamental for target recognition and for preventing an autoimmune response. In CRISPR-Cas systems of types I, II and V that target DNA, the presence of the PAM adjacent to the protospacer is mandatory for recognition and cleavage of the target by the interference complex. Thus, the PAM plays a key role during adaptation, but also in interference, since it is the first sequence sought by surveillance complexes. PAM motifs differ depending on the system type, or even subtype. Because repeats of the CRISPR array are devoid of PAM, the spacers are protected from self-targeting (Mojica et al. 2009). This conservative approach consisting of interacting only with non-self DNA, though efficient in preserving the integrity of the genome, makes these immune systems susceptible to rapid evolution of the phages (see section 8.5). Phages by accumulating mutations in the PAM can quickly no longer be recognized by the system (Deveau et al. 2010; Semenova et al. 2011).

In CRISPR-Cas systems of types III and VI that target RNA, discrimination between the self and the non-self does not depend on recognition of a PAM motif, but relies on recognition of a sequence, named *tag crRNA*, located at the 5' end of each mature crRNA (Marraffini and Sontheimer 2010; Meeske and Marraffini

2018). The transcription of the CRISPR array generates guide RNAs (crRNA) that are composed of the spacer sequence and are at their 5' end of a portion of the repeat sequence (tag crRNA). Self-targeting of these RNAs can only occur after a bidirectional transcription of the CRISPR array, a common phenomenon in prokaryotes (Lillestøl et al. 2009). To avoid an autoimmune response linked to the presence of antisense RNA (anti-crRNA), recognition of a pairing between the tag and its complementary sequence (i.e. the *antitag*) located at the 3' end of the targeted RNA makes it possible to identify this RNA as part of the self, and to suppress the activity of the interference complex. The tag-antitag crRNA pairing does not appear to affect RNA cleavage by type III effector complexes, but inhibits DNA cleavage (the ssDNase activity of Cas10), thereby protecting the host genome (Wang et al. 2019). On the other hand, it inhibits RNA cleavage by Cas13 (Wang et al. 2021). In the case of an infection by an exogenous agent, the lack of pairing between the crRNA tag and the 3' region that flanks the protospacer allows the target to be identified as foreign and a full and effective immune response to be triggered.

8.5. Phage escape mechanisms

Bacteria-phage interactions influence the structure, ecology and evolution of microbial communities and are therefore important determinants of the overall functioning of natural environments. Phages depend on specific hosts for their replication and are a major cause of bacterial cell death. Phages thus have an impact on the cycles of carbon and other nutrients, but also on the composition and diversity of the microbial population. In response to this pressure, prokaryotes have developed sophisticated defenses, including CRISPR-Cas systems, which allow targeted degradation of phages DNA limiting their spread. Needless to say, phages, in turn, have developed counterattack systems that allow them to bypass microbial defenses and prosper. Phages and their prey are thus engaged in a permanent arms race by reciprocally adapting their attack and defense systems to ensure their survival (Safari et al. 2020). To escape immunity conferred by CRISPR-Cas, phages use a variety of mechanisms involving either genomic modifications or specific proteins called anti-CRISPR.

8.5.1. *Genomic modifications*

Point mutations: Phages have a high mutation rate that allows them to adapt rapidly. Several co-evolution experiments between bacteria encoding a functional CRISPR-Cas system and phages have been conducted in *Streptococcus thermophilus* (Paez-Espino et al. 2015) and *Synechococcus* (Heidelberg et al. 2009).

These studies report the persistence, lesser or greater in length, of phages resistant to the CRISPR-Cas systems. Sequencing of these resistant phages reveals an accumulation of substitutions exclusively in regions of the phage genome targeted by the crRNAs of the CRISPR-Cas system. Thus, mutations are not randomly fixed in the phage genome, and substitutions are selected in the protospacers, particularly in the PAM motif. In several experiments, single mutations in protospacers rendered CRISPR-Cas inefficient. But, depending on the type of CRISPR-Cas systems, there is a certain tolerance when pairing the crRNA with its target, in that a crRNA is capable of targeting a set of partially complementary sequences. Thus, by accumulating substitutions preferentially in the PAM motif, phages prevent the interference stage, but also the adaptation stage, and thus ensure the complete abolition of CRISPR-Cas immunity.

Recombination: Phages are known to have a mosaic genomic organization resulting from frequent recombination events (Hendrix et al. 1999). In co-evolution experiments between bacteria encoding a CRISPR-Cas system and multiple phages, comparison of the genomes of evolved resistant phages with those of parental phages reveals that many recombination events occur between different phages (Paez-Espino et al. 2015). These studies show that the presence of multiple phages increases their persistence by enabling recombination-based formation of chimeric phage genomes, in which sequences highly targeted by CRISPR (protospacers) have been replaced.

Covalent modifications: Phages can also modify their genomes by attaching chemical groups to their DNA. Phage T4 is particularly known because the cytosines in its genome are modified by two covalent chemical reactions, 5-hydroxymethylation and glucosylation (ghmC modification). These covalent modifications render phage T4 resistant to most RM systems. Recently, it has been shown experimentally that the ghmC modification of the phage T4 genome also confers resistance to the two CRISPR-Cas systems tested (one type II, and one type V), although the level of resistance varies depending on the target (Liu et al. 2020). Thus, covalent DNA modifications constitute a counterattack mechanism common to RM and CRISPR-Cas systems. An important clarification: when two independent sites of the phage genome are targeted, a synergistic effect is observed and the resistance mediated by the ghmC modification is almost completely abolished.

Thus, when present and active, CRISPR-Cas-mediated immunity influences the accumulation of point mutations, covalent modifications and selects beneficial homologous recombination events between genetically related phages. CRISPR-Cas systems are therefore important drivers of the evolution and diversification of phages. The selective advantage of reduced sensitivity to CRISPR targeting may explain why recombination is a common evolutionary phenomenon in natural phage populations.

8.5.2. *Anti-CRISPR proteins*

MGEs encode small *anti-CRISPR* proteins, called *Acr* proteins, which are capable of inhibiting CRISPR-Cas systems. Recently discovered in phages and prophages, they are also present in other MGEs, such as plasmids, integrative and conjugative elements, and some pathogenicity islands. In many MGEs multiple *arc* genes are clustered in a single operon, suggesting that they can be exchanged between MGEs. These Acr proteins represent a true arsenal of *natural antagonists* of CRISPR-Cas systems. To date, some 50 non-homologous Acr proteins (24 for class 1 systems, and 21 for class 2) have been discovered, with very diverse mechanisms and structures, and no significant sequence similarity between them (Marino et al. 2020). Despite this diversity, characterized Arc proteins block the activity of CRISPR-Cas systems by interacting directly with certain Cas proteins, in three distinct ways: either by preventing the formation of the interference complex, inhibiting target recognition, or inhibiting target cleavage. Acr proteins are named according to the system they inhibit in the order in which they were discovered. Therefore, the AcrIIC2 protein corresponds to the second Acr protein discovered that specifically inhibits type II-C systems. Thus, by interacting with Cas9, the AcrIIC2 proteins interfere and prevent correct assembly of the interference complex (Zhu et al. 2019). AcrIIA4 proteins, by mimicking the PAM site recognition in the interference complex, or AcrIIC3, by inducing the dimerization of the interference complexes, block the PAM motif recognition site, and thus any possible target recognition (Harrington et al. 2017). Lastly, by specifically binding to Cas9, AcrIIC1 prevents cleavage of the target DNA (Harrington et al. 2017). This list is not exhaustive, but illustrates the diversity of mechanisms employed by the Arc proteins (Marino et al. 2020).

The Acr proteins identified so far only target components of interference complexes and show high specificity for particular subtypes. The existence of these Arc proteins in MGEs may explain why certain bacteria encode several distinct systems, and why exchanges of effector modules are frequently observed. The presence of different interference complexes with distinct molecular requirements would limit Acr-mediated phage immune evasion. The presence of these Arc proteins in MGEs would also explain why most genomes, despite the presence of CRISPR-Cas systems, contain MGEs integrated in chromosomes (e.g. prophages, integrative and conjugative elements, and transposons), and therefore why *arc* genes are widespread in bacterial genomes. Interestingly, this Acr-mediated immunosuppression may require cooperation between several MGEs. Indeed, such cooperation appears necessary in order to reach a sufficient concentration of Arcs in the infected cell to hinder CRISPR immunity and enable a successful infection (Landsberger et al. 2018). Thus, the interactions between MGE-encoded Arc proteins and the CRISPR-Cas systems that they target shape the evolution of MGEs and their hosts, but also the dynamics of their respective populations.

8.6. Biological cost of CRISPR-Cas systems

Despite the obvious advantages of immunity conferred by CRISPR-Cas when parasites are present in the environment, most bacterial genomes lack it. Given the specificity and adaptability of this immune mechanism, it is somewhat surprising that they are not more widespread. This is all the more striking because RM systems are ubiquitous, with an average of two copies per genome. Understanding the low prevalence of CRISPR-Cas systems in bacterial genomes is crucial and raises the question of the direct or indirect costs associated with their presence.

8.6.1. *Cost of expression*

When selection pressure linked to a parasite has disappeared, modifications to the genetic material of the bacterium that led to resistance can result in a *biological cost* to the bacterium, i.e. a reduction in its competitiveness, its *fitness*. The acquisition of a CRISPR-Cas system implies that part of the bacterium's metabolism is diverted to its maintenance, at the expense of its reproductive success. Analysis of the cost of expression of CRISPR-Cas systems in vitro reported contrasting results in *S. thermophilus* containing an II-A system, and in *P. aeruginosa*, containing an I-F system. In the case of *S. thermophilus*, the expression of the system has proven to be *constitutive* (always active) and particularly costly in the absence of infection (Vale et al. 2015), whereas in *P. aeruginosa*, the system only becomes costly upon infection, with an *inducible* cost that nevertheless increases with the frequency of infections. The increase in the inducible cost does not appear to be linked solely to the cost of expression, but would probably be due to an increase in the mortality rate owing to the cost linked to autoimmunity (Westra et al. 2015) (see section 8.6.2). These examples illustrate the difficulty of estimating the cost of expression of CRISPR-Cas systems, as they are so diverse, both in terms of proteins and regulatory mechanisms. Some systems prove to be constitutive, while others are inducible or repressed by very diverse and complex regulatory mechanisms, most of which remain to be characterized (Patterson et al. 2017). Understanding how each system is regulated and determining under what conditions they are activated are the challenges of future research. We can assume that constitutive CRISPR-Cas systems, being associated with a fixed cost of expression, can be easily lost in the absence of parasites. The regulation of CRISPR-Cas systems, and more specifically their inducible or repressed expression, by limiting the cost of constitutive expression, could promote their maintenance in the genome.

8.6.2. *Cost of autoimmunity*

While most of the identified protospacers correspond to sequences derived from exogenous agents, a small fraction corresponding to endogenous sequences (present in the host chromosome outside the CRISPR array) has been identified in a large number of studies. It is important to note that, in the vast majority of cases, breaks caused by Cas nucleases in the bacterial chromosome are deleterious, even lethal. This property is used to develop new antimicrobial strategies, both specific and programmable, in order to eliminate pathogenic or antibiotic-resistant bacteria (Bikard and Barrangou 2017). A comprehensive study of 330 organisms encoding a CRISPR revealed that one in five genomes contained at least one spacer targeting a host chromosomal sequence (Stern et al. 2010). Half of these spacers target MGEs integrated into the chromosome, suggesting that these elements have found a way to escape the immune system (see section 8.5), and the other half target housekeeping genes essential for bacterial survival. These self-targeting spacers are never conserved in different genomes, and are frequently flanked by degenerated repeats or associated with degraded *cas* genes, thus they are unlikely to have a role in the regulation of targeted genes. In addition, their location close to the leader sequence and their non-conservation suggest that they are mostly recent and result from either (i) the accidental incorporation of host nucleic acids into the CRISPR during the adaptation phase, or (ii) the incorporation of homologous sequences to those of the host carried by MGEs (common in transducer phages or in plasmids). Thus, the presence of a CRISPR-Cas system in a microorganism is associated with *an autoimmunity-related cost*.

This cost of autoimmunity is likely to vary depending on (1) the CRISPR-Cas system involved, with a difference between DNA- or RNA-targeting systems expected; (2) its level of expression, a difference between constitutive or regulated systems expected; and (3) the frequency of HGTs in the environment where the bacterium evolves. The incorporation rate of self-targeting spacers is estimated to be around 0.2 % (Stern et al. 2010). This rate is probably underestimated, since the calculation does not take into account all the deleterious spacers that have led to cell death or that have been counter-selected at population level. Recently, mathematical modeling has shown that spacer acquisition rates would be limited by the cost of autoimmunity (Bradde et al. 2019). The model assumes that higher acquisition rates increase not only the benefits of phage resistance, but also the cost of autoimmunity due to the more frequent acquisition of self-targeting spacers. This results in the selection of an optimal acquisition rate that maximizes bacterial survival (a trade-off between phage resistance and the risk of an autoimmune response), which in turn depends on ecological factors that influence the risk of infection, such as the size of bacterial and phage populations (see section 8.7). Thus, the cost of autoimmunity associated with the CRISPR system could explain the existence of many degenerated CRISPR-Cas systems observed in genomes. It could also explain the frequent losses of systems (or spacers) observed within the same species. However, it remains to be determined whether the relatively small sizes of the CRISPRs observed are favored to limit this cost, or whether they are the result of this constraint.

8.6.3. *The genetic background of the host*

Using in silico and experimental approaches, it has been shown that the benefits of carrying a CRISPR-Cas depend on the genetic background of the host. The most striking example is the absence of type II systems in archaea, due to the absence of RNase III (which is an exclusively bacterial protein). Microbes are very diverse and encode different functions with possibly different degrees of compatibility with CRISPR-Cas systems. Multiple positive and negative epistatic interactions appear to exist between different subtypes and different systems specialized in the repair of double-strand breaks (DSB) in bacteria, which act on DNA and can compete for the same substrate (Bernheim et al. 2019). On the one hand, some systems depend on specific repair pathways to be fully functional. For example, the subtype I-E system depends on RecBCD activity during the adaptation phase (Radovcic et al. 2018). Other systems, on the other hand, appear to be incompatible, such as subtype II-A system and the NHEJ (non-homologous end-joining) repair pathway. Thus, the Csn2 protein (specific to subtype II-A), which can affect the ability of NHEJ to repair DSB, would explain why these two systems do not coexist in the same genome (Bernheim et al. 2017). Upon transfer of a new CRISPR-Cas system to a bacterium, its interactions with the different existing DNA repair pathways would condition its loss or fixation in the lineage. Since the proteins involved in DNA repair mechanisms are under strong purifying selection, incompatibilities will most often lead to the loss of the CRISPR-Cas system and conservation of the extant DNA repair system. As a result of this process, bacteria with different repair systems will end up encoding different CRISPR-Cas systems. Hence, these positive and negative interactions, whose underlying mechanisms remain to be discovered, would shape the scattered distribution and the diversity of these immune systems in bacterial species.

8.6.4. *Limiting horizontal gene transfer*

Another explanation for the absence of CRISPR-Cas systems in many genomes is that they limit horizontal gene transfer, a key driver of the rapid evolution and adaptation of microorganisms. Experimental works have shown that CRISPR-Cas systems constitute a barrier to natural transformation (Bikard et al. 2012; Zhang et al. 2013), conjugation (Marraffini and Sontheimer 2008) and also transduction (Barrangou et al. 2007). This limitation of gene flows would explain in part why genomes encoding a CRISPR-Cas system are often smaller than those devoid of it within the same species (Wheatley and MacLean 2020). While this may be protective in some cases, it may also prevent the acquisition of genetic information that is potentially beneficial to the host, or even indispensable for its survival under certain conditions. It has been shown that CRISPR-Cas systems impede the acquisition of virulence genes in *E. coli* (García-Gutiérrez et al. 2015), or antibiotic resistance genes in *Enterococci faecalis* (Palmer and Gilmore 2010). In *Legionella pneumophila*, they lead to a specialization of the host by improving its fitness for an amoeba, but reducing its ability to reproduce in other hosts. Several studies have examined the relative benefit of encoding a CRISPR-Cas system or acquiring new genes by HGT. They showed that under certain conditions, with those where immunity directly harms the host, the loss or inactivation of the CRISPR-Cas system was adaptive (Bikard et al. 2012; Jiang et al. 2013). Although these studies conducted on specific species tend to show that CRISPR-Cas systems have a negative impact on short-term HGTs at population level, this does not appear to be generalizable to all microorganisms. For example, in *E. coli*, CRISPR-Cas systems do not appear to have any impact on the spread of antibiotic-resistance plasmids, and no association with the different lifestyles of strains of this species has yet been observed. More broadly, when HGTs were studied at the evolutionary scale of species by considering 1,400 genomes, no correlation was found between the number of HGT events and the length of CRISPR arrays. Thus, the impact of CRISPR-Cas on HGTs no longer seems detectable at this evolutionary scale, which is the likely consequence of the high turn-over of these systems during the course of evolution (Gophna et al. 2015).

Thus, it appears that there is a trade-off between active acquired immunity and the ability to acquire new genes and functions via HGTs at the population scale. This may partly explain why strains of the same species have a CRISPR-Cas system while others do not. The impact of these systems on HGTs appears to be variable from species to species, and even from genome to genome. It still remains to be determined whether the presence of Arc proteins in MGEs (upon infection or already integrated in the host genome) could provide a better understanding of the observed variability.

8.6.5. *Naïve and primed adaptation*

The acquisition of spacers from phage genomes has raised several questions, given the speed at which phages take control of the cell. Despite the limited number of studies focusing on the adaptation phase, there appears to exist at least two routes to spacer acquisition, known as *naïve* and *primed* adaptation. Naïve adaptation involves the de novo formation of immunity, independent of preexisting spacers. It is rarely observed under laboratory conditions and remains enigmatic for several CRISPR-Cas subtypes. Naïve adaptation takes time to become operational, and as a result is probably not efficient against highly virulent, fast-replicating phages. In contrast, primed adaptation uses existing spacers to enhance the acquisition of new spacers and therefore improve the efficiency of CRISPR-Cas-mediated immunity.

Recently, it has been shown that naïve adaptation can occur more efficiently when the bacteria are infected with defective phages (Hynes et al. 2014). Unlike active phages, defective phages will inject their DNA into the host cell without causing its death, thus giving it as much time as needed to acquire new spacers. This stage is surprisingly reminiscent of the vaccination process, which involves inoculating a host with defective (or attenuated) viruses to trigger protective immune responses. It has also been shown that the spacer acquisition rate increases with the number of defense systems present in the genome. For example, RM and CRISPR-Cas systems can, in some cases, work in synergy: with the RM system, which neutralizes the invader and supplies the CRISPR-Cas system with exogenous genome cleavage products that will be integrated, as spacers, more rapidly by the adaptation complex (Dupuis et al. 2013). A significant improvement has also been detected in the spacer acquisition rate as soon as preexisting spacers are already targeting the invasive agent. Indeed, primed adaptation is a positive feedback loop where target recognition, facilitated by preexisting (priming) spacers, enhances the acquisition of new spacers, thereby reinforcing immunity (Fineran et al. 2014). During the interference phase, the Cas3 protein in type I systems generates substrates (target cleavage products) that can be integrated into the CRISPR by the adaptation module. Thus, the new spacers are not acquired randomly, but generally come from protospacers clustered on a single strand around the region targeted by the priming spacer. Primed adaptation has been observed experimentally for several type I CRISPR-Cas systems and deduced by in silico approaches for type II systems (Nicholson et al. 2019). Note that this phenomenon can occur even with priming spacers that contain several mismatches, allowing hosts to update their immunity to counter escape mutants or highly divergent targets.

The primed adaptation allows the diversity of spacers in the population to be increased and the population's overall resistance to thus be increased. It therefore appears crucial in the efficiency of CRISPR-Cas-mediated immunity, as naïve adaptation is very limited. This priming requirement, on the one hand, imposes a significant constraint on the evolution of CRISPRs, and, on the other, probably limits the cost of autoimmunity by ensuring that only MGEs that resemble preexisting spacers (and therefore known parasites) are targeted.

8.7. Importance in nature: impact of ecological factors

The arms race between CRISPR-Cas systems and MGEs, the frequent HGT-gain and loss events of these CRISPR-Cas systems, and their direct and indirect costs, combine to produce a complex evolutionary dynamic. Despite this, the scarcity of CRISPR-Cas systems in bacteria remains surprising in view of the high prevalence in archaea. Various ecological factors, which could influence the presence of these systems in nature, have been proposed to explain this difference (Westra et al. 2016).

8.7.1. *Phage diversity – mutation rate*

Most sequenced archaea are thermophilic, while most sequenced bacteria are mesophilic. The ubiquity of CRISPR-Cas in archaea, which contrasts with low prevalence in bacteria, may depend more on environmental temperature than on bacterial-archaeal taxonomy (Weinberger et al. 2012). Biophysical models show that mutations are more likely to be lethal in thermophilic environments because high temperatures reduce protein stability (Wylie and Shakhnovich 2011). With an increased cost of mutation, thermophilic genomes have lower mutation rates than mesophilic genomes (Zeldovich et al. 2007). Similarly, phages infecting thermophiles are less likely to mutate and so escape CRISPR-Cas systems. Thus, the immunity conferred by CRISPR-Cas systems would be more efficient, more beneficial, and thus maintained in microorganisms evolving in thermophilic environments. Conversely, in mesophilic microorganisms where the mutation rate is higher, the CRISPR-Cas system would be quickly overwhelmed, as it would no longer be able to contain enough spacers to target the entire rapidly evolving phage population. The CRISPR-Cas system, unable to provide sufficient immunity, and due to the associated costs, would ultimately be lost (Weinberger et al. 2012). Supporting this idea, it has been shown in vitro that hypermutator bacteria (common in the environment) were much more likely to develop resistance by phage receptor mutation to the detriment of immunity mediated by the CRISPR-Cas system, which would probably be lost in the long term (Chevallereau et al. 2019). Thus, it has been proposed that the benefit of the CRISPR-Cas system would decrease as the mutation rate or phage diversity increases, or in the face of rapidly evolving phages.

8.7.2. *Phage diversity – population size*

Along the same lines, it has been suggested that the benefit of CRISPR-Cas-mediated immunity would decrease as the size of the microbial population increases (Iranzo et al. 2013). Because genetic diversity varies with mutation rate and population size, it is all the more significant, the larger the population. Thus, in a small microbial population, phages would be limited in number and not very diverse, whereas a high microbial density would be associated with a large and highly diversified viral population. As a result, the opportunities for phages to escape CRISPR-Cas systems (by mutation, recombination, or through the action of Arc proteins, see section 8.5) would be all the more common, the larger the microbial population. Theoretical models predict a threshold value for total viral diversity, proportional to the mutation rate and size of the population (viral and host), according to which CRISPR-Cas immunity would either be beneficial and thus conserved, or inefficient and costly and therefore lost (Iranzo et al. 2013; Weinberger et al. 2012). Thus, the high prevalence of CRISPR-Cas in thermophiles, which contrasts with that of mesophiles, would be related to the smaller population sizes observed in thermophilic environments.

8.7.3. *Infectious risk – alternative strategies*

A significant factor in understanding the distribution of CRISPR-Cas systems is probably that of their performance compared to other alternative immune strategies (Dy et al. 2014). These alternative mechanisms include (1) resistance by surface modification (results from mutation, masking, or loss of host receptor proteins which serve as entry points for the phage); (2) resistance by blocking the injection of phage DNA (a system encoded by prophages, thus preventing superinfection of the host by a similar phage); (3) resistance by degradation of the viral genome by intracellular nucleases (e.g. by RM and prokaryotic Argonaute (pAgo) systems); (4) resistance by abortive infection (where the bacterium triggers a suicidal reaction upon phage infection, thereby blocking viral spread, to the benefit of the rest of the population). Note that phages have found multiple and varied ways to escape from each of them. These many mechanisms, which act at different stages of infection, probably constitute different lines of defense and help to ensure efficient immunity. They are not mutually exclusive and some probably work together. Consistent with this idea, RM and CRISPR-Cas systems co-occur frequently in genomes, and certain combinations bring about an increase in immunity level and faster spacers acquisition (Dupuis et al. 2013). In the same vein, the pAgo expression stimulates the level of expression of the adaptation module, thus also promoting the adaptation phase of CRISPR-Cas systems.

But if it is beneficial to have multiple defense strategies, why do all bacteria not have all the mechanisms? According to the most likely hypothesis, the costs associated with different strategies (varying according to the conditions under which the microbe evolves) would condition their maintenance or loss in a given genome (van Houte et al. 2016). As a result, the strength of infection should be a key ecological factor in the evolution of these different immune strategies, since the overall cost of an inducible resistance strategy will depend on the frequency of infections. In accordance with this hypothesis, it has been shown in vitro that the strength of the infection can tilt the balance of an inducible immunity mediated by CRISPR-Cas toward constitutive immunity via surface modification (Westra et al. 2015). Inducible CRISPR-Cas systems would thus be more likely to confer an advantage in ecosystems where the risk of infection is low, such as high-temperature or resource-poor environments (because microbial and phage densities are lower), and would be disadvantaged because they are too costly or inefficient, when the force of infection is high. Recently, theoretical modeling has shown that the conditions under which CRISPR-Cas systems are favored in relation to these alternative defenses are very restricted (Gurney et al. 2019). Thus, microbes have many options for defending themselves against MGEs, and CRISPR-Cas systems, given their costs, quite simply may not always be the best solution, which would explain why they are so frequently lost. It should be noted, however, that our understanding of how different defense mechanisms evolve, their possible interactions, their associated costs and the ecological conditions in which they are or are not favored, is still very limited and will be the subject of much attention in the coming years.

8.8. Conclusions and perspectives

Several decades have passed since the fortuitous discovery of a CRISPR array. Several research groups have uncovered many of its mysteries, and the CRISPR-Cas system is probably one of the most fascinating discoveries in microbiology. Currently, a large number of CRISPR-Cas system types and subtypes have been described and many additional variants are probably waiting to be discovered. The development of ever greater numbers of CRISPR-Cas tools for genome editing is due, in part, to a better understanding of the stages of expression and interference that prove to be of remarkable diversity and complexity. The adaptation stage, which is indispensable for the efficiency of the immunity conferred by these systems, still remains enigmatic for most systems. The presence of inhibitors (in microbial genomes or in their MGEs), whose diversity remains to be explored, and the existence of multiple regulatory mechanisms, most of which are unknown, illustrate some of the difficulties to be overcome when studying these systems and represent future lines of research.

As studies on CRISPR-Cas systems have focused mainly on mechanistic details, our understanding of the evolutionary ecology of these immune systems is still relatively limited. Further studies conducted on various microbial and MGE populations in different natural environments will be necessary to gain a better understanding of the functioning and impact of these various systems in nature. It seems important to understand not only how often, but also when, where, why, and which CRISPR-Cas systems play a role in the defense against MGEs. Furthermore, it is still not known to what extent the selection induced by the different types of MGE is responsible for the existence, maintenance, loss or transfer of these CRISPR-Cas systems in microbial populations. If we wish to use CRISPR-Cas as antimicrobial or antiviral agents, it is imperative to determine their impact on the evolution and adaptation of microbes and their phages, and more generally on bacteria-phage interactions.

In most of the studies conducted, CRISPR-Cas systems are considered isolated from each other, but this is not the norm in nature, where they coexist with each other, as well as with other defense systems (often within genomic defense islands). How do these different defense systems work together in vivo in their natural environment? How do they interact, communicate with each other, and in what genetic backgrounds or environments are they favored? These are questions that remain unanswered at present, and are crucial to understanding the overall impact of CRISPR-Cas-mediated immunity in the microbial world.

CRISPR-Cas systems are considered to be mainly involved in defense against MGEs. However, relations between these systems and MGEs are much more complex. On the one hand, CRISPR-Cas systems are frequently transferred via HGTs, even though they are involved in the destruction of the main vectors providing these transfers. Understanding how these systems are able to invade a population, and how they integrate the host chromosome and adapt to the host genetic background, is of great importance in shedding light on this apparent paradox. On the other hand, several types of MGEs have contributed to the emergence and diversification of CRISPR-Cas systems and, conversely, CRISPR-Cas systems and their components have been recruited by various MGEs for functions that remain largely uncharacterized (Faure et al. 2019). It is therefore imperative to study these complex relationships in order to better understand the impact on the joint evolution of these systems, MGEs and their hosts.

While their role in protecting the host against MGEs is now well established, there is growing evidence that these systems modulate other processes, such as DNA repair, genome evolution and gene-expression regulation (Westra et al. 2014). Thus, it has been shown that Cas1 increases the resistance of *Mycobacterium smegmatis* to several antituberculosis drugs, that Cas2 is an essential factor in facilitating infection of amoebae by *Legionella pneumophila* and that Cas9 is involved in the regulation of the expression of genes modulating the virulence of several pathogenic bacteria including *Francisella novicida*, *Campylobacter jejuni* and *Neisseria meningitidis*. It should be noted, however, that the mechanisms involved in these different unconventional functions largely remain to be elucidated. A central question relating to the evolution of these non-canonical functions is whether they have actually been selected or whether they are simply by-products of their role in immunity. Understanding the role of CRISPR-Cas in these processes should improve our understanding of the evolution and maintenance of CRISPR-Cas systems in prokaryotic genomes.

Lastly, it is important to emphasize that CRISPR-Cas systems remain relatively rare for an immune system. Different ecological factors and associated costs have been proposed to explain their absence in more than half of the bacterial genomes sequenced. However, none of these proposals are fully satisfactory. Future research will likely focus on this fundamental question, which currently remains open.

8.9. References

Amitai, G. and Sorek, R. (2016). CRISPR-Cas adaptation: Insights into the mechanism of action. *Nat. Rev. Microbiol.*, 14(2), 67–76.

Anderson, R.E., Brazelton, W.J., Baross, J.A. (2011). Using CRISPRs as a metagenomic tool to identify microbial hosts of a diffuse flow hydrothermal vent viral assemblage. *FEMS Microbiol. Ecol.*, 77(1), 120–133.

Andersson, A.F. and Banfield, J.F. (2008). Virus population dynamics and acquired virus resistance in natural microbial communities. *Science*, 320(5879), 1047–1050.

Barrangou, R., Fremaux, C., Deveau, H., Richards, M., Boyaval, P., Moineau, S., Romero, D.A., Horvath, P. (2007). CRISPR provides acquired resistance against viruses in prokaryotes. *Science*, 315(5819), 1709–1712.

Bernheim, A., Calvo-Villamañán, A., Basier, C., Cui, L., Rocha, E.P.C., Touchon, M., Bikard, D. (2017). Inhibition of NHEJ repair by type II-A CRISPR-Cas systems in bacteria. *Nat. Commun.*, 8(1), 2094.

Bernheim, A., Bikard, D., Touchon, M., Rocha, E.P.C. (2019). A matter of background: DNA repair pathways as a possible cause for the sparse distribution of CRISPR-Cas systems in bacteria. *Philos. Trans. R Soc. Lond. B Biol. Sci.*, 374(1772), 20180088.

Bernheim, A., Bikard, D., Touchon, M., Rocha, E.P.C. (2020). Atypical organizations and epistatic interactions of CRISPRs and cas clusters in genomes and their mobile genetic elements. *Nucleic Acids Res.*, 48(2), 748–760.

Bikard, D. and Barrangou, R. (2017). Using CRISPR-Cas systems as antimicrobials. *Curr. Opin. Microbiol.*, 37, 155–160.

Bikard, D., Hatoum-Aslan, A., Mucida, D., Marraffini, L.A. (2012). CRISPR interference can prevent natural transformation and virulence acquisition during in vivo bacterial infection. *Cell Host Microbe*, 12(2), 177–186.

Biswas, A., Staals, R.H., Morales, S.E., Fineran, P.C., Brown, C.M. (2016). CRISPRDetect: A flexible algorithm to define CRISPR arrays. *BMC Genomics*, 17, 356.

Bolotin, A., Quinquis, B., Sorokin, A., Ehrlich, S.D. (2005). Clustered regularly interspaced short palindrome repeats (CRISPRs) have spacers of extrachromosomal origin. *Microbiology (Reading)*, 151(Pt 8), 2551–2561.

Bradde, S., Mora, T., Walczak, A.M. (2019). Cost and benefits of clustered regularly interspaced short palindromic repeats spacer acquisition. *Philos. Trans. R. Soc. Lond. B Biol. Sci.*, 374(1772), 20180095.

Burstein, D., Sun, C.L., Brown, C.T., Sharon, I., Anantharaman, K., Probst, A.J., Thomas, B.C., Banfield, J.F. (2016). Major bacterial lineages are essentially devoid of CRISPR-Cas viral defence systems. *Nat. Commun.*, 7, 10613.

Charpentier, E., Richter, H., Van Der Oost, J., White, M.F. (2015). Biogenesis pathways of RNA guides in archaeal and bacterial CRISPR-Cas adaptive immunity. *FEMS Microbiol. Rev.*, 39(3), 428–441.

Chevallereau, A., Meaden, S., Van Houte, S., Westra, E.R., Rollie, C. (2019). The effect of bacterial mutation rate on the evolution of CRISPR-Cas adaptive immunity. *Philos. Trans. R. Soc. Lond. B Biol. Sci.*, 374(1772), 20180094.

Chylinski, K., Le Rhun, A., Charpentier, E. (2013). The tracrRNA and Cas9 families of type II CRISPR-Cas immunity systems. *RNA Biol.*, 10(5), 726–737.

Cornuault, J.K., Moncaut, E., Loux, V., Mathieu, A., Sokol, H., Petit, M.A., De Paepe, M. (2020). The enemy from within: A prophage of Roseburia intestinalis systematically turns lytic in the mouse gut, driving bacterial adaptation by CRISPR spacer acquisition. *ISME J.*, 14(3), 771–787.

Couvin, D., Bernheim, A., Toffano-Nioche, C., Touchon, M., Michalik, J., Néron, B., Rocha, E.P.C., Vergnaud, G., Gautheret, D., Pourcel, C. (2018). CRISPRCasFinder, an update of CRISRFinder, includes a portable version, enhanced performance and integrates search for Cas proteins. *Nucleic Acids Res.*, 46(W1), W246–W251.

Delaney, N.F., Balenger, S., Bonneaud, C., Marx, C.J., Hill, G.E., Ferguson-Noel, N., Tsai, P., Rodrigo, A., Edwards, S.V. (2012). Ultrafast evolution and loss of CRISPRs following a host shift in a novel wildlife pathogen, *Mycoplasma gallisepticum*. *PLoS Genet,* 8(2), e1002511.

Deveau, H., Garneau, J.E., Moineau, S. (2010). CRISPR/Cas system and its role in phage-bacteria interactions. *Annu. Rev. Microbiol,* 64, 475–493.

Dupuis, M.E., Villion, M., Magadan, A.H., Moineau, S. (2013). CRISPR-Cas and restriction-modification systems are compatible and increase phage resistance. *Nat. Commun.,* 4, 2087.

Dy, R.L., Richter, C., Salmond, G.P., Fineran, P.C. (2014). Remarkable mechanisms in microbes to resist phage infections. *Annu. Rev. Virol.,* 1(1), 307–331.

Erdmann, S. and Garrett, R.A. (2012). Selective and hyperactive uptake of foreign DNA by adaptive immune systems of an archaeon via two distinct mechanisms. *Mol. Microbiol.,* 85(6), 1044–1056.

Faure, G., Shmakov, S.A., Yan, W.X., Cheng, D.R., Scott, D.A., Peters, J.E., Makarova, K.S., Koonin, E.V. (2019). CRISPR-Cas in mobile genetic elements: Counter-defence and beyond. *Nat. Rev. Microbiol.,* 17(8), 513–525.

Fineran, P.C., Gerritzen, M.J., Suarez-Diez, M., Kunne, T., Boekhorst, J., Van Hijum, S.A., Staals, R.H., Brouns, S.J. (2014). Degenerate target sites mediate rapid primed CRISPR adaptation. *Proc. Natl. Acad. Sci. USA*, 111(16), E1629–E1638.

García-Gutiérrez, E., Almendros, C., Mojica, F.J., Guzmán, N.M., García-Martínez, J. (2015). CRISPR content correlates with the pathogenic potential of *Escherichia coli*. *PLoS One,* 10(7), e0131935.

Gomaa, A.A., Klumpe, H.E., Luo, M.L., Selle, K., Barrangou, R., Beisel, C.L. (2014). Programmable removal of bacterial strains by use of genome-targeting CRISPR-Cas systems. *mBio,* 5(1), e00928–e00913.

Gophna, U., Kristensen, D.M., Wolf, Y.I., Popa, O., Drevet, C., Koonin, E.V. (2015). No evidence of inhibition of horizontal gene transfer by CRISPR-Cas on evolutionary timescales. *ISME J.,* 9(9), 2021–2027.

Gurney, J., Pleska, M., Levin, B.R. (2019). Why put up with immunity when there is resistance: An excursion into the population and evolutionary dynamics of restriction-modification and CRISPR-Cas. *Philos. Trans. R. Soc. Lond. B Biol. Sci.,* 374(1772), 20180096.

Haft, D.H., Selengut, J., Mongodin, E.F., Nelson, K.E. (2005). A guild of 45 CRISPR-associated (Cas) protein families and multiple CRISPR/Cas subtypes exist in prokaryotic genomes. *PLoS Comput. Biol.,* 1(6), e60.

Harrington, L.B., Doxzen, K.W., Ma, E., Liu, J.J., Knott, G.J., Edraki, A., Garcia, B., Amrani, N., Chen, J.S., Cofsky, J.C. et al. (2017). A broad-spectrum inhibitor of CRISPR-Cas9. *Cell,* 170(6), 1224–1233.e1215.

Heidelberg, J.F., Nelson, W.C., Schoenfeld, T., Bhaya, D. (2009). Germ warfare in a microbial mat community: CRISPRs provide insights into the co-evolution of host and viral genomes. *PLoS One,* 4(1), e4169.

Heler, R., Samai, P., Modell, J.W., Weiner, C., Goldberg, G.W., Bikard, D., Marraffini, L.A. (2015). Cas9 specifies functional viral targets during CRISPR-Cas adaptation. *Nature,* 519(7542), 199–202.

Hendrix, R.W., Smith, M.C., Burns, R.N., Ford, M.E., Hatfull, G.F. (1999). Evolutionary relationships among diverse bacteriophages and prophages: All the world's a phage. *Proc. Natl. Acad. Sci. USA,* 96(5), 2192–2197.

Horn, H., Slaby, B.M., Jahn, M.T., Bayer, K., Moitinho-Silva, L., Förster, F., Abdelmohsen, U.R., Hentschel, U. (2016). An enrichment of CRISPR and other defense-related features in marine sponge-associated microbial metagenomes. *Front. Microbiol.,* 7, 1751.

Hynes, A.P., Villion, M., Moineau, S. (2014). Adaptation in bacterial CRISPR-Cas immunity can be driven by defective phages. *Nat. Commun.,* 5, 4399.

Iranzo, J., Lobkovsky, A.E., Wolf, Y.I., Koonin, E.V. (2013). Evolutionary dynamics of the prokaryotic adaptive immunity system CRISPR-Cas in an explicit ecological context. *J. Bacteriol.,* 195(17), 3834–3844.

Jansen, R., Embden, J.D., Gaastra, W., Schouls, L.M. (2002). Identification of genes that are associated with DNA repeats in prokaryotes. *Mol. Microbiol.,* 43(6), 1565–1575.

Jiang, W., Maniv, I., Arain, F., Wang, Y., Levin, B.R., Marraffini, L.A. (2013). Dealing with the evolutionary downside of CRISPR immunity: Bacteria and beneficial plasmids. *PLoS Genet.,* 9(9), e1003844.

Koonin, E.V. and Makarova, K.S. (2019). Origins and evolution of CRISPR-Cas systems. *Philos. Trans. R. Soc. Lond. B Biol. Sci.,* 374(1772), 20180087.

Koonin, E.V. and Wolf, Y.I. (2009). Is evolution Darwinian or/and Lamarckian? *Biol. Direct,* 4, 42.

Krupovic, M., Béguin, P., Koonin, E.V. (2017). Casposons: Mobile genetic elements that gave rise to the CRISPR-Cas adaptation machinery. *Curr. Opin. Microbiol.,* 38, 36–43.

Kunin, V., Sorek, R., Hugenholtz, P. (2007). Evolutionary conservation of sequence and secondary structures in CRISPR repeats. *Genome Biol.,* 8(4), R61.

Kuo, C.H. and Ochman, H. (2009). Deletional bias across the three domains of life. *Genome Biol. Evol.,* 1, 145–152.

Landsberger, M., Gandon, S., Meaden, S., Rollie, C., Chevallereau, A., Chabas, H., Buckling, A., Westra, E.R., Van Houte, S. (2018). Anti-CRISPR phages cooperate to overcome CRISPR-Cas immunity. *Cell,* 174(4), 908–916.e912.

Lillestøl, R.K., Shah, S.A., Brügger, K., Redder, P., Phan, H., Christiansen, J., Garrett, R.A. (2009). CRISPR families of the crenarchaeal genus Sulfolobus: Bidirectional transcription and dynamic properties. *Mol. Microbiol.,* 72(1), 259–272.

Liu, Y., Dai, L., Dong, J., Chen, C., Zhu, J., Rao, V.B., Tao, P. (2020). Covalent modifications of the bacteriophage genome confer a degree of resistance to bacterial CRISPR systems. *J. Virol.,* 94(23), e01630-20.

Makarova, K.S., Wolf, Y.I., Iranzo, J., Shmakov, S.A., Alkhnbashi, O.S., Brouns, S.J.J., Charpentier, E., Cheng, D., Haft, D.H., Horvath, P. et al. (2020). Evolutionary classification of CRISPR-Cas systems: A burst of class 2 and derived variants. *Nat. Rev. Microbiol.,* 18(2), 67–83.

Marino, N.D., Pinilla-Redondo, R., Csörgő, B., Bondy-Denomy, J. (2020). Anti-CRISPR protein applications: Natural brakes for CRISPR-Cas technologies. *Nat. Methods,* 17(5), 471–479.

Marraffini, L.A. and Sontheimer, E.J. (2008). CRISPR interference limits horizontal gene transfer in staphylococci by targeting DNA. *Science,* 322(5909), 1843–1845.

Marraffini, L.A. and Sontheimer, E.J. (2010). Self versus non-self discrimination during CRISPR RNA-directed immunity. *Nature,* 463(7280), 568–571.

Meeske, A.J. and Marraffini, L.A. (2018). RNA guide complementarity prevents self-targeting in type VI CRISPR systems. *Mol. Cell,* 71(5), 791–801.e793.

Mitrofanov, A., Alkhnbashi, O.S., Shmakov, S.A., Makarova, K.S., Koonin, E.V., Backofen, R. (2021). CRISPRidentify: Identification of CRISPR arrays using machine learning approach. *Nucleic Acids Res.,* 49(4), e20.

Modell, J.W., Jiang, W., Marraffini, L.A. (2017). CRISPR-Cas systems exploit viral DNA injection to establish and maintain adaptive immunity. *Nature,* 544(7648), 101–104.

Mohanraju, P., Makarova, K.S., Zetsche, B., Zhang, F., Koonin, E.V., Van Der Oost, J. (2016). Diverse evolutionary roots and mechanistic variations of the CRISPR-Cas systems. *Science,* 353(6299), aad5147.

Mojica, F.J., Ferrer, C., Juez, G., Rodríguez-Valera, F. (1995). Long stretches of short tandem repeats are present in the largest replicons of the Archaea *Haloferax mediterranei* and *Haloferax volcanii* and could be involved in replicon partitioning. *Mol. Microbiol.,* 17(1), 85–93.

Mojica, F.J., Díez-Villaseñor, C., García-Martínez, J., Soria, E. (2005). Intervening sequences of regularly spaced prokaryotic repeats derive from foreign genetic elements. *J. Mol. Evol.,* 60(2), 174–182.

Mojica, F.J., Díez-Villaseñor, C., García-Martínez, J., Almendros, C. (2009). Short motif sequences determine the targets of the prokaryotic CRISPR defence system. *Microbiology (Reading),* 155(Pt 3), 733–740.

Nicholson, T.J., Jackson, S.A., Croft, B.I., Staals, R.H.J., Fineran, P.C., Brown, C.M. (2019). Bioinformatic evidence of widespread priming in type I and II CRISPR-Cas systems. *RNA Biol.,* 16(4), 566–576.

Nussenzweig, P.M. and Marraffini, L.A. (2020). Molecular mechanisms of CRISPR-Cas immunity in bacteria. *Annu. Rev. Genet.,* 54, 93–120.

Paez-Espino, D., Sharon, I., Morovic, W., Stahl, B., Thomas, B.C., Barrangou, R., Banfield, J.F. (2015). CRISPR immunity drives rapid phage genome evolution in Streptococcus thermophilus. *mBio,* 6(2), e00262-15.

Palmer, K.L. and Gilmore, M.S. (2010). Multidrug-resistant enterococci lack CRISPR-cas. *mBio,* 1(4), e00227-10.

Patterson, A.G., Yevstigneyeva, M.S., Fineran, P.C. (2017). Regulation of CRISPR-Cas adaptive immune systems. *Curr. Opin. Microbiol.,* 37, 1–7.

Pickar-Oliver, A. and Gersbach, C.A. (2019). The next generation of CRISPR-Cas technologies and applications. *Nat. Rev. Mol. Cell Biol.,* 20(8), 490–507.

Pourcel, C., Salvignol, G., Vergnaud, G. (2005). CRISPR elements in *Yersinia pestis* acquire new repeats by preferential uptake of bacteriophage DNA, and provide additional tools for evolutionary studies. *Microbiology (Reading),* 151(Pt 3), 653–663.

Pourcel, C., Touchon, M., Villeriot, N., Vernadet, J.P., Couvin, D., Toffano-Nioche, C., Vergnaud, G. (2020). CRISPRCasdb a successor of CRISPRdb containing CRISPR arrays and cas genes from complete genome sequences, and tools to download and query lists of repeats and spacers. *Nucleic Acids Res.,* 48(D1), D535–D544.

Pride, D.T., Sun, C.L., Salzman, J., Rao, N., Loomer, P., Armitage, G.C., Banfield, J.F., Relman, D.A. (2011). Analysis of streptococcal CRISPRs from human saliva reveals substantial sequence diversity within and between subjects over time. *Genome Res.,* 21(1), 126–136.

Radovcic, M., Killelea, T., Savitskaya, E., Wettstein, L., Bolt, E.L., Ivancic-Bace, I. (2018). CRISPR-Cas adaptation in *Escherichia coli* requires RecBCD helicase but not nuclease activity, is independent of homologous recombination, and is antagonized by 5' ssDNA exonucleases. *Nucleic Acids Res.,* 46(19), 10173–10183.

Rho, M., Wu, Y.W., Tang, H., Doak, T.G., Ye, Y. (2012). Diverse CRISPRs evolving in human microbiomes. *PLoS Genet.,* 8(6), e1002441.

Rollie, C., Chevallereau, A., Watson, B.N.J., Chyou, T.Y., Fradet, O., Mcleod, I., Fineran, P.C., Brown, C.M., Gandon, S., Westra, E.R. (2020). Targeting of temperate phages drives loss of type I CRISPR-Cas systems. *Nature,* 578(7793), 149–153.

Safari, F., Sharifi, M., Farajnia, S., Akbari, B., Karimi Baba Ahmadi, M., Negahdaripour, M., Ghasemi, Y. (2020). The interaction of phages and bacteria: The co-evolutionary arms race. *Crit. Rev. Biotechnol.,* 40(2), 119–137.

Semenova, E., Jore, M.M., Datsenko, K.A., Semenova, A., Westra, E.R., Wanner, B., Van Der Oost, J., Brouns, S.J., Severinov, K. (2011). Interference by clustered regularly interspaced short palindromic repeat (CRISPR) RNA is governed by a seed sequence. *Proc. Natl. Acad. Sci. USA,* 108(25), 10098–10103.

Shmakov, S.A., Sitnik, V., Makarova, K.S., Wolf, Y.I., Severinov, K.V., Koonin, E.V. (2017). The CRISPR spacer space is dominated by sequences from species-specific mobilomes. *mBio,* 8(5), e01397-17

Stern, A., Keren, L., Wurtzel, O., Amitai, G., Sorek, R. (2010). Self-targeting by CRISPR: Gene regulation or autoimmunity? *Trends Genet.,* 26(8), 335–340.

Stern, A., Mick, E., Tirosh, I., Sagy, O., Sorek, R. (2012). CRISPR targeting reveals a reservoir of common phages associated with the human gut microbiome. *Genome Res.,* 22(10), 1985–1994.

Vale, P.F., Lafforgue, G., Gatchitch, F., Gardan, R., Moineau, S., Gandon, S. (2015). Costs of CRISPR-Cas-mediated resistance in *Streptococcus thermophilus. Proc. Biol. Sci.,* 282(1812), 20151270.

Van Der Oost, J., Westra, E.R., Jackson, R.N., Wiedenheft, B. (2014). Unravelling the structural and mechanistic basis of CRISPR-Cas systems. *Nat. Rev. Microbiol.,* 12(7), 479–492.

Van Der Ploeg, J.R. (2009). Analysis of CRISPR in *Streptococcus mutans* suggests frequent occurrence of acquired immunity against infection by M102-like bacteriophages. *Microbiology (Reading),* 155(Pt 6), 1966–1976.

Van Houte, S., Buckling, A., Westra, E.R. (2016). Evolutionary ecology of prokaryotic immune mechanisms. *Microbiol. Mol. Biol. Rev.,* 80(3), 745–763.

Wang, L., Mo, C.Y., Wasserman, M.R., Rostøl, J.T., Marraffini, L.A., Liu, S. (2019). Dynamics of Cas10 govern discrimination between self and non-self in type III CRISPR-Cas immunity. *Mol. Cell,* 73(2), 278–290.e274.

Wang, B., Zhang, T., Yin, J., Yu, Y., Xu, W., Ding, J., Patel, D.J., Yang, H. (2021). Structural basis for self-cleavage prevention by tag:anti-tag pairing complementarity in type VI Cas13 CRISPR systems. *Mol. Cell,* 81(5), 1100–1115.e1105.

Wei, Y., Chesne, M.T., Terns, R.M., Terns, M.P. (2015). Sequences spanning the leader-repeat junction mediate CRISPR adaptation to phage in *Streptococcus thermophilus. Nucleic Acids Res,* 43(3), 1749–1758.

Weinberger, A.D., Wolf, Y.I., Lobkovsky, A.E., Gilmore, M.S., Koonin, E.V. (2012). Viral diversity threshold for adaptive immunity in prokaryotes. *mBio,* 3(6), e00456–00412.

Westra, E.R., Buckling, A., Fineran, P.C. (2014). CRISPR-Cas systems: Beyond adaptive immunity. *Nat. Rev. Microbiol.*, 12(5), 317–326.

Westra, E.R., Van Houte, S., Oyesiku-Blakemore, S., Makin, B., Broniewski, J.M., Best, A., Bondy-Denomy, J., Davidson, A., Boots, M., Buckling, A. (2015). Parasite exposure drives selective evolution of constitutive versus inducible defense. *Curr. Biol.*, 25(8), 1043–1049.

Westra, E.R., Dowling, A.J., Broniewski, J.M., Houte, S.V. (2016). Evolution and ecology of CRISPR. *Annual Review of Ecology, Evolution, and Systematics*, 47(1), 307–331.

Wheatley, R.M. and Maclean, R.C. (2020). CRISPR-Cas systems restrict horizontal gene transfer in *Pseudomonas aeruginosa*. *ISME J.*, 15(5), 1420–1433.

Wylie, C.S. and Shakhnovich, E.I. (2011). A biophysical protein folding model accounts for most mutational fitness effects in viruses. *Proc. Natl. Acad. Sci. USA*, 108(24), 9916–9921.

Yosef, I., Goren, M.G., Qimron, U. (2012). Proteins and DNA elements essential for the CRISPR adaptation process in *Escherichia coli*. *Nucleic Acids Res.*, 40(12), 5569–5576.

Zeldovich, K.B., Chen, P., Shakhnovich, E.I. (2007). Protein stability imposes limits on organism complexity and speed of molecular evolution. *Proc. Natl. Acad. Sci. USA*, 104(41), 16152–16157.

Zhang, Y., Heidrich, N., Ampattu, B.J., Gunderson, C.W., Seifert, H.S., Schoen, C., Vogel, J., Sontheimer, E.J. (2013). Processing-independent CRISPR RNAs limit natural transformation in *Neisseria meningitidis*. *Mol. Cell*, 50(4), 488–503.

Zhu, Y., Gao, A., Zhan, Q., Wang, Y., Feng, H., Liu, S., Gao, G., Serganov, A., Gao, P. (2019). Diverse mechanisms of CRISPR-Cas9 inhibition by type IIC Anti-CRISPR proteins. *Mol. Cell*, 74(2), 296–309.e297.

List of Authors

Patricia BALARESQUE
Diversification écologique et évolution des populations
Laboratoire d'Evolution et Diversité Biologique
Université Paul Sabatier
CNRS UMR5174
Toulouse
France

Camille BERTHELOT
Institut de biologie de l'École Normale Supérieure
CNRS UMR8197
Inserm U1024
and
Génomique fonctionnelle comparative
Institut Pasteur
CNRS UMR3525
Paris
France

Amandine BONNET
Génomes, biologie cellulaire et thérapeutique
Université Paris Cité
CNRS UMR7212
Inserm U944
France

Clément CARRÉ
Epigénétique transgénérationnelle et biologie des petits ARN
Institut de Biologie Paris Seine
Sorbonne Université
CNRS UMR7622
France

Karine CASIER
Epigénétique transgénérationnelle et biologie des petits ARN
Institut de Biologie Paris Seine
Sorbonne Université
CNRS UMR7622
France

Nuria CORTES-SILVA
Evolution des centromères et ségrégation des chromosomes
Institut Curie
CNRS UMR3664
Paris
France

Franklin DELEHELLE
Institut de Biologie de l'École Normale Supérieure
CNRS UMR8197
Université Paris Sciences & Lettres
France

Bernard DUJON
Sorbonne Université
and
Institut Pasteur
Institut de France
Académie des Sciences
Paris
France

Ines A. DRINNENBERG
Evolution des centromères et ségrégation des chromosomes
Institut Curie
CNRS UMR3664
Paris
France

Pascale LESAGE
Génomes, biologie cellulaire et thérapeutique
Université Paris Cité
CNRS UMR7212
Inserm U944
France

Emilia Puig LOMBARDI
SFR Necker
CNRS UAR3633
Inserm US24
Paris
France

Arturo LONDOÑO-VALLEJO
Télomères et cancer
Institut Curie
CNRS UMR3347
Paris
France

Elise PAREY
Institut de biologie de l'École Normale Supérieure
CNRS UMR8197
INSERM U1024
Paris
France

Guy-Franck RICHARD
Instabilités naturelles & synthétiques des génomes
Institut Pasteur
CNRS UMR3525
Paris
France

Aruni P. SENARATNE
Evolution des centromères et ségrégation des chromosomes
Institut Curie
CNRS UMR3664
Paris
France

Laure TEYSSET
Epigénétique transgénérationnelle et biologie des petits ARN
Institut de Biologie Paris Seine
Sorbonne Université
CNRS UMR7622
France

Marie TOUCHON
Génomique évolutive des microbes
Institut Pasteur
CNRS UMR3525
Paris
France

Wilhelm VAYSSE–ZINKHÖFER
Instabilités naturelles & synthétiques des génomes
Institut Pasteur
CNRS UMR3525
Paris
France

Index

I, L

M, N

P

R, S

T, V

Printed and bound by CPI Group (UK) Ltd, Croydon, CR0 4YY
13/09/2026
14960999-0007